METHODS IN MOLECULAR BIOLOGY™

AF181023

For other titles published in this series, go to
www.springer.com/series/7651

Protein Misfolding and Cellular Stress in Disease and Aging

Concepts and Protocols

Edited by

Peter Bross and Niels Gregersen

Research Unit for Molecular Medicine, Aarhus University Hospital, Skejby, Århus, Denmark; Department of Clinical Medicine, Faculty of Health Sciences, Aarhus University, Århus, Denmark

Humana Press

Editors
Peter Bross
Research Unit for Molecular Medicine
Aarhus University Hospital
Skejby, Århus
Denmark
and
Department of Clinical Medicine
Faculty of Health Sciences
Aarhus University
Århus
Denmark
peter.bross@ki.au.dk

Niels Gregersen, Dr.
Research Unit for Molecular Medicine
Aarhus University Hospital
Skejby, Århus
Denmark
and
Department of Clinical Medicine
Faculty of Health Sciences
Aarhus University
Århus
Denmark
NIG@ki.au.dk

ISSN 1064-3745 e-ISSN 1940-6029
ISBN 978-1-4939-5759-0 ISBN 978-1-60761-756-3 (eBook)
DOI 10.1007/978-1-60761-756-3
Springer New York Dordrecht Heidelberg London

Cover illustration: The inset shows a scheme from chapter 1.

Printed on acid-free paper

Humana Press is part of Springer Science+Business Media (www.springer.com)

Preface

This volume of Methods in Molecular Biology is intended as a follow-up to the MiMB volume on "PROTEIN MISFOLDING AND DISEASE". It is thus not an update of the previous volume but rather an addition containing new content that may be used in combination with the previous content. The major focus of the previous volume was on how and why certain proteins misfold and how this is linked to many disease processes. In this current volume, prime emphasis is given to concepts and methods to determine the molecular effects of protein misfolding at a cellular level and to delineate the impacts and cellular reactions that play a role in pathogenetic mechanisms, and possible manipulations and treatment strategies that can counteract, modify, or delay the consequences of misfolding.

In the first part of the new volume (Chaps. 1–8), concepts and approaches that have been developed in the recent past are discussed in overview chapters. We have also included a connection to the research fields of aging, where the concepts of protein misfolding disease mechanisms have been shown to be relevant because misfolding diseases can be conceived as premature aging processes. The second part of the volume (Chaps. 9–22) contains detailed descriptions of protocols for relevant experimental approaches. We believe that the current volume can be of value for researchers working in the field, as well as for medical professionals and molecular biologists, who wish to get an overview of the recent developments in this active research area.

We hope that the chapters gathered in this volume prove useful for shaping and performing research related to protein misfolding and that they encourage many new researchers to work in this intriguing research field.

Århus, Denmark

Peter Bross
Niels Gregersen

Contents

Contributors

RAJIV VAID BASAIAWMOIT • *Department of Molecular Biology, Aarhus University, Århus, Denmark*

PETER BROSS • *Research Unit for Molecular Medicine, Aarhus University Hospital, Skejby, Århus, Denmark; Department of Clinical Medicine, Faculty of Health Sciences, Aarhus University, Århus, Denmark*

JOHN H. BRUMELL • *Cell Biology Program, The Hospital for Sick Children, Toronto, ON, Canada*

MONICA BUCCIANTINI • *Research Centre on the Molecular Basis of Neurodegeneration (CIMN) and Department of Biochemical Sciences, University of Florence, Italy*

D. ALLAN BUTTERFIELD • *Department of Chemistry, Center of Membrane Sciences and Sanders-Brown Center on Aging, University of Kentucky, Lexington, KY, USA*

ORAZIO CANTONI • *Institute of Pharmacology and Pharmacognosy, University of Urbino "Carlo Bo", Urbino, Italy*

GIORGIO CASARI • *Neurogenomics Unit, Center for Genomics, Bioinformatics and Biostatistics, San Raffaele Scientific Institute and San Raffaele University, Milan, Italy*

CRISTINA CECCHI • *Research Centre on the Molecular Basis of Neurodegeneration (CIMN) and Department of Biochemical Sciences, University of Florence, Italy*

LIANA CERIONI • *Institute of Pharmacology and Pharmacognosy, University of Urbino "Carlo Bo", Urbino, Italy*

RODNEY J. DEVENISH • *Department of Biochemistry and Molecular Biology and the ARC Centre of Excellence in Structural and Functional Microbial Genomics, Monash University, Clayton, VIC, Australia*

NIELS GREGERSEN • *Research Unit for Molecular Medicine, Aarhus University Hospital, SkejbyÅrhus, Denmark; Department of Clinical Medicine, Faculty of Health Sciences, Aarhus University, Århus, Denmark*

JAKOB HANSEN • *Research Unit for Molecular Medicine, Aarhus University Hospital Skejby, Århus, Denmark; Department of Clinical Medicine, Faculty of Health Sciences, Aarhus University, Århus, Denmark*

TOSHIHISA ISHIKAWA • *Department of Biomolecular Engineering, Graduate School of Bioscience and Biotechnology, Tokyo Institute of Technology, Yokohama, Japan; Omics Science Center, RIKEN Yokohama Institute, Yokohama, Japan*

NATALIA A. KANIUK • *Cell Biology Program, The Hospital for Sick Children, Toronto, ON, Canada*

SHOKO KOSHIBA • *Department of Biomolecular Engineering, Graduate School of Bioscience and Biotechnology, Tokyo Institute of Technology, Yokohama, Japan*

CRISTOVAO F. LIMA • *CITAB, Department of Biology, School of Sciences, University of Minho, Braga, Portugal*

Francesca Maltecca • *Neurogenomics Unit, Center for Genomics, Bioinformatics and Biostatistics, San Raffaele Scientific Institute and San Raffaele University, Milan, Italy*

Aurora Martinez • *Department of Biomedicine, University of Bergen, Bergen, Norway*

Dalibor Mijaljica • *Department of Biochemistry and Molecular Biology, Monash University, Clayton, VIC, Australia*

Carlos Torres Moraes • *The Neuroscience Program and Department of Neurology, Miller School of Medicine, University of Miami, Miami, FL, USA*

Hiroshi Nakagawa • *Department of Biomolecular Engineering, Graduate School of Bioscience and Biotechnology, Tokyo Institute of Technology, Yokohama, Japan*

Nobuyuki Nukina • *Laboratory for Structural Neuropathology, RIKEN Brain Science Institute, Saitama, Japan*

Joshua B. Owen • *Department of Chemistry, University of Kentucky, Lexington, KY, USA*

Johan Palmfeldt • *Research Unit for Molecular Medicine, Aarhus University Hospital, Skejby, Århus, Denmark; Department of Clinical Medicine, Faculty of Health Sciences, Aarhus University, Århus, Denmark*

Christina Bak *Pedersen* • *Research Unit for Molecular Medicine, Aarhus University Hospital, Skejby, Århus, Denmark; Department of Clinical Medicine, Faculty of Health Sciences, Aarhus University, Århus, Denmark*

Alicia Mae Pickrell • *The Neuroscience Program, Miller School of Medicine, University of Miami, Miami, FL, USA*

Mark Prescott • *Department of Biochemistry and Molecular Biology, Monash University, Clayton, VIC, Australia*

Suresh I. S. Rattan • *Department of Molecular Biology, Aarhus University, Århus, Denmark*

Javier Sancho • *Departamento de Bioquimica y Biologia Molecular y Celular, Facultad de Ciencias, Universidad de Zaragoza, Zaragoza 50009, Spain*

Shoichi Sasaki • *Department of Neurology, Tokyo Women's Medical University, Tokyo, Japan*

Martin Schröder • *School of Biological and Biomedical Sciences, Durham University, Durham, UK*

Anatoly A. Starkov • *Department of Neurology and Neuroscience, Weill Medical College of Cornell University, New York, NY, USA*

Massimo Stefani • *Department of Biochemical Sciences and Research Centre on the Molecular Basis of Neurodegeneration, University of Florence, Florence, Italy*

Raymond C. Stevens • *Department of Molecular Biology, The Scripps Research Institute, La Jolla, CA, USA*

Louise Sutcliffe • *School of Biological and Biomedical Sciences, Durham University, Durham, UK*

Ai Tamura • *Department of Biomolecular Engineering, Graduate School of Bioscience and Biotechnology, Tokyo Institute of Technology, Yokohama, Japan*

Isei Tanida • *Department of Biochemistry and Cell Biology, National Institute of Infectious Diseases, Tokyo, Japan*

YU TOYODA • *Department of Biomolecular Engineering, Graduate School of Bioscience and Biotechnology, Tokyo Institute of Technology, Yokohama, Japan*
SØREN VANG • *Faculty of Health Sciences; Research Unit for Molecular Medicine, Aarhus University Hospital, Skejby, Århus, Denmark; Department of Clinical Medicine, Aarhus University, Århus, Denmark*
SATOSHI WAGURI • *Department of Anatomy and Histology, Fukushima Medical University School of Medicine, Fukushima, Japan*
KANAKO WAKABAYASHI-NAKAO • *Medical Genetics Division, Shizuoka Cancer Center Research Institute, Shizuoka, Japan*
TOMOYUKI YAMANAKA • *Laboratory for Structural Neuropathology, RIKEN Brain Science Institute, Saitama, Japan; Special Postdoctoral Researchers Program, RIKEN, Saitama, Japan*

Part I

Concepts

Chapter 1

Protein Misfolding and Cellular Stress: An Overview

Niels Gregersen and Peter Bross

Abstract

Cell survival and death are complex matters. Too much survival may lead to cancer and too much cell death may result in tissue degeneration. In this chapter, we will first of all focus on the cellular survival mechanisms that promote correct folding and maintenance of protein function. These mechanisms include protein quality control (PQC) systems comprising molecular chaperones and intracellular proteases in the cytosol, endoplasmatic reticulum (ER) and in the mitochondria. In addition to the PQC systems, mechanisms elicited by misfolded proteins, known as unfolded protein responses (UPRs), including induction/activation of antioxidant systems are also present in the three compartments of the cell. Second, we will discuss the mechanisms by which misfolded proteins lead to the generation of oxidative stress in the form of reactive oxygen species (ROS) and reactive nitrogen species (RNS). These species are produced mainly from superoxide (O_2^-) generated in the mitochondrial respiratory chain and from nitrogen oxide (NO) produced by the mitochondrial nitrogen oxide synthetase (mtNOS). Third, the effects of oxidative stress will be discussed, both with respect to mitochondrial dynamics, i.e., fission and fusion, and the related elimination of dysfunctional mitochondria by cellular cleaning systems, i.e., mitophagy or mitoptosis, and related to the generation and cellular effects of oxidatively modified proteins, which closes a vicious cycle of protein misfolding and oxidative stress.

Key words: Protein misfolding, Unfolded protein response, Heat shock response, Oxidative stress, Reactive oxygen species, Reactive nitrogen species, Protein quality control system, Antioxidative defense, Reactive carbonyl compounds

1. Introduction

All living cells contain molecular systems that are vital for proper function and survival. Among these, organelle specific networks of protein quality control (PQC) systems, comprising molecular chaperones and intracellular proteases, as well as antioxidant systems, maintain appropriate protein folding, structure, and function. Although an important function of the chaperone components

Peter Bross and Niels Gregersen (eds.), *Protein Misfolding and Cellular Stress in Disease and Aging: Concepts and Protocols*, Methods in Molecular Biology, vol. 648, DOI 10.1007/978-1-60761-756-3_1, © Springer Science+Business Media, LLC 2010

of the PQC systems is assistance during folding of nascent proteins, a main function is, together with the proteases and antioxidant systems, to be the first line of cellular protection against misfolded and damaged proteins due to internal or external stressors. In young and healthy cells, the function and possible induction of these systems are sufficient to maintain homeostasis and restore it if disturbed. However, in cells carrying genetic defects and in stressed and ageing cells these "damage protection" systems may be overloaded, and damaged and misfolded proteins may accumulate in various forms. The result is loss as well as gain of protein function, which may disturb a variety of cellular functions, including oxidative balances, energy production, and mitochondrial integrity, which may activate a second line of defense, i.e., the elimination of accumulated proteins and dysfunctional organelles, eventually initiating the cell death mechanisms.

2. Cytosolic Protein Quality Control, Unfolded Protein Response, and Misfolded Protein Overload

Proteins with functions in the cytosol, nucleus, and peroxisomes emerge from the ribosome and fold in a concerted action (1). In order to shield the hydrophobic parts from interacting with other components of the crowded cellular environment, the emerging polypeptides interact with a variety of chaperones and assistant factors, i.e., nascent-polypeptide-associated complex (NAC), prefoldin, Hsp40, and Hsp70 (2). A subset also binds to the chaperonin TCP-1 ring complex (TRiC) and is held in a folding-competent state until released from the ribosome (3). The majority of small cytosolic proteins complete their folding without further assistance, whereas a fraction requires further assistance from the chaperone Hsp90 and the chaperonin TRiC (2). If the native structure is not easily achieved, the interaction with TRiC may be prolonged, and the polypeptide, which may be arrested in an intermediate conformation (3), is presented for the degradation components of the cytosolic PQC system: the ubiquitin-proteasome system (UPS), which functions as a safeguard for acquisition of functional conformers of nascent proteins by eliminating potentially toxic conformers (4, 5). This is only one of the mechanisms exerted by the PQC system. Another is "damage protection," which was first described for the cytosol as the cytosolic heat-shock response (6). This is a dynamic and very complex system, which could be coined the cytosolic unfolded protein response (cytUPR) to indicate the functional similarity to the endoplasmatic reticulum UPR (erUPR) and the mitochondria UPR (mitUPR) (7).

The working mechanism of cytUPR has been intensively studied (8–10). At non-stressed cellular conditions, components of the PQC system are constitutively expressed in amounts sufficient to counteract sporadic misfolding and accumulation of newly synthesized proteins including defective translation products (DRiPs) (11, 12), as well as accumulation of moderately damaged and oxidatively modified proteins.

The master regulator of cytUPR is the heat-shock factor 1 (HSF1). In non-stressed situations, HSF1 is present in an inactive monomeric form in complex with chaperones, such as hsp90, hsp70, and hsp40 (10, 13). During stress, unfolded proteins are present in the cell, and HSF1 is activated. The mechanism of activation is conceptionally simple, but in reality complicated. Accumulated unfolded proteins, either variant proteins caused by gene variations or oxidatively modified proteins, which are not easily degraded, may lead to sequestration of chaperones, especially hsp70 and hsp40 as well as hsp90, which also are involved in the degradation process (14). Initially, the unfolded proteins thereby compete with HSF1 for binding of chaperones, thus releasing HSF1. Released HSF1 undergoes a conformational change, trimerisation, becomes phosphorylated, and translocates to the nucleus, where it binds to heat-shock elements (HSE) of a large number of genes, including those coding for chaperones (10). In this way, the amounts of chaperones capable of occupying HSF1 increase. The activation is alleviated at the same time as the unfolded proteins, assisted by the chaperones, may be refolded or presented for the degradation apparatus. In addition to the induction of chaperones, the HSF1 orchestrated stress response also includes the important cytosolic antioxidant system, heme oxygenase (HO-1), which converts heme to biliverdin that is further metabolized to the powerful antioxidant bilirubin (9). This reflects that oxidative stress is generated in stressed cells and that combating it is important for cell survival. Indeed, recent results have added significant knowledge to the mechanisms that may alleviate the consequences of oxidative stress (15). Westerheide and coworkers showed that HSF1 binding to and release from the hsp70 promoter is dependent on acetylation of HSF1, which is regulated by the metabolically controlled activity of the deacetylase SIRT1. Activation of SIRT1 by a high NAD^+/NADH ratio, which is present in metabolically active and healthy cells, hold HSF1 in the binding competent form, thereby enhancing the stress response and cell survival. Thus, the interplay between HSF1 and SIRT1 together with other SIRT1 regulated transcription factors, such as FOXO3, p53, and NF-κB, are thus involved in the stress response, including combating of oxidative stress. Indeed, the involvement of these cellular mechanisms may be seen as a reflection of the importance of oxidative stress, perhaps the most important, patophysiological manifestation in many diseases (16).

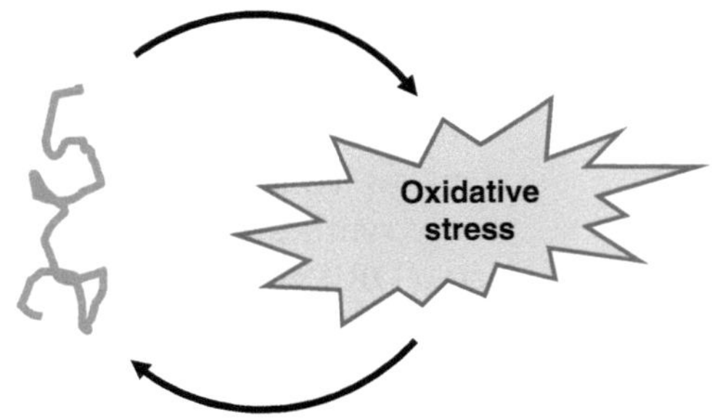

Fig. 1. Simplified overview of the conception of the vicious cycle, comprising on one side misfolded protein, either due to inherited genetic defects or oxidative damage, and on the other, oxidative stress that – by oxidative damage – cause protein misfolding.

Accordingly, there exists a wealth of knowledge about the antioxidant systems, such as glutathione, glutathione peroxidases and reductases as well as superoxide dismutases, catalase and the mentioned heme oxygenase, as the first line of defense, but also about oxidatively modified proteins, which – if not eliminated – feed into a vicious cycle of protein misfolding and further oxidation (Fig. 1), eventually leading to cytotoxicity and cell death (9, 17–21).

In contrast to this large amount of knowledge about the generation, regulation, and effects of oxidative stress, the knowledge about the mechanism(s), by which misfolded proteins in the cytosol mediate the generation of oxidative stress, is currently very limited. One mechanism, discussed in Chap. 2, concerns misfolded proteins, which form fibrils that may permeabilise membranes and increase intracellular free calcium ions, which in turn activate mitochondrial metabolic processes and production of reactive oxygen species (ROS) as well as nitrogen oxide synthetase (mtNOS), producing reactive nitrogen species (RNS) (22, 23). This direct effect, which is associated with an effect on the mitochondria dynamics (see below and Chaps. 4 and 6), is probably involved in the mechanism by which expanded CAG trinucleotide repeats create oxidative stress in neuronal cells from patients with Huntington's disease (24). Pertubation of the mitochondrial dynamics seems also to be involved in a neuronal model of Alzheimer's disease, where oxidative stress – paradoxically and as yet unexplained – was associated with down-regulation of the mitochondrial antioxidant superoxide dismutase (MnSOD) (25).

Another oxidative stress mediating mechanism, which may be active in cases of fibril-forming misfolded proteins as well as in cases of accumulated misfolded proteins in general, is overloading and inhibition of the proteasome, which constitutes the main cytosolic protease containing system. The proteasome is important, both in normally functioning cells and in cells accumulating misfolded/damaged proteins. It is noteworthy that oxidatively modified proteins can be degraded directly by the proteasome, thus circumventing the usual ubiquitinylation as a signal for degradation (21). This feature may reflect that an evolutionary important system that has evolved as a safeguard if the cellular antioxidant systems, including the cytosolic, may be overloaded or defective. In this respect, it is interesting to note that mild inhibition of the proteasome, which could be affected by misfolded/damaged proteins, is able to elicit oxidative stress due to mitochondrial dysfunction secondary to decrease of the capacity of the respiratory chain (26). It is further noteworthy that this mild proteasome inhibition does not result in apoptosis, but rather in selective elimination of dysfunctional mitochondria by mitophagy or mitoptosis (27, 28), thus functioning as a cell survival mechanism. These mechanisms are discussed in Chaps. 5 and 6 and will be touched again below. It is sufficient to say that these mechanisms are induced by oxidative stress and function as a second line of defense, subsequent to or in parallel with the PQC and antioxidant systems mentioned above and as a last survival mechanism before apoptotic cell death (29).

In conclusion, the cytosolic stress responses comprise components which counteract protein misfolding and oxidative damage, and the proteasome, which can eliminate damaged proteins as such, as well as mechanistic links to the autophagic/mitoptotic systems, which can eliminate dysfunctional cell elements and promote survival. Ultimately, if restoration of homeostasis cannot be achieved by these mechanisms, cell death by apoptosis (or necrosis) may occur.

3. Endoplasmic Reticulum Protein Quality Control, Unfolded Protein Response, and Oxidative Stress

The endoplasmic reticulum (ER) is the site of processing secretory and plasma membrane-associated proteins including membrane receptors, as well as proteins destined for cellular vesicles, such as the lysosomes, and proteins involved in the processing of all these proteins. All ER-processed proteins are nuclear encoded, co-translational translocated through the ER membrane, guided by leader peptides, co-translocational N-glycosylated and leader peptide deleted before folding (30) (see also Chap. 3). Folding is accomplished by a specialized set of chaperones, including BiP (GRP78), an ER equivalent to hsp70, and GRP94, as well as folding enzymes, for example, protein disulfide isomerase (PDI), and the lectin chaperones, calnexin and calreticulin (22, 31). Indeed, in addition to interacting with these chaperones and enzymes, most ER-processed proteins further interact with lectin chaperones, which supervise the folding process. The initial N-glycosyl moiety, N-acetylglucosamine$_2$-mannose$_9$-glucose, is trimmed by glucosidase I and II to yield a folding intermediate with one unit of glucose left to bind to calnexin and calreticulin. Subsequently, the last glucose is removed by glucosidase I, which prevents further binding to calnexin and calreticulin, and if the folding is completed, the native protein is transport competent and leave the ER for further processing in Golgi and trafficking to its destination. Thus, this terminal glucose functions as an ER retention signal. If the folding is not completed, due to amino acid alterations or other damage, glucose is reintroduced by UDP-glucose:glycoprotein glucosyltranferase (UGT1) and the folding process is repeated. Eventually, after continuous failures, the misfolded intermediate is retrogradly translocated through the membrane into the cytosol and degraded by the UPS (32, 33).

As discussed above for the cytosolic PQC system, the ER PQC system should be sufficiently effective to cope with erroneous and oxidatively damaged proteins in the young and unstressed cell. Indeed, it has been estimated that only about 30% of newly synthesized proteins processed through the ER reaches the native transport competent state (11). The remaining is degraded as fast as possible. However, in the stressed and old cells, the ER PQC system may be overloaded with misfolded conformers, producing proteotoxic stress, which elicits the endoplasmatic reticulum unfolded protein response (erUPR) (31, 34) and oxidative stress (22, 31).

The various elements of the erUPR are well documented and comprise three signaling pathways: IRA1, ATF6, and PERK (22, 31, 34) (see also Chap. 3). Without going into details, these three signaling pathways work in concert to enhance the folding capacity as well as the defense against stress (35, 36). IRA1, ATF6, and PERK are all membrane spanning proteins that are held in inactive states in complexes with BiP. When challenged with misfolded polypeptides, BiP is competitively released, leading to activation of the three factors. IRA1 undergoes dimerization to an active endonuclease which removes an intron from the X-box binding protein 1 (XBP1) mRNA, rendering it translation competent to initiate synthesis of XBP1 protein, which translocates to the nucleus and contributes to the transcription of the cytoprotective genes. When released from BiP ATF6 is translocated to the Golgi, where two proteases, S1P and S2P, cleave and activate ATF6, which is then trafficked to the nucleus as a co-activator of the cytoprotective genes. PERK also dimerises and subsequently autophosphorylates and associates with eukaryotic initiation factor 2α (eIF2α), which is phosphorylated and thereby inactivated. By decreasing the translation, activation of PERK alleviates the burden of newly synthesized polypeptides, thus leaving the folding and degradation mechanisms to their cleaning function. In addition to the inhibition of general translation, PERK activates the translation of some mRNAs, most notably *Atf4* mRNA, that produces the transcription factor ATF4, which is also involved in the activation of cytoprotective genes.

In addition to these protective responses, which are set into action by acute "mild" challenges, elimination signals may also be initiated, most probably after severe misfolding and oxidative stress. The proapoptotic transcription activator C/EBP homologous protein (CHOP) can be activated by ATF4, and caspases may by cleaved and activated by IRA1, thus shifting the balance from protection to execution. If, at the same time, generation of ROS exceeds the antioxidant capacity, apoptosis may also be initiated.

ROS is usually created by leakage of superoxide from the mitochondrial respiratory chain due to perturbation of mitochondrial structure and/or function. Indeed, this seems to be the case during ER stress, since Ca^{++} redistribution from ER to mitochondria and activation of mitochondrial nitrogen oxide synthetase (mtNOS) creating RNS is promoted by severe ER stress (22, 23). These effects will be discussed below in connection with mitochondrial stress. Suffice here to say that ROS and RNS from mitochondria may aggravate Ca^{++} leakage from ER and increase the production of ROS/RNS, thus creating a vicious cycle. If the production of ROS/SNS is elevated beyond a certain level, the mitochondrial permeability transition pores eventually open and give rise to liberation of cytochrome c and execution of the apoptosis programme (29, 37, 38).

ROS may in addition be created inside ER by hyper-activity of the so-called oxidative protein folding. Since ER constitutes an oxidative milieu and many ER processed proteins contain disulfide bonds, this compartment, in contrast to the cytosol and mitochondria, comprise a number of oxidoreductases by which disulfide bonds can be created as well as redistributed and broken if wrongly synthesized. PDI is the most important and abundant. Although glutathione is able to reduce misplaced disulfide bonds in polypeptides, PDI in conjunction with ER oxidoreductin 1 (Ero1p) is considered a major pathway since Ero1p is induced by ER stress (39). Interestingly, Ero1p is a flavoenzyme which may interact directly with molecular oxygen, producing ROS in the form of hydrogen peroxide and thereby contribute to the burden of oxidative stress. Further, it has been implicated, at least in neuronal cells containing a nitrogen oxide synthetase, that NO may inactivate PDI by nitrosylation, thereby inhibiting reduction and/or redistribution of protein disulfides and aggravating the ER stress (19, 40). Since mitochondria contain an NO synthetase (mtNOS) (23), which is stimulated by the above discussed redistribution of Ca^{++} from ER, the mitochondria may generate NO. NO is free diffusible and may migrate to the ER for nitrosylation of proteins, such as PDI.

In conclusion, ER contains a unique set of chaperones, including lectins, which in the young and healthy cell can cope with newly imported and damaged proteins generated after "mild" stress. However, in case of "severe" stress a number of misfolded protein sensors in concert induce a range of protective genes, including folding chaperones as well as genes coding for antioxidants. At the same time, signals initiating apoptosis are activated, and the balance between restoration and cell death depends on the severity and duration of the insult.

4. Mitochondrial Protein Quality Control, Unfolded Protein Response and Oxidative Stress

Mitochondria contain proteins coded from at least a thousand different genes (41, 42), nearly all of which are nuclear encoded and contain a leader sequence used for targeting to the mitochondrial import machinery (43). In addition, 22 tRNAs, two ribosomal RNAs as well as 13 components of the respiratory chain are coded from the mitochondrial DNA (mtDNA), which is a reminiscence from the aerobic protobacteria which invaded pre-eukaryotic cells more than a billion years ago. Eukaryotic mitochondria are formed by an outer membrane (OM), an inter-membrane space (IMS), inner membrane (IM), and a matrix (M). Relevant for the presence context is that most of the metabolic activity takes place in the matrix, and the production of energy-rich ATP as a result of the respiratory chain activity, is associated with IM and IMS (see also Chap. 6) and (16).

Although it may not be correct to talk about numbers, since there is a dynamic balance between small distinct and large reticulum formed mitochondria, the "number" of active units varies drastically, depending on the metabolic activity needed (44). Especially muscle, heart, and brain cells that are highly metabolically active may contain several hundred active units, each containing many mtDNAs. This large number and the dynamic nature of the mitochondria network make the cell robust to insults, such as oxidative stress. This will be discussed in details below; suffice here to say that "mildly" damaged mitochondrial units can fuse and be repaired by "healthy" ones, and that "severely" dysfunctional units may be removed without immediate serious consequences for the survival of the cell (44, 45). Before discussing this and other consequences of mitochondrial stress, it is appropriate shortly to recapitulate the current knowledge about protein folding and misfolding inside mitochondria as well as the stress response through the mitochondrial unfolded protein response (mtUPR).

Nuclear encoded polypeptides destined for mitochondria are translated by cytosolic ribosomes, shielded by cytosolic chaperones, and trafficked to the mitochondria, where they dock into receptors of the very complex import machinery (43). Guided by the import signal, which may be an N-terminal leader sequence or an internal amino acid sequence, and by other protein–protein interaction signals, the final destination may be OM, IMS, IM, or M. Matrix proteins as well as many IM embedded proteins are transferred to the matrix assisted by chaperones, especially the mitochondrial hsp70, and the leader is removed by peptidases. Subsequently, small proteins may fold to the active structure without more assistance, while larger, multi-domain and multimeric proteins may need further assistance from the mitochondrial chaperonin hsp60, which functions in concert with the co-chaperone hsp10 (46). Proteins with folding difficulties, either due to inherited or acquired amino acid alterations or damage, such as oxidative modifications, may eventually be eliminated by proteases in the matrix, i.e., LON, or IM, i.e., AAA-proteases, or they may accumulate (44, 47, 48). Together, the network of chaperones and proteases comprise the mitochondrial protein quality control system (mtPQC). The balance between folding, degradation, and accumulation depends on the nature of the amino acid alteration/damage and the efficiency of the PQC, which is governed by genetic, cellular and environmental factors. Despite the fact that loss of protein function can be disease-causing, as seen in many inborn errors of metabolism (49, 50), the accumulation of misfolded protein inside the mitochondria may further add to the pathology, especially in active tissues, like brain, heart, and muscles, as well as in ageing cells where the efficiency of the various PQC components, for example, the Lon protease, is declining (47).

As discussed above, mitochondria are implicated in cellular dysfunction, especially in the generation of oxidative stress, when the unfolded protein responses (UPRs) in the cytosol and ER are not able to cope with misfolded and damaged proteins. The same seems to be the case for accumulated proteins inside the mitochondria themselves. As discussed for protein accumulation in the cytosol and ER, mitochondria have their own defense mechanisms, of which the first layer of protection is a mtUPR.

The first indication of the existence of an mtUPR stems from over-expression of a misfolding deletion mutant of ornitine transcarbamylase (OTC) in COS cells (51). These cells showed induction of Hsp60 and Hsp10 as well as of the mtDnaJ chaperone and of the ClpP protease. Although the induction pathway was not elucidated at that time, it was shown that the common promoter of Hsp60/Hsp10 (52) as well as the promoters of mtDnaJ and ClpP contain a binding element for CHOP, which was up-regulated in the stressed cells. More recent results from C. elegans (7, 53) has partly elucidated the signal pathway. Upon unfolding stress, the cytosolic DVE-1, which is homologous to the mammalian SATB1 and SATB2, relocate to the nucleus and interact with the hsp60/10 promoter. Simultaneously, UBL-5, which is up-regulated during the unfolding stress, interacts with DVE-1, thus possibly enhancing the stress response. Despite these results, the stress signal(s) inside and from the mitochondria have not yet been found. However, it seems that ClpP activity is necessary for the response, since down-regulation of the expression of ClpP in C. elegans alleviated the induction of DVE-1 and UBL-5. So, proteolytic digestion of some mitochondrial proteins that are able to sense the accumulation of misfolded proteins is probably needed. It could be speculated that such proteins are normally stabilized by chaperones, for example, hsp70, like PERK, IRA1, and ATF6 in the ER. Upon loading with unfolded proteins, hsp70 is released from the protein, leaving it exposed and subject to degradation. It has been speculated that the degradation products constitute the active signalling substances or that the putative effecter protein is a repressor of the stress response (7). But how is the signal then transmitted to DVE-1? Whatever the exact mechanism, the important point here is that the mitochondrion is able to sense misfolded proteins and promote induction of protective chaperones and proteases. Interestingly, as mentioned, CHOP, which is the same pro-apoptotic transcription factor as activated in the ER stress pathway (22, 34), is also implicated in the mtUPR (51, 54). However, in the case of mtUPR, CHOP is activated through an AP-1 binding site, and probably through the pro-apoptotic JNK pathway (54), whereas CHOP in the erUPR is activated through the ER stress element (ERSE) binding motive as well as through the PERK-ATF4 pathways (55). Thus, like the erUPR, the mtUPR at the same time as giving signal to inducing protective genes, also initiates cell death signals. Whether these signals may take over must depend on both the duration and severity of the stressor(s).

As discussed above, the stressors in the cytosol and in the ER may be misfolded proteins, which through more or less elucidated mechanisms, for example, fibril formation, Ca^{++} redistribution, and proteasome inhibition, disturb the mitochondrial function, giving rise to oxidative stress and cell injury, which is combated by the various defense systems, and if not possible leading to cell death. Surprisingly, although the mentioned over-expression of delta-OTC in COS cells induced expression of chaperones and proteases, the effect on the respiratory chain and the possible creation of oxidative stress was not investigated (51). However, in experiments where a misfolding variant of the short-chain acyl-CoA dehydrogenase (SCAD) protein was transiently over-expressed in an astrocytic cell line we have observed disturbance of the mitochondria dynamics, probably due to oxidative stress (87). Additionally, in another astrocytic cell line, stably expressing the same variant SCAD protein, we observe induction of the mitochondrial superoxide dismutase (MnSOD or SOD2), most pronounced after exposure to mild heat-stress (40°C) for 24 h (87).

These experiments indicate that regardless the origin of misfolded proteins, the perturbation of mitochondria function and creation of oxidative stress most probably provide the first steps of the patophysiological manifestations of protein misfolding diseases. Despite the sparse knowledge about the exact mechanisms by which misfolded proteins create oxidative stress, it is today common wisdom that oxidative stress is implicated in a large number of diseases of which the neurodegenerative diseases as well as disorders of the respiratory chain are the best known (16). Thus, notwithstanding the eliciting factors, oxidative stress is important, especially because – once created – it propagates the vicious cycle of oxidatively modified/damaged proteins, which enhance misfolding and increase the oxidative stress, etc.

Since the main source of ROS production is the mitochondria, it is appropriate to discuss the creation, regulation, and general molecular effects of mitochondria-mediated oxidative stress.

5. Oxidative Stress: Creation, Regulation, and Molecular Effects

Creation of mitochondrial oxidative stress is due to an imbalance between production and elimination of ROS and RNS. ROS comprises the primary produced superoxide ($O_2^{\cdot-}$) as well as secondary products, such as hydrogen peroxide (H_2O_2) and hydroxyl radical ($HO^{\cdot}$) (29), all of which can modify/damage DNA, lipids, and proteins. The main source of RNS in the form of NO is nitrogen oxide synthetases (NOSs) (22, 23), which may react with simultaneously produced $O_2^{\cdot-}$ to form the highly reactive peroxynitrite, $ONOO^-$. Both H_2O_2 and NO function as redox regulators of various transcription factors. Their production is therefore necessary for normal physiological function (56–59), requiring continuous survey of the amounts, especially by the ROS defense systems, comprising a multitude of antioxidant enzymes, including superoxide dismutases (SOD1 and 2), glutathione (GSH), glutathione reductases, and peroxidases as well as catalase, thioredoxins, and thioredoxin reductase (29, 60). However, these systems may be overwhelmed or down-regulated with consequential damaging oxidative stress.

As mentioned earlier, mtNOS, is stimulated by Ca^{++}, and overproduction of NO may thus be a consequence of calcium influx and activation of mtNOS (22, 23). Likewise, ROS production may be stimulated by Ca^{++}, which induces the activity of mitochondrial metabolic processes and inhibits proper action of the respiratory chain (22) (and Chap. 2). Moreover, mitochondrial overproduction of ROS may arise directly by perturbation of the respiratory chain, such as inherited or acquired respiratory deficiencies (61), as well as by a number of other sites, including the enzymes 2-oxo glutarate dehydrogenase and aconitase (60). Further, ROS may be produced by the electron transfer flavoprotein (ETF)/ETF quinone oxidoreductase (ETFQO) complex (62), and by still undefined sites in organic acidemias caused by mitochondrial enzyme deficiencies, including fatty acid oxidation deficiencies (63, 64). In mitochondrial enzyme deficiencies, including fatty acid oxidation deficiencies, glutaryl-CoA dehydrogenase deficiency, propionic and methylmalonic acidemias, it has been indicated that the observed oxidative stress was initiated by induction of NO, giving rise to lipid oxidation and membrane distortion. However, large numbers of patients with mitochondrial enzyme deficiencies harbor missense gene variations, coding for misfolding enzyme proteins, which – together with loss of enzyme function and toxic accumulation of metabolites – may contribute to the patophysiology, including oxidative stress (48, 65, 66).

Despite the fact that oxidative stress as a patophysiological consequence of intra- as well as extra-mitochondrial stressors in most cases is considered to be created by increased production of $o_2^{\cdot-}$ from the respiratory chain, a decrease in the antioxidant defense may also contribute in certain cases. Indeed, in the mentioned mitochondrial enzyme deficiencies the total radical-trapping antioxidant potential (TRAP) has been found decreased (63, 64). Further, and probably more relevant in the present context, SOD2/MnSOD, the mitochondrial superoxide dismutase, is down-regulated in the neuronal model of Alzheimer's Disease mentioned above (25). Additionally, we have observed reduced expression of SOD2 in cultured fibroblasts from patients with SCAD deficiency (88) and decreased resistance to menadione created O_2^- in fibroblasts from patients with this and other mitochondrial fatty acid oxidation defects (Zolkipli et al. 2009, unpublished).

Thus, down-regulation of the intra-mitochondrial enzymatic defense may be under-appreciated simply because it has not been investigated.

The regulatory mechanisms of the respiratory chain, including production of ROS, and of the antioxidant system, including enzymatic and non-enzymatic antioxidants, are extremely complex and a detailed discussion of them is outside the scope of the present text. What is interesting here is that small alterations in the ROS production and elimination rates are capable of shifting the balance from physiological healthy to pathological. As previously mentioned, overloading of the UPRs and perturbation of the redox balance in the cytosol, ER, and mitochondrion eventually result in the overproduction of respiratory chain created ROS – as well as NOS created RNS – and mitochondrial dysfunction, leading to cell death by apoptosis or necrosis.

However, as will be discussed next, depending on the oxidative stress load and duration, the effects and long-term consequences are quite different.

In recent years, it has been realized that mitochondria are not small round organelles, as seen after isolation from various tissues, but constitute in young and healthy cells a network of elongated filamentous structures. They are in constant movement and change shape and size through dynamic processes of fission and fusion, conducted by a number of dynamin-like GTPases (67–69). Important fusion inducers are: Mfn1 and Mnf2 and Opa1, while fission factors are Fis1, Drp1, and MTP18, which all are transcriptionally regulated.

Since each cell contains numerous of these mitochondrial structures and each mitochondrion carry many mtDNAs, the dynamic process is thought to function as rescue mechanism for dysfunctional/damaged structures by fusing with healthy ones (44, 70), or alternatively, when damage is excessive by isolating structures for subsequent engulfment into autophagosomes (69, 71). This process has been named mitophagy, which together with classical modes of autophagy (macroautophagy, microphagy, and chaperone-mediated (CMA) autophagy) have been studied intensively (72–74) (see also Chap. 5). Of relevance in the present context is also the so-called mitoptosis, a related and highly selective mechanism for eliminating dysfunctional mitochondria (28) (see also Chap. 6). All these modes for eliminating dysfunctional sections of mitochondria are intimately linked to mitochondrial fission. For instance, mitophagy is characterized by increased fission compared to fusion (75). Mitophagy, and possibly mitoptosis, apparently functions as intermediate but cell-saving steps on the road to apoptosis (see Fig. 2). Indeed, as shown in human endothelial cells (HUVEC) loaded for a short time with pathological, but realistic, amounts of H_2O_2, the mitochondria recover from grain-like fission product to thread-like structures after 24–48 h (45). Similarly, young primary cortical neuronal cultures treated with 50–200 μmol/lS-nitrosocystein (SNOC), yielding realistic amounts of NO, showed fragmented mitochondria after 20 min; this was recovered after 60 min (71), documenting that these types of acute stress is not lethal for the cells. The mechanism by which the fission is promoted in these neuronal cells was shown to involve nitrosylation of a Cys^{344} in the fission factor Drp1, increasing its activity (76). Thus, alleviating the stress loading, that is, NO, must dilute the nitrosylated Drp1 by new synthesis, since the level is transcriptionally regulated.

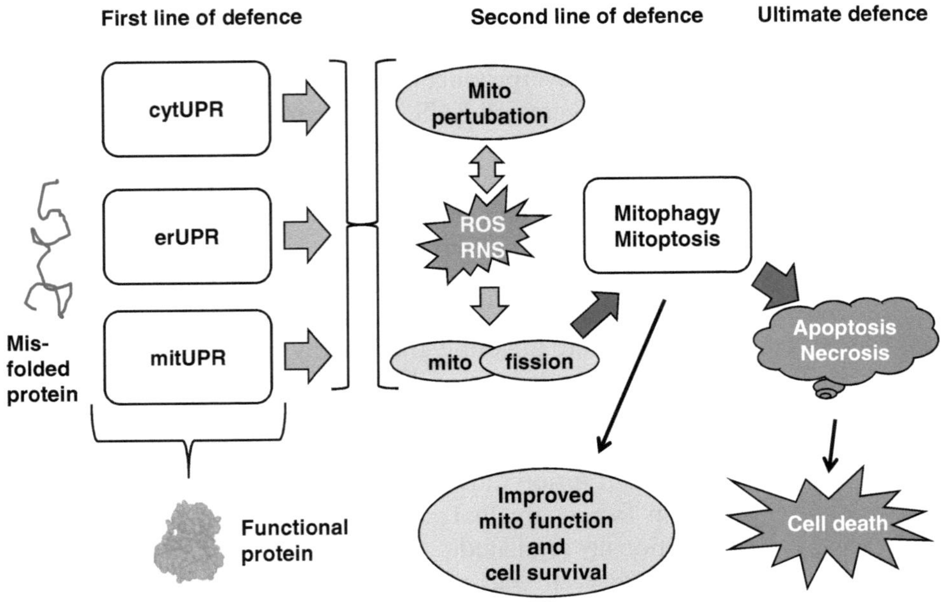

Fig. 2. Schematic overview of the mechanisms that may be elicited by misfolded proteins. Depending the nature, amount and localization of the misfolded protein as well as other genetic, cellular and environmental factors, the mechanisms may eliminate the misfolded protein or dysfunctional mitochondria in order to rescue the cell. If this is not possible, apoptosis or necrosis signal take over and execute cell death.

Interestingly, loading the cells with the Amyloidβ-25-35, the pathological peptide fragment in Alzheimer's disease, showed the same phenomenon, that is, mitochondrial fragmentation, also by the mechanism of Drp1 nitrosylation, probably by inducing neuronal NOS (nNOS) after activation of the NMDA receptor (76) and Ca^{++} in-flux. In this connection, it is noteworthy that nitrosylated Drp1 has been detected in AD brains.

Taking these results together with the ones obtained by Bereiter-Hahn's group (45) it is evident that oxidative stress elicited by NO and/or $O_2^{\cdot-}/H_2O_2$ results in mitochondrial fission. The mechanism by which this is affected by NO seems to be elucidated, but the effect of ROS still remains to be clarified.

Both ROS and RNS exert other – more traditionally appreciated– effects. We will focus on reactive carbonyl compounds and the resulting modified products, for which both traditionally cell biological and advanced protein chemical detection methods have been developed. Especially protein modifications/oxidations are interesting in connection with protein misfolding diseases, since misfolded proteins within the cell may be more prone to oxidative attacks due to their extended structure compared to the compact structured proteins.

6. Reactive Carbonyl Compounds: Creation and Protein Modification

As mentioned earlier, the proteolytic systems in the cytosol (the proteasome) as well as that in the mitochondria (the Lon protease) are able to degrade modified proteins (21, 47). The presence of these systems in young and healthy cells signifies a need to eliminate modified proteins produced continuously as byproducts from "normal" and stress-induced ROS/RNS production. In this connection, it is noteworthy that a protein mass spectrometric analysis of carbonylated proteins from a mitochondria enriched fraction from the hind limb muscle of young (18 month old) female adult rats revealed 243 carbonylated proteins of which 94 were localized to the mitochondria (77) (see below). To set this in perspective, it is of special interest in the present context that the activity of both the proteasome and the Lon protease decline with age (21, 47), indicating that accumulation of numerous modified proteins are important contributors to age-related diseases as well as to the general ageing. This view is compatible with the modern theories of ageing (78, 79) (see also Chap. 7).

Carbonylation of proteins may arise predominantly from two sources, lipid peroxidation and oxidative break-down of glycated proteins (18, 80, 81). Lipid peroxidation occurs when unsaturated fatty acid components of lipids are attacked by nucleophilic ROS and RNS molecules. The primary products are subsequently metabolized through metabolic and detoxification pathways to small molecular compounds, of which the best known are aldehydes, for example, malondialdehyde, hexanal, acrolein, glyoxal, crotonaldehyde, 2-nonenal, 4-oxo-2-nonenal, and 4-hydroxy-2-nonenal. Glycated proteins on the other hand are produced from sugars containing reducing aldehyde groups, such as glucose, which react with free amino groups (arginine and lysine) in proteins to form so-called Schiff bases. Subsequent rearrangements of these bases give rise to advanced glycation endproducts (AGE). If not accumulated or excreted, these AGEs may be oxidatively metabolized to small molecules, such as the dicarbonyls methylglyoxal, glyoxal and 3-deoxyglucosone, as well as the aldehydes diacetyl, acetol, pyruvaldehyde and acrolein (18). This probably unfinished list of carbonyl compounds, which may all react non-enzymatically with proteins, DNA as well as other small molecules, such as ketone bodies (Poulsen T and Johannsen M 2009, unpublished), emphasizes the size of the problem for the cell to cope with.

As mentioned above, a crudely mitochondrial enriched fraction of young rat muscle contained 243 identified carbonylated proteins, of which 94 could be assigned mitochondrial (77). The study was methodological in its nature and designed to evaluate the power of gel-free liquid chromatography coupled to tandem mass spectrometry, especially with respect to the detection of membrane and hydrophobic proteins, which may escape detection by 2-dimensional gel-electrophoresis combined with mass spectrometry (see also Chap. 9). It is interesting to note that many components of the respiratory chain as well as metabolic enzymes, including several components of the TCA cycle, the mitochondrial PQC, and the antioxidant systems, were found to be carbonylated. In a follow-up study, the same group investigated the age-dependent protein carbonylation in slow- and fast-twitch muscles from rats (82). Surprisingly, they found that the glucose oxidation-dependent fast-twitch muscles contained twice as many carbonylated proteins compared to fatty acid oxidation dependent slow-twich muscles, and that many enzymes of the fatty acid oxidation pathway, the respiratory chain as well as the chaperones hsp60 and hsp70 were more carbonylated in aged compared to young muscles. First of all, these results may indicate that small carbonyl compounds produced non-enzymatically from the glycolytic pathway and from AGE contributes to the carbonylation burden. Secondly, they support the notion that the efficiency of the PQC systems decline with age, and thirdly that, since carbonylation in most cases causes inactivation, this mechanism may contribute to the decline in ATP production with age.

A few other studies of carbonylation of multiple proteins have been performed (83). In one study, brain tissue from aged mice was investigated. However, much fewer and mostly high abundance proteins were found carbonylated, for example, actin and tubulin, but also carbonylation of a single TCA cycle enzyme, isocitrate dehydrogenase, as well as SOD2 and protein tyrosine phosphatases, were detected. SOD2 is noteworthy, since decreased activity has been observed in the neuronal model of Alzheimer's disease (25) and in skin fibroblasts from patients with SCAD deficiency (88). Likewise, protein tyrosine phosphatases are interesting since they are involved in the redox regulation (84) and in the regulation of para-catalytic reactions producing ROS (85).

Some other investigations have been designed to answer more focused questions. Of interest in this context is a study where brain tissue from patients with Parkinson's disease was investigated for carbonylated complex I components (86). The study identified carbonylation of 14 components, which constitutes the catalytic core of complex I.

In addition to the study in rat muscles, this study illustrates that carbonylation is a common modification, which – if not eliminated by the appropriate cellular mechanisms – may disturb vital processes and contribute to disease initiation, progression, and end stage as well as to the common ageing.

7. Conclusion and Perspective

The interest in misfolded proteins and cellular stress has emerged from the realization that these pathological mechanisms are involved in a wide variety of diseases, especially through perturbation of the mitochondrial function and creation of oxidative stress, as discussed above as well as in several of the chapters in this book.

In the present chapter, we have tried to give an overview of the diverse cellular systems/mechanisms that are involved in the production and protection of correctly folded proteins as well as in elimination of misfolded and damaged proteins, all orchestrated and executed by molecular chaperones and intracellular proteases, which are also involved in the various UPRs. The mechanisms, including the various components of the UPRs, are fairly well understood although the signalling pathway(s) from the mitochondria is (are) not yet elucidated. What is also not fully understood are the exact mechanisms by which misfolded proteins/aggregates lead to mitochondrial dysfunction, oxidative stress, fragmentation, and various mitophagic/mitoptotic mechanisms, which – as mentioned – seem to be common pathological events. A schematic overview of the mechanisms discussed is given in Fig. 2.

The next goal is to study the mechanisms and interventions that can alleviate the misfolding (Chap. 8). This may be achieved by stimulating the folding or stabilization of misfolded conformations by chemical/pharmacological chaperones, either synthetically produced or extracted, or directly consumed from various foodstuff which may also decrease the amounts of oxidative stress. More knowledge is needed about the mechanisms involved and the technologies used to measure such effects, and it can be foreseen that a continuation of this book will be concerned with these issues.

References

1. Komar AA (2009) A pause for thought along the co-translational folding pathway. Trends Biochem Sci 34:16–24
2. Hartl FU, Hayer-Hartl M (2002) Molecular chaperones in the cytosol: from nascent chain to folded protein. Science 295: 1852–1858
3. Albanese V, Yam AY, Baughman J, Parnot C, Frydman J (2006) Systems analyses reveal two chaperone networks with distinct functions in eukaryotic cells. Cell 124:75–88
4. Esser C, Alberti S, Hohfeld J (2004) Cooperation of molecular chaperones with the ubiquitin/proteasome system. Biochim Biophys Acta 1695:171–188
5. McClellan AJ, Tam S, Kaganovich D, Frydman J (2005) Protein quality control: chaperones culling corrupt conformations. Nat Cell Biol 7:736–741
6. Lindquist S (1986) The heat-shock response. Annu Rev Biochem 55:1151–1191
7. Broadley SA, Hartl FU (2008) Mitochondrial stress signaling: a pathway unfolds. Trends Cell Biol 18:1–4
8. Richter-Landsberg C, Goldbaum O (2003) Stress proteins in neural cells: functional roles

in health and disease. Cell Mol Life Sci 60:337–349
9. Calabrese V, Cornelius C, Mancuso C, Pennisi G, Calafato S, Bellia F, Bates TE, Giuffrida Stella AM, Schapira T, Dinkova Kostova AT, Rizzarelli E (2008) Cellular stress response: a novel target for chemoprevention and nutritional neuroprotection in aging, neurodegenerative disorders and longevity. Neurochem Res 33:2444–2471
10. Morimoto RI (2008) Proteotoxic stress and inducible chaperone networks in neurodegenerative disease and aging. Genes Dev 22: 1427–1438
11. Schubert U, Anton LC, Gibbs J, Norbury CC, Yewdell JW, Bennink JR (2000) Rapid degradation of a large fraction of newly synthesized proteins by proteasomes. Nature 404:770–774
12. Connolly JB (2005) Neurodegeneration caused by the translation of nonsense: does macromolecular misfolding impair the synchrony of gene expression? Med Hypotheses 64:968–972
13. Zou J, Guo Y, Guettouche T, Smith DF, Voellmy R (1998) Repression of heat shock transcription factor HSF1 activation by HSP90 (HSP90 complex) that forms a stress-sensitive complex with HSF1. Cell 94:471–480
14. McClellan AJ, Frydman J (2001) Molecular chaperones and the art of recognizing a lost cause. Nat Cell Biol 3:E51–E53
15. Westerheide SD, Anckar J, Stevens SM Jr, Sistonen L, Morimoto RI (2009) Stress-inducible regulation of heat shock factor 1 by the deacetylase SIRT1. Science 323:1063–1066
16. Pieczenik SR, Neustadt J (2007) Mitochondrial dysfunction and molecular pathways of disease. Exp Mol Pathol 83:84–92
17. Scandalios JG (2005) Oxidative stress: molecular perception and transduction of signals triggering antioxidant gene defenses. Braz J Med Biol Res 38:995–1014
18. Ellis EM (2007) Reactive carbonyls and oxidative stress: potential for therapeutic intervention. Pharmacol Ther 115:13–24
19. Nakamura T, Lipton SA (2007) Molecular mechanisms of nitrosative stress-mediated protein misfolding in neurodegenerative diseases. Cell Mol Life Sci 64:1609–1620
20. Grune T, Jung T, Merker K, Davies KJ (2004) Decreased proteolysis caused by protein aggregates, inclusion bodies, plaques, lipofuscin, ceroid, and "aggresomes" during oxidative stress, aging, and disease. Int J Biochem Cell Biol 36:2519–2530
21. Poppek D, Grune T (2006) Proteasomal defense of oxidative protein modifications. Antioxid Redox Signal 8:173–184
22. Malhotra JD, Kaufman RJ (2007) Endoplasmic reticulum stress and oxidative stress: a vicious cycle or a double-edged sword? Antioxid Redox Signal 9:2277–2293
23. Dedkova EN, Ji X, Lipsius SL, Blatter LA (2004) Mitochondrial calcium uptake stimulates nitric oxide production in mitochondria of bovine vascular endothelial cells. Am J Physiol Cell Physiol 286:C406–C415
24. Wang H, Lim PJ, Karbowski M, Monteiro MJ (2009) Effects of overexpression of huntingtin proteins on mitochondrial integrity. Hum Mol Genet 18:737–752
25. Sompol P, Ittarat W, Tangpong J, Chen Y, Doubinskaia I, Batinic-Haberle I, Abdul HM, Butterfield DA, St Clair DK (2008) A neuronal model of Alzheimer's disease: an insight into the mechanisms of oxidative stress-mediated mitochondrial injury. Neuroscience 153: 120–130
26. Sullivan PG, Dragicevic NB, Deng JH, Bai Y, Dimayuga E, Ding Q, Chen Q, Bruce-Keller AJ, Keller JN (2004) Proteasome inhibition alters neural mitochondrial homeostasis and mitochondria turnover. J Biol Chem 279: 20699–20707
27. Salminen A, Kaarniranta K (2009) Regulation of the aging process by autophagy. Trends Mol Med 15:217–224
28. Lyamzaev KG, Nepryakhina OK, Saprunova VB, Bakeeva LE, Pletjushkina OY, Chernyak BV, Skulachev VP (2008) Novel mechanism of elimination of malfunctioning mitochondria (mitoptosis): formation of mitoptotic bodies and extrusion of mitochondrial material from the cell. Biochim Biophys Acta 1777:817–825
29. Orrenius S, Gogvadze V, Zhivotovsky B (2007) Mitochondrial oxidative stress: implications for cell death. Annu Rev Pharmacol Toxicol 47:143–183
30. Wickner W, Schekman R (2005) Protein translocation across biological membranes. Science 310:1452–1456
31. Kincaid MM, Cooper AA (2007) ERADicate ER stress or die trying. Antioxid Redox Signal 9:2373–2387
32. Meusser B, Hirsch C, Jarosch E, Sommer T (2005) ERAD: the long road to destruction. Nat Cell Biol 7:766–772
33. Schroder M, Kaufman RJ (2005) The mammalian unfolded protein response. Annu Rev Biochem 74:739–789

34. Lin JH, Walter P, Yen TS (2008) Endoplasmic reticulum stress in disease pathogenesis. Annu Rev Pathol 3:399–425
35. Harding HP, Zhang Y, Zeng H, Novoa I, Lu PD, Calfon M, Sadri N, Yun C, Popko B, Paules R, Stojdl DF, Bell JC, Hettmann T, Leiden JM, Ron D (2003) An integrated stress response regulates amino acid metabolism and resistance to oxidative stress. Mol Cell 11:619–633
36. Rutkowski DT, Kaufman RJ (2007) That which does not kill me makes me stronger: adapting to chronic ER stress. Trends Biochem Sci 32:469–476
37. Brookes PS, rley-Usmar VM (2004) Role of calcium and superoxide dismutase in sensitizing mitochondria to peroxynitrite-induced permeability transition. Am J Physiol Heart Circ Physiol 286:H39–H46
38. Jacobson J, Duchen MR (2002) Mitochondrial oxidative stress and cell death in astrocytes – requirement for stored Ca2+ and sustained opening of the permeability transition pore. J Cell Sci 115:1175–1188
39. Pagani M, Fabbri M, Benedetti C, Fassio A, Pilati S, Bulleid NJ, Cabibbo A, Sitia R (2000) Endoplasmic reticulum oxidoreductin 1-lbeta (ERO1-Lbeta), a human gene induced in the course of the unfolded protein response. J Biol Chem 275:23685–23692
40. Uehara T (2007) Accumulation of misfolded protein through nitrosative stress linked to neurodegenerative disorders. Antioxid Redox Signal 9:597–601
41. Calvo S, Jain M, Xie X, Sheth SA, Chang B, Goldberger OA, Spinazzola A, Zeviani M, Carr SA, Mootha VK (2006) Systematic identification of human mitochondrial disease genes through integrative genomics. Nat Genet 38:576–582
42. Pagliarini DJ, Calvo SE, Chang B, Sheth SA, Vafai SB, Ong SE, Walford GA, Sugiana C, Boneh A, Chen WK, Hill DE, Vidal M, Evans JG, Thorburn DR, Carr SA, Mootha VK (2008) A mitochondrial protein compendium elucidates complex I disease biology. Cell 134:112–123
43. Mokranjac D, Neupert W (2009) Thirty years of protein translocation into mitochondria: unexpectedly complex and still puzzling. Biochim Biophys Acta 1793:33–41
44. Tatsuta T, Langer T (2008) Quality control of mitochondria: protection against neurodegeneration and ageing. EMBO J 27:306–314
45. Jendrach M, Mai S, Pohl S, Voth M, Bereiter-Hahn J (2008) Short- and long-term alterations of mitochondrial morphology dynamics and mtDNA after transient oxidative stress. Mitochondrion 8:293–304
46. Ostermann J, Horwich AL, Neupert W, Hartl FU (1989) Protein folding in mitochondria requires complex formation with hsp60 and ATP hydrolysis. Nature 341:125–130
47. Ngo JK, Davies KJ (2007) Importance of the lon protease in mitochondrial maintenance and the significance of declining lon in aging. Ann NY Acad Sci 1119:78–87
48. Gregersen N, Bross P, Vang S, Christensen JH (2006) Protein misfolding and Human disease. Annu Rev Genomics Hum Genet 7:103–124
49. Gregersen N, Bross P, Andresen BS (2004) Genetic defects in fatty acid beta-oxidation and acyl-CoA dehydrogenases. Eur J Biochem 271:470–482
50. Gregersen N, Bolund L, Bross P (2005) Protein misfolding, aggregation, and degradation in disease. Mol Biotechnol 31: 141–150
51. Zhao Q, Wang J, Levichkin IV, Stasinopoulos S, Ryan MT, Hoogenraad NJ (2002) A mitochondrial specific stress response in mammalian cells. EMBO J 21:4411–4419
52. Corydon MJ, Andresen BS, Bross P, Kjeldsen M, Andreasen PH, Eiberg H, Kølvraa S, Gregersen N (1997) Structural organization of the human short-chain acyl-CoA dehydrogenase gene. Mamm Genome 8:922–926
53. Haynes CM, Petrova K, Benedetti C, Yang Y, Ron D (2007) ClpP mediates activation of a mitochondrial unfolded protein response in C. elegans. Dev Cell 13:467–480
54. Horibe T, Hoogenraad NJ (2007) The chop gene contains an element for the positive regulation of the mitochondrial unfolded protein response. PLoS One 2:e835
55. Ma Y, Brewer JW, Diehl JA, Hendershot LM (2002) Two distinct stress signaling pathways converge upon the CHOP promoter during the mammalian unfolded protein response. J Mol Biol 318:1351–1365
56. Sun Y, Oberley LW (1996) Redox regulation of transcriptional activators. Free Radic Biol Med 21:335–348
57. D'Autreaux B, Toledano MB (2007) ROS as signalling molecules: mechanisms that generate specificity in ROS homeostasis. Nat Rev Mol Cell Biol 8:813–824
58. Miao L, St Clair DK (2009) Regulation of superoxide dismutase genes: Implications in disease. Free Radic Biol Med 47:344–56

59. Murphy MP (2009) How mitochondria produce reactive oxygen species. Biochem J 417:1–13
60. Starkov AA (2008) The role of mitochondria in reactive oxygen species metabolism and signaling. Ann NY Acad Sci 1147:37–52
61. Reinecke F, Smeitink JA, van der Westhuizen FH (2009) OXPHOS gene expression and control in mitochondrial disorders. Biochim Biophys Acta 1792:1113–1121
62. St-Pierre J, Buckingham JA, Roebuck SJ, Brand MD (2002) Topology of superoxide production from different sites in the mitochondrial electron transport chain. J Biol Chem 277:44784–44790
63. Wajner M, Latini A, Wyse AT, Dutra-Filho CS (2004) The role of oxidative damage in the neuropathology of organic acidurias: insights from animal studies. J Inherit Metab Dis 27:427–448
64. Schuck PF, Ferreira GC, Moura AP, Busanello EN, Tonin AM, Dutra-Filho CS, Wajner M (2009) Medium-chain fatty acids accumulating in MCAD deficiency elicit lipid and protein oxidative damage and decrease non-enzymatic antioxidant defenses in rat brain. Neurochem Int 54:519–525
65. Keyser B, Muhlhausen C, Dickmanns A, Christensen E, Muschol N, Ullrich K, Braulke T (2008) Disease-causing missense mutations affect enzymatic activity, stability and oligomerization of glutaryl-CoA dehydrogenase (GCDH). Hum Mol Genet 17:3854–3863
66. Gregersen N, Andresen BS, Pedersen CB, Olsen RK, Corydon TJ, Bross P (2008) Mitochondrial fatty acid oxidation defects–remaining challenges. J Inherit Metab Dis 31:643–657
67. Honda S, Hirose S (2003) Stage-specific enhanced expression of mitochondrial fusion and fission factors during spermatogenesis in rat testis. Biochem Biophys Res Commun 311:424–432
68. Okamoto K, Shaw JM (2005) Mitochondrial morphology and dynamics in yeast and multicellular eukaryotes. Annu Rev Genet 39:503–536
69. Bereiter-Hahn J, Voth M, Mai S, Jendrach M (2008) Structural implications of mitochondrial dynamics. Biotechnol J 3:765–780
70. Busch KB, Bereiter-Hahn J, Wittig I, Schagger H, Jendrach M (2006) Mitochondrial dynamics generate equal distribution but patchwork localization of respiratory complex I. Mol Membr Biol 23:509–520
71. Barsoum MJ, Yuan H, Gerencser AA, Liot G, Kushnareva Y, Graber S, Kovacs I, Lee WD, Waggoner J, Cui J, White AD, Bossy B, Martinou JC, Youle RJ, Lipton SA, Ellisman MH, Perkins GA, Bossy-Wetzel E (2006) Nitric oxide-induced mitochondrial fission is regulated by dynamin-related GTPases in neurons. EMBO J 25:3900–3911
72. Kim I, Rodriguez-Enriquez S, Lemasters JJ (2007) Selective degradation of mitochondria by mitophagy. Arch Biochem Biophys 462:245–253
73. Mizushima N, Levine B, Cuervo AM, Klionsky DJ (2008) Autophagy fights disease through cellular self-digestion. Nature 451:1069–1075
74. Scherz-Shouval R, Elazar Z (2007) ROS, mitochondria and the regulation of autophagy. Trends Cell Biol 17:422–427
75. Priault M, Salin B, Schaeffer J, Vallette FM, di Rago JP, Martinou JC (2005) Impairing the bioenergetic status and the biogenesis of mitochondria triggers mitophagy in yeast. Cell Death Differ 12:1613–1621
76. Cho DH, Nakamura T, Fang J, Cieplak P, Godzik A, Gu Z, Lipton SA (2009) S-nitrosylation of Drp1 mediates beta-amyloid-related mitochondrial fission and neuronal injury. Science 324:102–105
77. Meany DL, Xie H, Thompson LV, Arriaga EA, Griffin TJ (2007) Identification of carbonylated proteins from enriched rat skeletal muscle mitochondria using affinity chromatography-stable isotope labeling and tandem mass spectrometry. Proteomics 7:1150–1163
78. Lane N (2003) A unifying view of ageing and disease: the double-agent theory. J Theor Biol 225:531–540
79. Rattan SI (2006) Theories of biological aging: genes, proteins, and free radicals. Free Radic Res 40:1230–1238
80. Esterbauer H, Cheeseman KH, Dianzani MU, Poli G, Slater TF (1982) Separation and characterization of the aldehydic products of lipid peroxidation stimulated by ADP-Fe2+ in rat liver microsomes. Biochem J 208:129–140
81. Thornalley PJ, Langborg A, Minhas HS (1999) Formation of glyoxal, methylglyoxal and 3-deoxyglucosone in the glycation of proteins by glucose. Biochem J 344(Pt 1):109–116
82. Feng J, Xie H, Meany DL, Thompson LV, Arriaga EA, Griffin TJ (2008) Quantitative proteomic profiling of muscle type-dependent and age-dependent protein carbonylation in rat skeletal muscle mitochondria. J Gerontol A Biol Sci Med Sci 63:1137–1152
83. Soreghan BA, Yang F, Thomas SN, Hsu J, Yang AJ (2003) High-throughput proteomic-based

identification of oxidatively induced protein carbonylation in mouse brain. Pharm Res 20: 1713–1720

84. Tonks NK (2005) Redox redux: revisiting PTPs and the control of cell signaling. Cell 121:667–670

85. Bunik VI, Schloss JV, Pinto JT, Gibson GE, Cooper AJ (2007) Enzyme-catalyzed side reactions with molecular oxygen may contribute to cell signaling and neurodegenerative diseases. Neurochem Res 32:871–891

86. Keeney PM, Xie J, Capaldi RA, Bennett JP Jr (2006) Parkinson's disease brain mitochondrial complex I has oxidatively damaged subunits and is functionally impaired and misassembled. J Neurosci 26:5256–5264

87. Schmidt SP, Corydon TJ, Pedersen CB, Bross P, & Gregersen N. (2010) Misfolding of shortchain acyl-CoA dehydrogenase leads to mitochondrial fission and oxidative stress. Mol Genet Metab 100:155–162

88. Pedersen CB, Zolkipli Z, Vang S, Palmfeldt J, Kjeldsen M, Stenbroen V, Schmidt SP, Wanders RJA, Ruiter J, Wibrand F, Tein I, & Gregersen N. (2010) Antioxidant dysfunction – potential risk for neurotoxicity in ethylmalonic aciduria. J Inherit Metab Dis 33:211–222

Chapter 2

Protein Aggregation Diseases: Toxicity of Soluble Prefibrillar Aggregates and Their Clinical Significance

Massimo Stefani

Abstract

Amyloid diseases, the most clinically relevant protein misfolding pathologies due to the high prevalence of some of them in the population, are characterized by the presence, in specific tissues and organs, of fibrillar deposits of specific peptides or proteins. Increasing efforts are presently dedicated at investigating the structural features and the structure-toxicity relation of the soluble oligomeric precursors arising in the path of fibril formation. In fact, it is increasingly recognised that these unstable, dynamic assemblies are remarkably toxic to cells thus featuring these as the main factor responsible for cell impairment in amyloid diseases. This chapter will review shortly the data presently available on the structural and biochemical features of these assemblies, as well as on their biological and clinical significance.

Key words: Amyloid, Amyloid oligomers, Amyloid fibrils, Amyloid cytotoxicity

1. Introduction

Amyloid diseases are, by far, the most clinically relevant protein misfolding pathologies due to the high prevalence of some of them in the population, including type II diabetes mellitus, Alzheimer's, and Parkinson's diseases. They are characterized by the presence, either localized or spread in specific tissues and organs, of fibrillar deposits of specific peptides or proteins (reviewed in (1)). Amyloid fibrils are polymeric assemblies of 1 out of around 20 peptides or proteins, each characteristic of a specific disease or of a group of strictly similar pathological conditions associated with peculiar clinical signs (reviewed in (2)). The presently described amyloid diseases include over 20 familial, sporadic, or transmissible conditions, although other degenerative pathologies with amyloid deposition are increasingly being

Peter Bross and Niels Gregersen (eds.), *Protein Misfolding and Cellular Stress in Disease and Aging: Concepts and Protocols*, Methods in Molecular Biology, vol. 648, DOI 10.1007/978-1-60761-756-3_2, © Springer Science+Business Media, LLC 2010

reported (reviewed in (3)). The best known amyloid diseases affect the brain and the central nervous system (e.g., Alzheimer's, Parkinson's, Huntington, Creutzfeldt-Jakob diseases, and other neurodegenerative conditions such as several ataxias) or peripheral tissues and organs (type-2 diabetes mellitus, several systemic amyloidoses, dialysis-related amyloidosis). In the latter, the deposits are predominantly extracellular, whereas in most neurodegenerative diseases protein fibrillar polymers are found inside the cells, either in the cytoplasm or in the nucleus (4).

The presence in tissue of fibrillar proteinaceous deposits of a specific protein/peptide is a recognized hallmark of any peculiar amyloid condition leading to suggest that a causative link must exist between aggregate deposition and clinical symptoms (the amyloid hypothesis). The latter is presently supported by many biochemical and genetic studies (5–7) although the structural features of the pathogenic aggregated species (mature fibrils, their oligomeric precursors, or both) and the molecular basis of their cytotoxicity are still under intense investigation (8–12). The polypeptides found aggregated in the differing amyloid diseases can display either wild type sequences, as in the sporadic diseases, or be variants resulting from genetic mutations associated with early-onset, familial forms. The existence of the latter has provided significant clues as to the origins of these pathologies indicating the existence of a link between aggregation propensity of the mutant protein and the time of appearance and severity of the clinical signs of a specific disease (13, 14).

In the past, the idea was generally accepted that protein aggregation into amyloid assemblies resulted from unusual conformational changes inherently related to some specific structural features of the peptides and proteins associated with amyloid diseases. This view was challenged in 1998, when it was first reported that, under mild denaturing conditions, two different proteins unrelated to any amyloid disease were able to aggregate in vitro into ordered polymers comparable to the amyloid fibrils grown from disease-associated peptides and proteins (15, 16). Soon after that, a similar behaviour was described for other proteins (17) and subsequently confirmed for an increasing number of natural proteins and peptides as well as for amino acid homopolymers and very short synthetic peptides ((18) and references therein, (19)). These data led some authors to propose that amyloid aggregation is associated with physicochemical properties inherent to the shared covalent peptide backbone of any protein/peptide, at variance with the sequences of their side-chains, whose interactions primarily dictate a protein's specific fold (20). Nevertheless, the properties of the side-chains affect profoundly the propensities of proteins and peptides to generate amyloid assemblies under given conditions and influence the structural details of the latter.

In spite of the inherent propensity to aggregate of any polypeptide chain, only a very limited number of proteins and peptides are found aggregated in specific diseases under normal conditions. Such an apparent paradox can be explained considering the highly cooperative nature of the process by which a protein gains its native structure, and the structural evolutive adaptations aimed at increasing the resistance of natural proteins against aggregation (21). The evolution of the complex mechanisms and quality control of folding also allows natural proteins to escape the tendency to aggregate for significant lengths of time (22). However, the build-up of the precursor, aggregation-prone, species that nucleate rapid aggregate growth, can be triggered by any alteration of a protein's levels or structure (increased synthesis or reduced degradation, presence of specific mutations) or by minor, even subtle, changes in the environmental conditions. The latter can include heat shock, oxidative stress or chemical modifications, alterations of the intracellular macromolecular crowding, presence of suitable surfaces, absence of stabilizing ligands, any impairment of the quality control of protein folding in the cell, and others (23).

Finally, increasing efforts are presently dedicated at investigating the structural features and the structure-toxicity relation of the soluble oligomeric precursors arising in the path of fibril formation. In fact, it is increasingly recognised that these unstable, dynamic assemblies are endowed with the highest toxicity to cells thus featuring these as the main factor responsible for cell impairment in amyloid diseases. This chapter will focus on the structural and biochemical features, as well as the biological and clinical significance, of these assemblies.

2. Soluble Pre-fibrillar Aggregates Are Generated in the Path of Protein Fibrillization In Vitro and In Vivo

Amyloid assemblies are presently considered the main culprits of cell impairment in affected tissues in a number of degenerative diseases including several systemic amyloidoses, type II diabetes mellitus, Alzheimer's, Parkinson's and prion diseases (1–4). The study of the amyloid structure has therefore been a main focus in the investigation of the molecular basis of amyloid diseases. However, although considerable information has recently been gained on the structural features of the ordered β-sheet-rich core of amyloid fibrils and their supramolecular organization (24–26) there is still a severe lack of knowledge on the structural features of fibril precursors. Therefore, a key issue in the investigation of the amyloid structures is the description of either the growth mechanism from their monomeric precursors and the structural features of their intermediates.

Many studies have highlighted the morphological modifications occurring in amyloid assemblies made from different disease-related and disease-unrelated peptides and proteins in the path of fibrillization from unfolded monomers to aggregation nuclei and mature fibrils. Several more-or-less defined steps are involved in such an assembly process (27, 28). EM and AFM studies have imaged the initial presence of transient, unstable, roundish or tubular particles 2.5–5.0 nm in diameter generally enriched in β-structure often called "amorphous aggregates" or sometimes "micelles" (27–30). These species, characterized as protein/peptide oligomers (see later), frequently associate into bead-like chains, small annular rings ("doughnuts" or "pores"), or curvy protofibrils that appear to be precursors of longer protofilaments and mature fibrils (31). Other common and highly organised amyloids appear as large closed rings and ribbons (28–32). Such pre-fibrillar assemblies have been reported for many proteins and peptides associated with disease, including Aβ peptides, α-synuclein, superoxide dismutase, huntingtin, the androgen receptor, amylin, tranthyretin, serum amyloid A, and others (27, 29–32). Similar pre-fibrillar assemblies have also been imaged in the aggregation path of several proteins not associated with amyloid disease, including HypF-N, an SH3 domain, apomyoglobin, stefin-B and others (28, 33–35). Most of these assemblies are considered intermediates in the path of fibrillization, although some of them, such as the small annular oligomers, could be "dead end" products of the process. Finally, soluble oligomers grown in the fibrillization path of several peptides and proteins have been repeatedly imaged also in vivo in cultured cells and in tissues where the monomeric precursors are expressed. These species are presently considered the main responsible for cell/tissue impairment (see later).

A better knowledge of the "early aggregates" preceding the appearance of mature fibrils is considered very important to understand the nature and origins of the pathological properties of amyloid structures associated with disease, particularly with neurodegenerative conditions. However, the intrinsic disordered nature of these assemblies makes it very difficult to get solid data on their structural features. A previous study reporting the generation of a conformational antibody recognising a generic amyloid fibril epitope highlighted for the first time the presence of shared conformational features in amyloid fibrils grown from different peptides and proteins (36). This information was complemented and substantiated by a later report describing antibodies able to recognise specifically amyloid oligomers (37) and amyloid pores (38), but not mature fibrils, grown from differing peptides and proteins. These findings indicate that pre-fibrillar aggregates of different proteins and peptides share common structural features that are different from those displayed by the folded monomers or their fibrillar counterparts; such features can be the result of the exposure to the aqueous environment of hydrophobic patches normally buried into the protein compact structure, thus explaining the intrinsic instability of these assemblies. They also give clues to explain the tendency of these assemblies to interact inappropriately with other cellular components and, accordingly, their toxic potential (see later).

An issue of growing interest is the effect of protein chemical modifications and environmental conditions not only in promoting/hindering protein misfolding and aggregation but also in affecting the structural features of the aggregation nuclei thus targeting them to give rise to fibril precursors and mature fibrils with different structures, stabilities, and cytotoxicities (39–41). Assessing the nature and structural features of the aggregated species directly involved in the impairment of cell physiology and viability is of the outmost importance to decide which amyloid assembly (mature fibrils or their precursors) should be the target of pharmacological research.

3. Protein Pre-fibrillar Aggregates Are Endowed with High Cytotoxicity

Presently, the amyloid hypothesis is supported by a large number of data on many in vivo and in vitro amyloidogenic peptides and proteins indicating a direct cytotoxic effect of amyloid aggregates (4). Until recently, mature amyloid fibrils were considered the key factor responsible for cell damage and tissue impairment since they were the form of the aggregates commonly found in the pathological deposits (extracellular plaques or intracellular inclusions). Therefore, it appeared that the pathogenic features of amyloid diseases resulted from the interaction with cell components of the deposits of the aggregated material. By providing a theoretical frame to understand the molecular basis of these diseases, such a scenario stimulated the exploration of therapeutic approaches to amyloidoses mainly focused at searching molecules able to impair the growth and deposition of fibrillar aggregates.

However, the idea that mature fibrils are the main factor responsible for cell demise in the affected tissues has been recently challenged by an increasing body of experimental data. Actually, it has been repeatedly reported that although in some cases mature amyloid fibrils can impair cell viability (42, 43), most often the pre-fibrillar assemblies grown from Aβ peptides, α-synuclein, amylin, β2-microglobulin, transthyretin, and others are the main or even the sole cytotoxic species (8–12, 33, 44–46). There is growing awareness that the unstable, oligomeric assemblies arising early in the fibrillization path are the most potent toxins to cells stems in particular from a large number of in vitro and in vivo studies on Aβ peptide aggregation and Aβ aggregate neurotoxicity (reviewed in (47)). The idea that neuronal cells do really produce Aβ oligomers and that the latter are responsible for neuronal impairment was first proposed over 10 years ago by the William Klein group (8). However, one of the first evidences that Aβ oligomers are really produced in cultured neuronal cells impairing cell physiology and viability was provided in 2002 by the Dennis Selkoe group working on rat hippocampal cells overexpressing the V717F mutant of APP (10). The Selkoe results have been subsequently confirmed by many studies, progressively leading to assess the nature of those Aβ oligomers as spherical trimer-hexamer to 24mer aggregates (ADDLs, amylospheroids) (8, 48–50), their presence inside AD brains (51, 52), their formation within neuronal cells (53–55), their involvement in physical degeneration of synapses (56) and their ability to impair memory and cognitive function (48, 52).

The idea that pre-fibrillar aggregates are the most highly toxic species leads to consider mature fibrils as inert, harmless deposits of the toxic precursors and hence their growth could be interpreted as a cell defense mechanism; it can also explain the lack of direct correlation between density of fibrillar plaques in the brains of Alzheimer's disease patients and the severity of their clinical symptoms (57). However, in spite of the growing information on the effects of amyloids on cell biochemical and functional features, a unifying model for protein aggregation and aggregate damage under physiological conditions has not yet been proposed for all forms of amyloidoses; moreover, for many of these, no information is currently available on either the identity of the supramolecular assemblies responsible for tissue damage or the molecular mechanism(s) of cell impairment.

A nice confirmation of the role of Aβ oligomers in synaptotoxicity has came very recently from a high resolution array tomography study. The authors imaged a sharp reduction of dendritic spine density in neuronal cells surrounded by amyloid plaques in tissue slices from AD transgenic mice and demonstrated that such a reduction depended strictly on the gradient of Aβ oligomers irradiating from the plaques, strongly suggesting that the latter can be a source of the toxic species (58). This study highlights the importance of amyloid fibrils in amyloid synaptotoxicity not directly but, rather, as providers of toxic species. It also supports other considerations; in fact, recent evidence suggests that amyloid fibrils exhibit molecular recycling with dissociation and re-association of the monomeric and, possibly, oligomeric components (59). In addition, fibrils of the same monomer arising under different conditions can differ in stability and hence in resistance against breakage (60), a process of possible importance in fibril proliferation. Actually, a frequent breakage of the fibrils results in increased free ends favouring not only further protein misfolding but also binding of incoming misfolded monomers, in competition with their oligomerization into toxic assemblies. Accordingly, fibril stability can affect the rate of progression of systemic amyloidoses and the transmission of specific prion strains (60). Finally, it has also been reported that Aβ fibrils grown in vitro can disassemble into oligomers or protofibrils upon interaction with membranes (61). Taken together, these data support the idea that amyloid fibrils, though in general harmless by themselves, can be a source of toxic oligomers; such a behaviour can depend on the type and stability of the fibrils and hence on their growth conditions, as well as on the environment where they are settled. Even though more information on a wider population of fibrils grown from a variety of peptides and proteins is needed to explore the generality of these considerations, as a whole they can provide additional clues to explain the above mentioned lack of relation between amyloid load and severity of the clinical symptoms in AD (57).

The increasing knowledge supporting the cytotoxicity of early aggregates of peptides and proteins either associated or not associated with amyloid disease supports the idea that any amyloid aggregate in its pre-fibrillar organization can be intrinsically toxic to living cells (33, 34). Moreover, it further implies that amyloid cytotoxicity arises from shared characteristics of the supramolecular structure of the aggregates rather than from any specific feature of the amino acid sequences of their parent soluble polypeptides. This concept appears in contrast to the properties of functional proteins, whose native structures and biological functions are specifically determined by their amino acid sequences. Actually, such a remarkable result arises from the conclusion discussed above that the structure, and hence the properties, of conformational species other than the native state of a protein are not directly determined by the specific interactions among side-chains. Several reports indicate that similar aggregates of different peptides and proteins elicit comparable biochemical modifications in the same exposed cells, thus supporting the idea that the cytotoxicity of amyloids grown from different peptides and proteins relies on shared structural features of the amyloid assemblies (37, 62).

Overall, the above considerations led to conclude that any future design of therapeutic interventions against the differing amyloid diseases should be primarily targeted at avoiding the appearance of early aggregates; such a goal can be accomplished either directly or by reducing the load of misfolded molecules populated at equilibrium. In this view, hindering the formation of mature fibrils without reducing protein misfolding or increasing the efficiency of the mechanisms aimed at clearing misfolded proteins and their oligomeric assemblies could be detrimental rather than beneficial.

4. Mechanisms of Toxicity of Protein Soluble Aggregates

In spite of the increasing number of studies appeared in recent years, much must still be learnt on the molecular, biochemical, and biological fundamentals underlying the effects of amyloids on living systems. Although in systemic amyloidoses tissue impairment is believed to arise mainly from the diffusion barrier created by the huge amount of deposited fibrillar material hindering secretion and gas/nutrient absorption (63), it is generally recognised that, at least in most cases, protein-folding variants and their early aggregates are endowed with intrinsic and generic cytotoxicity (33) (see also above). Indeed, it is likely that amyloid cytotoxicity arises from the misfolded nature of the aggregated species and their precursors and from the exposure, in them, of aggregation-prone regions normally buried in the compact native state, including patches of hydrophobic residues and the peptide backbone. Such an idea is supported by the intrinsic instability of pre-fibrillar species enabling them to further organise into more ordered and stable structures and/or to interact with synthetic phospholipid bilayers (64–67) and cell components; the latter include the plasma (32, 45, 68), the ER (69) and the mitochondria (70) membranes and the microtubules (71), where they can accumulate causing dysfunction (72). Any increased load of misfolded molecules and/or their early aggregates can also engulf the ubiquitin/proteasome path of protein degradation with severe outcomes for cell viability (reviewed in (73)).

Increasing information on similar toxic effects raised by early aggregates of peptides and proteins not associated with amyloid disease strongly supports the idea that toxicity may be inherent to some shared structural feature of amyloids (37, 74, 75). Several lines of evidence suggest that such interactions result in membrane disassembly with non-specific permeabilization, impairment of specific membrane-bound protein function (45, 64–68, 76) and penetration of the aggregates inside the cell (46). In particular, the suggestion appears very intriguing that a subpopulation of protofibrils, also referred to as amyloid pores may account for non-specific cell permeabilization by amyloids. Indeed, since 1993, the "channel hypothesis" of amyloid toxicity was proposed suggesting that the toxic aggregated species form non-specific pore-like channels in the membranes of the exposed cells (64).

Early aggregates featuring small annular rings with a central pore have been imaged by AM and AFM in the aggregation paths of many proteins and peptides including the Aβ and ABri peptides, α-synuclein, transthyretin, β2-microglobulin and others (27, 29, 66, 77–80). However, it is not clear whether the above mentioned doughnut-shaped pre-fibrillar aggregates are directly responsible for ion homeostasis impairment in the exposed cells. In fact, modifications of membrane permeability can occur in cells exposed to misfolded monomers or their pre-fibrillar aggregates with no pore-like appearance. In general, it is believed that the interaction of the aggregates or their precursors with the cell membranes is, by itself, able to destabilize the phospholipid bilayer creating disordered areas allowing reduction or destruction of the ion gradients (37, 74, 75). The pore-or-oligomer toxicity dilemma can be overcome in the light of a recent study showing that homogeneous populations of preformed annular Aβ and α-synuclein protofibrils are stable and do not permeabilize phospholipid bilayers at variance with similar assemblies grown from pre-fibrillar oligomers on the bilayer surface (44). It could well be that often oligomer toxicity is mediated by their assembly into pores at the membrane surface, similarly to what happens with bacterial pore-forming toxins (44). These findings also support the idea that annular protofibrils are off-pathway intermediates of fibril growth and that the toxicity of amyloid assemblies is related to their relative stabilities.

The intra- or extracellular presence of toxic aggregates can impair a number of cell functions eventually leading to cell death by apoptosis or necrosis (81–83). However, in most cases initial alterations of fundamental cellular processes associated with membrane perturbation appear to underlie subsequent cell impairment. Increasing information points to a central role of early alterations of the intracellular redox state and free Ca^{2+} levels (see below) following membrane permeabilization by toxic aggregates. Yet, other causes have been proposed that are not necessarily alternative to the Ca^{2+}/ROS dyshomeostasis; these include transcriptional derangement in poly(Q) extension diseases, microtubular transport alterations in poly(Q), SOD, AD and tauopathies, excitotoxicity through deregulation of the NMDA or AMPA receptors, and the cytotoxic effect of pro-inflammatory factors secreted by microglia in AD (81, 85–88).

In general, cells exposed to toxic amyloids display a remarkable increase in the levels of reactive oxygen and nitrogen species (ROS, RNS); the modification of the intracellular redox state can result in overstimulation of excitatory glutamate receptors (89), lipid peroxidation, deregulation of NO metabolism, protein nitrosylation, and up-regulation of heme oxygenase-1 (90). The key role performed by the oxidative stress in amyloid aggregate cytotoxicity is supported by many experimental data and by the protection provided to exposed cells by antioxidants such as tocopherol, lipoic acid, or reduced glutathione (46, 91). This is also supported by data from prion-infected mice showing that increased free radical production with reduced efficacy of the anti-oxidant defenses in the mitochondria results in brain damage (92). Furthermore, recent data point to direct effects of ageing and oxidative stress on the activity and expression levels of the proteasome in the nervous tissue (93). Actually, oxidative proteasome inhibition or its engulfment with oxidised or otherwise damaged or misfolded proteins can result in the accumulation of the latter thus reinforcing the detrimental effects of ROS.

It has been suggested that, in cells exposed to amyloid aggregates, intracellular ROS elevation can be an immediate consequence of the intracellular Ca^{2+} increase following non-specific membrane permeabilization or the activation of glutamate-gated calcium channels. Increased intracellular levels of free Ca^{2+} can stimulate the oxidative metabolism by activating the dehydrogenases of the citric acid cycle in order to provide the ATP needed to the ion pumps to clear the excess Ca^{2+}. The resulting oxidative stress can increase free Ca^{2+} levels by modifying the physicochemical (lipid peroxidation) and/or functional (ion pump inactivation) features of the cell membrane (94, 95) with further Ca^{2+} increase in a self-reinforcing loop. Such a scenario can explain the relationship between ROS, intracellular Ca^{2+} increase, mitochondrial damage and apoptosis described in cells exposed to toxic amyloid aggregates (95–97). A similar chain of events could also occur in the old age, where cells are more susceptible to oxidative stress and their energy load is basically reduced. Actually, many studies support a close link between Alzheimer's, Parkinson's, and prion diseases and calcium homeostasis deregulation. Recent data show that cells exposed to early aggregates of proteins unrelated to amyloid disease display similarly increased ROS and free Ca^{2+} levels with an apoptotic or a necrotic outcome (34, 45, 46, 83) further underscoring the generality of these effects.

It is known that different cell types in tissue are variously affected by amyloids; for instance, AD preferentially affects neurons whereas glial cells remain intact, supporting the idea that synaptic dysfunction can be at the basis of the disease (98). This and other findings strengthen the need to investigate the biochemical features underlying the different vulnerability of varying cell types to the same toxic pre-fibrillar aggregates. A recent report shows significant correlations between cell vulnerability, aggregate interaction with the plasma membrane, cholesterol content, total antioxidant capacity and basal Ca^{2+}-ATPase activity in a panel of cultured cells (99). These data support the importance of the aggregate-cell membrane interaction and the subsequent early modifications of free Ca^{2+} and redox state in triggering the chain of events culminating with cell death; they also highlight the importance of the cell defences against any modification of the free Ca^{2+} and ROS levels provided by the interaction with toxic aggregates. Overall, these data agree with the idea that in most cases aggregate cytotoxicity and its early manifestations are generic features of amyloids.

5. Solution Conditions Can Affect the Structural and Toxicity Features of Protein Soluble Aggregates

Most of the above reported cytotoxicity studies have been carried out mainly by exposing cells to aggregates added to their culture media whereas much less information is currently available on the toxicity of similar aggregates found inside the cell, with the exception of poly(Q) aggregates (38, 84). What is best known is that increasing the intracellular load of misfolded and aggregated proteins besides favouring aggregate nucleation can saturate the ER and cytosolic quality control of protein folding resulting in the activation of the unfolded protein response or the heat-shock response, respectively (73).

Macromolecular crowding is a key, yet poorly considered, factor affecting protein folding, stability, aggregation propensity, and aggregate interaction with cell surfaces (100). The excluded volume effect in the cytosol and the ER can modify the rate of protein folding/aggregation as well as aggregate stability. Actually, it can increase not only the folding of polypeptide chains into compact, globular states but also their aggregation rate (101) possibly favouring closely associated states (compactly folded molecules or mature fibrils) over less structured, loosely folded ones (early aggregates). Furthermore, the crowded intracellular milieu is very different from the cell culture media where toxic amyloid oligomers are delivered and the outer surface of the plasma membrane can be reached more easily than the cytosolic surface. It must also be considered that the leaflets of the cell membranes display a characteristic asymmetry in protein and lipid composition. For instance, lipid rafts are found only in the outer leaflet of the cell membrane whereas anionic phospholipids are present in the inner plasma membrane leaflet in all cells other than cancer and apoptotic cells (102). Accordingly, the inner leaflet displays a remarkable density of negative charge whereas the outer leaflet is more rigid. These and other biochemical and physicochemical asymmetries can modulate monomer generation (in the case of peptides arising from processing of parental membrane proteins), recruitment, and aggregation at the cell membrane; they can also affect aggregate interaction with the latter and their structural reorganization resulting in bilayer disassembly (reviewed in (103)). In addition, the mechanisms and severity of cell impairment can vary in the presence of aggregates displaying different conformations (40, 41, 44, 104, 105) possibly as a consequence of their growth under slightly different cell/ tissue conditions.

6. Conclusions

Significant information has been reported in recent years on the biophysical and molecular determinants of protein folding, misfolding, and aggregation, the structural features of amyloid aggregates and the biochemical and physiologic effects they produce into exposed cells. In particular, the molecular basis of the toxicity of small aggregates appearing during the early stages of protein/peptide fibrillization appears to display shared characteristics in the different systems that have been studied so far, including aggregates grown from disease-associated and disease-unrelated proteins and peptides. Present knowledge depicts a scenario whereby, in most cases, the misfolded monomers or their pre-fibrillar oligomers appear to be the main factor responsible for amyloid cytotoxicity that often can be traced back to a few shared, early biochemical modifications. Presently, amyloid cytotoxicity appears, at least in most cases, to require the interaction of misfolded proteins and their early aggregates with the cell membranes, with alterations of physical, chemical, and biochemical features of the latter. Such interaction triggers complex cascades of biochemical modifications starting with the alteration of the intracellular ion content, redox state and energy load eventually culminating with cell death.

Much must still be learnt about these key points. For example, increased information on the site(s) of a cell where those interactions occur as well as on the existence of amyloid receptors or preferential interaction sites is needed. Nonetheless, the idea is steadily gaining support that the most toxic species are the oligomers arising in the initial steps of protein aggregation. The key role played in most cases by biological surfaces, particularly membranes, in the onset of the chain of events starting with protein misfolding and aggregation and culminating with cell death is also emerging.

Acknowledgements

The author gratefully acknowledges financial support from the Ente Cassa di Risparmio di Firenze and Italian MURST (PRIN Project 2007XY59ZJ_001).

References

1. Stefani M, Dobson CM (2003) Protein aggregation and aggregate toxicity, new insights into protein folding, misfolding diseases and biological evolution. J Mol Med 81:678–699
2. Chiti F, Dobson CM (2006) Protein misfolding, functional amyloid, and human disease. Annu Rev Biochem 75:333–366
3. Stefani M (2004) Protein misfolding and aggregation: new examples in medicine and biology of the dark side of the protein world. Biochim Biophys Acta 1739:5–25
4. Selkoe DJ (2003) Folding proteins in fatal ways. Nature 426:900–904
5. Reilly MM (1998) Genetically determined neuropathies. J Neurol 245:6–13
6. Kelly J (1998) Alternative conformation of amyloidogenic proteins and their multi-step assembly pathways. Curr Opin Struct Biol 8:101–106
7. Dobson CM (2001) The structural basis of protein folding and its links with human disease. Phil Trans R Soc Lond B 356:133–145
8. Lambert MP, Barlow AK, Chromy BA, Edwards C, Freed R, Liosatos M, Morgan TE, Rozovsky I, Trommer B, Viola KL, Wals P, Zhang C, Finch CE, Krafft GA, Klein WL (1998) Diffusible nonfibrillar ligands derived from Aβ-42 are potent central nervous system neurotoxins. Proc Natl Acad Sci USA 95:6448–6453
9. Walsh DM, Hartley DM, Kusumoto Y, Fezoui Y, Condron MM, Lomakin A, Benedek GB, Selkoe DJ, Teplow DB (1999) Amyloid β-protein fibrillogenesis. Structure and biological activity of protofibrillar intermediates. J Biol Chem 274:25945–25952
10. Walsh DM, Klyubin I, Fadeeva JV, Cullen WK, Anwyl R, Wolfe MS, Rowan MJ, Selkoe DJ (2002) Naturally secreted oligomers of amyloid β protein potently inhibit hippocampal long-term potentiation in vivo. Nature 416:535–539
11. Conway KA, Lee S-J, Rochet JC, Ding TT, Williamson RE, Lansbury PT (2000) Acceleration of oligomerization not fibrillization is a shared property of both alpha-synuclein mutations linked to early-onset Parkinson's disease. Implication for pathogenesis and therapy. Proc Natl Acad Sci USA 97:571–576
12. Reixach N, Deechingkit S, Jiang X, Kelly JW, Buxbaum JN (2004) Tissue damage in the amyloidoses: transthyretin monomers and nonnative oligomers are the major cytotoxic species in tissue culture. Proc Natl Acad Sci USA 101:2817–2822
13. Clarke G, Collins RA, Leavitt BR, Andrews DF, Hayden MR, Lumsden CJ, McInnes RR (2000) A one-hit model of cell death in inherited neuronal degeneration. Nature 406: 195–199
14. Perutz MF, Windle AH (2001) Cause of neuronal death in neurodegenerative diseases attributable to expansion of glutamine repeats. Nature 412:143–144
15. Litvinovich SV, Brew SA, Aota S, Akiyama SK, Haudenschild C, Ingham KC (1998) Formation of amyloid-like fibrils by self-association of a partially unfolded fibronectin type III module. J Mol Biol 280:245–258
16. Gujiarro JI, Sunde M, Jones JA, Campbell ID, Dobson CM (1998) Amyloid fibril formation by an SH3 domain. Proc Natl Acad Sci USA 95:4224–4228
17. Chiti F, Webster P, Taddei N, Clark A, Stefani M, Ramponi G, Dobson CM (1999) Designing conditions for in vitro formation of amyloid protofilaments and fibrils. Proc Natl Acad Sci USA 96:3590–3594
18. Chiti F, Bucciantini M, Capanni C, Taddei N, Dobson CM, Stefani M (2001) Solution conditions can promote formation of either amyloid protofilaments or mature fibrils from the HypF N-terminal domain. Protein Sci 10:2541–2547
19. Fändrich M, Dobson CM (2002) The behaviour of polyamino acids reveals an inverse side chain effect in amyloid structure formation. EMBO J 21:5682–5690
20. Dobson CM (2003) Protein folding and misfolding. Nature 426:884–890
21. Monsellier E, Chiti F (2007) Prevention of amyloid-like aggregation as a driving force of protein evolution. EMBO Rep 8:737–742
22. Dobson CM (1999) Protein misfolding, evolution and disease. Trends Biochem Sci 24:329–332
23. Chiti F, Stefani M, Taddei N, Ramponi G, Dobson CM (2003) Rationalization of the effects of mutations on peptide and protein aggregation rates. Nature 424:805–808
24. Serpell LC, Sunde M, Benson MD, Tennent GA, Pepys MB, Fraser PE (2000) The protofilament substructure of amyloid fibrils. J Mol Biol 300:1033–1039
25. Jiménez JL, Nettleton EJ, Bouchard M, Robinson CV, Dobson CM, Saibil HR

(2002) The protofilament structure of insulin amyloid fibrils. Proc Natl Acad Sci USA 99:9196–9201
26. Lührs T, Ri C, Adrian M, Riek-Loher D, Bohrmann B, Döbeli H, Schubert D, Riek R (2005) 3D structure of Alzheimer's amyloid-β(1-42) fibrils. Proc Natl Acad Sci USA 102:17342–17347
27. Quintas A, Vaz DC, Cardoso I, Saraiva MJM, Brito RMM (2001) Tetramer dissociation and monomer partial unfolding precedes protofibril formation in amyloidogenic transthyretin variants. J Biol Chem 276: 27207–27213
28. Relini A, Torrassa S, Rolandi R, Ghiozzi A, Rosano C, Canale C, Bolognesi M, Plakoutsi G, Bucciantini M, Chiti F, Stefani M (2004) Monitoring the process of HypF fibrillization and liposome permeabilization by protofibrils. J Mol Biol 338:943–957
29. Lashuel HA, Petre BM, Wall J, Simon M, Nowak RJ, Walz T, Lansbury PT (2002) α-Synuclein, especially the Parkinson's disease-associated mutants, forms pore-like annular and tubular protofibrils. J Mol Biol 322: 1089–1102
30. Poirier MA, Li H, Macosko J, Cail S, Amzel M, Ross CA (2002) Huntingtin spheroids and protofibrils as precursors in polyglutamine fibrillization. J Biol Chem 277:41032–41037
31. Caughey B, Lansbury PT (2003) Protofibrils, pores, fibrils and neurodegeneration: separating the responsible protein aggregates from the innocent bystanders. Annu Rev Neurosci 26:267–298
32. Lin H, Bhatia R, Lal R (2001) Amyloid β protein forms ion channels: implications for Alzheimer's disease pathophysiology. FASEB J 15:2433–2444
33. Bucciantini M, Giannoni E, Chiti F, Baroni F, Formigli L, Zurdo J, Taddei N, Ramponi G, Dobson CM, Stefani M (2002) Inherent toxicity of aggregates implies a common mechanism for protein misfolding diseases. Nature 416:507–511
34. Sirangelo I, Malmo C, Iannuzzi C, Mezzogiorno A, Bianco MR, Papa M, Irace G (2004) Fibrillogenesis and cytotoxic activity of the amyloid-forming apomyoglobin mutant W7FW14F. J Biol Chem 279:13183–13189
35. Ceru S, Kokalj SJ, Rabzelj S, Skarabot M, Gutierrez-Aguirre I, Kopitar-Jerala N, Anderluh G, Turk D, Turk V, Zerovnik E (2008) Size and morphology of toxic oligomers of amyloidogenic proteins: a case study of human stefin B. Amyloid 15:47–59
36. O'Nuallain B, Wetzel R (2002) Conformational Abs recognizing a generic amyloid fibril epitope. Proc Natl Acad Sci USA 99: 1485–1490
37. Kayed R, Head E, Thompson JL, McIntire TM, Milton SC, Cotman CW, Glabe CG (2003) Common structure of soluble amyloid oligomers implies common mechanisms of pathogenesis. Science 300:486–489
38. Ren P-H, Laucker JE, Kachirskaia I, Heuser JE, Melki R, Kopito RR (2009) Cytoplasmic penetration and persistent infection of mammalian cells by polyglutamine aggregates. Nat Cell Biol 11:219–225
39. Chen Y, Kokholyan N (2005) A single disulfide bond differentiates aggregation pathways of β2-microglobulin. J Mol Biol 354: 473–482
40. Danzer KM, Haasen D, Karow AR, Moussaud S, Habeck M, Giese A, Kretzschmar H, Hengerer B, Kostka M (2007) Different species of α-synuclein oligomers indice calcium influx and seeding. J Neurosci 271: 9220–9232
41. Bravo R, Arimon M, Valle-Delgado JJ, Garcia R, Durany N, Castel S, Cruz M, Ventura S, Fernandez-Busquets X (2008) Sulfated polysaccharides promote the assembly of amyloid $\beta_{1\text{-}42}$ peptide into stable fibrils of reduced cytotoxicity. J Biol Chem 283: 32471–32783
42. Gharibyan AL, Zamotin V, Yanamandra K, Moskaleva OS, Margulis BA, Kostanyan IA, Morozova-Roche LA (2007) Lysozyme amyloid oligomers and fibrils induce cellular death via different apoptotic/necrotic pathways. J Mol Biol 365:1337–13349
43. Novitskaya V, Bocharova OV, Bronstein I, Baskakov IV (2006) Amyloid fibrils of mammalian prion protein are highly toxic to cultured cells and primary neurons. J Biol Chem 281:13828–13836
44. Kayed R, Pensalfini A, Margol L, Sokolov Y, Sarsoza F, Head E, Hall J, Glabe C (2009) Annular protofibrils are a structurally and functionally distinct type of amyloid oligomer. J Biol Chem 284:4230–4237
45. Hirakura Y, Kagan BL (2001) Pore formation by beta-2-microglobulin: a mechanism for the pathogenesis of dialysis-associated amyloidosis. Amyloid 8:94–100
46. Bucciantini M, Calloni G, Chiti F, Formigli L, Nosi D, Dobson CM, Stefani M (2004) Pre-fibrillar amyloid protein aggregates share common features of cytotoxicity. J Biol Chem 279:31374–31382
47. Walsh DM, Selkoe DJ (2004) Oligomers on the brain, the emerging role of soluble

protein aggregates in neurodegeneration. Protein Peptide Lett 11:1–16

48. Cleary JP, Walsh DM, Hofmeister JJ, Shankar GM, Kuskowski MA, Selkoe DJ, Ashe KH (2005) Natural oligomers of the amyloid-β protein specifically disrupt cognitive function. Nat Neurosci 8:79–84
49. Townsend M, Shankar GM, Mehta T, Walsh DM, Selkoe DJ (2006) Effects of secreted oligomers on amyloid β-protein on hippocampal synaptic plasticity: a potent role for trimers. J Physiol 572(2):477–492
50. Chromy BA, Nowak RJ, Lambert MP, Viola KI, Chang L, Velasco PT, Jones BW, Fernandez SJ, Lacor PN, Horowitz P, Finch CE, Krafft GA, Klein WL (2003) Self-assembly of Aβ1-42 into globular neurotoxins. Biochemistry 42:12749–12760
51. Gong Y, Chang I, Viola KI, Lacor PN, Lambert MP, Finch CE, Krafft GA, Klein WI (2003) Alzheimer's disease-affected brain: presence of oligomeric Aβ ligands (ADDLs) suggests a molecular basis for reversible memory loss. Proc Natl Acad Sci USA 100:10417–10422
52. Lesné S, Koh MT, Kotlinek L, Kayed R, Glabe CG, Yang A, Gallagher M, Ashe KH (2006) A specific amyloid-beta protein assembly in the brain impairs memory. Nature 440:352–357
53. Gouras GK, Tsai J, Nasslund J, Vincent B, Edgar M, Checler F, Greefiels JP, Haroutunian V, Buxbaum JD, Xu H, Greengard P, Relkin NR (2000) Intraneuronal Aβ accumulation in human brain. Am J Pathol 156:15–20
54. Hoshi M, Sato M, Matsumoto S, Noguchi A, Yasutake K, Yoshida N, Sato K (2003) Spherical aggregates of β-amyloid (amylospheroid show high neurotoxicity and activate tau protein kinase I/glycogen synthase kinase-3β. Proc Natl Acad Sci USA 100:6370–6375
55. Walsh DM, Tseng BP, Rydel RE, Podlisny MB, Selkoe DJ (2000) The oligomerization of amyloid β-protein begins intracellularly in cell derived from human brain. Biochemistry 39:10831–10839
56. Billings LM, Oddo S, Green KN, McGaugh JL, LaFerla F (2005) Intraneuronal Aβ causes the onset of early Alzheimer's disease-related cognitive deficits in transgenic mice. Neuron 45:675–688
57. Dickson DW (1995) Correlation of synaptic and pathological markers with cognition of the elderly. Neurobiol Aging 16:285–298
58. Koffie RM, Meyer-Luehmann M, Hashimoto T, Adams KW, Mielke ML, Garcia-Alloza M, Micheva KD, Smith SJ, Kim ML, Lee VM, Hyman BT, Spires-Jones TL (2009) Oligomeric amyloid beta associates with postsynaptic densities and correlates with excitatory synapse loss near senile plaques. Proc Natl Acad Sci USA 106:4012–4027
59. Carulla N, Caddy GL, Hall DR, Zurdo J, Gairi M, Feliz M, Giralt E, Robinson C, Dobson CM (2005) Molecular recycling within amyloid fibrils. Nature 436:554–558
60. Smith JF, Knowles TPJ, Dobson CM, MacPhee CE, Welland ME (2006) Characterization of the nanoscale properties of individual amyloid fibrils. Proc Natl Acad Sci USA 103:15806–15811
61. Martins IC, Kuperstein I, Wilkinson H, Maes E, Vambrabant M, Jonckheere W, Van Gelder P, Hartmann D, D'Hooge R, De Strooper B, Schymkowitz J, Rousseau F (2008) Lipids revert inert Abeta amyloid fibrils to neurotoxic protofibrils that affect learning in mice. EMBO J 27:224–233
62. Pellistri F, Bucciantini M, Relini A, Gliozzi A, Robello M, Stefani M (2008) Generic interaction of pre-fibrillar amyloid aggregates with NMDA and AMPA receptors results in free Ca^{2+} increase in primary neuronal cells. J Biol Chem 283:29950–29960
63. Pepys MB (1995) In: Wheaterall DJ, Ledingham JG, Warrel DA (eds) Oxford textbook of medicine, 3rd Edition, Oxford University Press, Oxford, pp 1512–1524.
64. Arispe N, Rojas E, Pollard HD (1993) Alzheimer's disease amyloid beta protein forms calcium channels in bilayer membranes: blockade by tromethamine and aluminium. Proc Natl Acad Sci USA 89: 10940–10944
65. Mirzabekov TA, Lin MC, Kagan BL (1996) Pore formation by the cytotoxic islet amyloid peptide amylin. J Biol Chem 271:1988–1992
66. Kourie JI (1999) Synthetic C-type mammalian natriuretic peptide forms large cation selective channels. FEBS Lett 445:57–62
67. Volles MJ, Lansbury PT (2001) Vesicle permeabilization by protofibrillar α-synuclein: comparison of wild-type with Parkinson's disease linked mutants and insights in the mechanisms. Biochemistry 40:7812–7819
68. Zhu YJ, Lin H, Lal R (2000) Fresh and nonfibrillar amyloid β protein (1-40) induces rapid cellular degeneration in aged human fibroblasts: evidence for AβP-channel-mediated cellular toxicity. FASEB J 14:1244–1254
69. Ferreiro E, Resende R, Costa R, Oliveira C, Pereira CMF (2006) An endoplasmic-reticulum-specific apoptotic pathway is involved in prion and amyloid-beta peptides neurotoxicity. Neurobiol Dis 23:669–678

70. Aleardi AM, Bernard G, Augereau O, Malgat M, Talbot JC, Mazat JP, Letellier T, Dachary-Prigent J, Solaini GC, Rossignol R (2005) Gradual alteration of mitochondrial structure and function by beta-amyloids: importance of membrane viscosity changes, energy deprivation, reactive oxygen species production, and cytochrome c release. J Bioenerg Biomem 37:207–225
71. King ME, Kan H-M, Baas PW, Erisir A, Glabe CG, Bloom S (2006) Tau-dependent microtubule disassembly initiated by prefibrillar β-amyloid. J Cell Biol 175:541–546
72. Lustbader JW, Cirilli M, Lin C, Xu HW, Takuma K, Wang N, Caspersen C, Chen X, Pollak S, Chaney M, Trinchese F, Liu S, Gunn-Moore F, Lue LF, Walker DG, Kuppusamy P, Zewier ZL, Arancio O, Stern D, Yan SS, Wu H (2004) ABAD directly links Abeta to mitochondrial toxicity in Alzheimer's disease. Science 304:448–452
73. Sherman MY, Goldberg AL (2001) Cellular defences against unfolded proteins: a cell biologist thinks about neurodegenerative diseases. Neuron 29:15–32
74. Kayed R, Sokolow Y, Edmonds B, McIntire TM, Milton SC, Hall JE, Glabe CG (2004) Permeabilization of lipid bilayers is a common conformation-dependent activity of soluble amyloid oligomers in protein misfolding diseases. J Biol Chem 279:46363–46366
75. Demuro A, Mina E, Kayed R, Milton SC, Parker I, Glabe CG (2005) Calcium dysregulation and membrane disruption as a ubiquitous neurotoxic mechanism of soluble amyloid oligomers. J Biol Chem 280:17294–17300
76. Kourie JI, Shorthouse AA (2000) Properties of cytotoxic peptide-induced ion channels. Am J Physiol Cell Physiol 278:C1063–C1087
77. Wang L, Lashuel HA, Walz T, Colòn W (2002) Murine apolipoprotein serum amyloid A in solution forms a hexamer containing a central channel. Proc Natl Acad Sci USA 99:15947–15952
78. Chung J, Yang H, de Beus MD, Ryu CY, Cho K, Colòn W (2003) Cu/Zn superoxide dismutase can form pore-like structures. Biochem Biophys Res Commun 312:873–876
79. Srinivasan R, Marchant RE, Zagorski MG (2004) ABri peptide associated with familial British dementia forms annular and ring-like protofibrillar structures. Amyloid 11:10–13
80. Vendrely C, Valadie H, Bednarova L, Cardin L, Pasdeloup M, Cappadoro J, Bednar J, Rinaudo M, Jamin M (2005) Assembly of the full-length recombinant mouse prion protein I. Formation of soluble oligomers. Biochim Biophys Acta 1724:355–366
81. Morishima Y, Gotoh Y, Zieg J, Barrett T, Takano H, Flavell R, Davis RJ, Shirasaki Y, Greenberg ME (2001) Beta-amyloid induces neuronal apoptosis via a mechanism that involves the c-Jun N-terminal kinase pathway and the induction of Fas ligand. J Neurosci 21:7551–7560
82. Velez-Pardo C, Arroyave ST, Lopera F, Castano AD, Jimenez Del Rio M (2001) Ultrastructure evidence of necrotic neural cell death in familial Alzheimer's disease brains bearing presenilin-1 E280A mutation. J Alzheimer's Dis 3:409–415
83. Bucciantini M, Rigacci S, Berti A, Pieri L, Cecchi C, Nosi D, Formigli L, Chiti F, Stefani M (2005) Patterns of cell death triggered in two different cell lines by HypF-N pre-fibrillar aggregates. FASEB J 19:437–439
84. Ross CA (2002) Polyglutamine pathogenesis: emergence of unifying mechanisms for Huntington's disease and related disorders. Neuron 35:819–822
85. Hsieh H, Boehm J, Sato C, Iwatsubo T, Tomita T, Sisodia S, Malinow R (2006) AMPAR removal underlies Aβ-induced synaptic depression and dendritic spine loss. Neuron 52:831–843
86. De Felice FG, Velasco PT, Lambert MP, Viola K, Fernandez SJ, Ferreira ST, Klein WL (2007) Aβ oligomers induce neuronal oxidative stress through an N-methyl-d-aspartate receptor-dependent mechanism that is blocked by the Alzheimer drug memantine. J Biol Chem 282:11590–11601
87. Chevalier-Larsen E, Holzbaur ELF (2006) Axonal transport and neurodegenerative disease. Biochim Biophys Acta 1762:1094–1108
88. Brandt R, Hundelt M, Shahani N (2004) Tau alteration and neuronal degeneration in tauopathies: mechanisms and models. Biochim Biophys Acta 1739:331–354
89. Shen Y, He P, Zhong Z, McAllister C, Lindholm K (2006) Distinct destructive signal pathways of neuronal death in Alzheimer's disease. Trends Mol Med 12:574–579
90. Choi YG, Kim JL, Lee HP, Jin JK, Choi EK, Carp RI, Kim YS (2000) Induction of heme oxygenase-1 in the brain of scrapie-infected mice. Neurosci Lett 11:173–176
91. Zhang L, Xing Gq, Barker JL, Chang Y, Maric D, Ma W, Li B-s, Rubinow DR (2001) α-Lipoic acid protects rat cortical neurons against cell death induced by amyloid and hydrogen peroxide through the Akt signalling pathway. Neurosci Lett 312:125–128
92. Lee DW, Sohn HO, Lim HB, Lee YG, Kim YS, Carp RJ, Wisnievski HM (1999)

Alteration of free radical metabolism in the brain of mice infected with scrapie agent. Free Rad Res 30:499–507

93. Keller JN, Huang FF, Markesbery WR (2002) Decreased levels of proteasome activity and proteasome expression in aging spinal cord. Neuroscience 98:149–156
94. Butterfield AD, Drake J, Pocernich C, Castegna A (2001) Evidence of oxidative damage in Alzeimer's disease brain: central role for amyloid β-peptide. Trends Mol Med 7:548–554
95. Varadarajan S, Yatin S, Aksenova M, Butterfield DA (2000) Alzheimer's amyloid β-peptide-associated free radical oxidative stress and neurotoxicity. J Struct Biol 130:184–208
96. Squier TC (2001) Oxidative stress and protein aggregation during biological aging. Exp Gerontol 36:1539–1550
97. Kawahara M (2004) Disruption of calcium homeostasis in the pathogenesis of Alzheimer's disease and other conformational diseases. Curr Alz Res 1:87–95
98. Lacor PN, Buniel MC, Chang L, Fernandez SJ, Gong Y, Viola KL, Lambert M, Velasco PT, Bigio EH, Finch CE, Krafft G, Klein WI (2004) Synaptic targeting by Alzheimer's-related amyloid β oligomers. J Neurosci 24:10191–10200
99. Cecchi C, Baglioni S, Fiorillo C, Pensalfini A, Liguri G, Nosi D, Rigacci S, Bucciantini M, Stefani M (2005) Insights into the molecular basis of the differing susceptibility of varying cell types to the toxicity of amyloid aggregates. J Cell Sci 118:3459–3470
100. Ellis RJ (2001) Macromolecular crowding: an important but neglected aspect of the intracellular environment. Curr Opin Struct Biol 11:114–119
101. van den Berg B, Ellis J, Dobson CM (1999) Effects of macromolecular crowding on protein folding and aggregation. EMBO J 18:6927–6933
102. Ran S, Thorpe PE (2002) Phosphatidylserine is a marker of tumour vasculature and a potential target for cancer imaging and therapy. Int J Radiat Oncol Biol Phys 54: 1479–1484
103. Stefani M (2008) Protein folding, misfolding and aggregation on surfaces. Int J Mol Sci 9:2515–2542
104. Deshpande A, Mina E, Glabe C, Busciglio J (2006) Different conformations of amyloid β induce neurotoxicity by distinct mechanisms in human cortical neurons. J Neurosci 26: 6011–6018
105. Nekooki-Machida Y, Kurosawa M, Nukina N, Ito K, Tanaka M (2009) Distinct conformations of in vitro and vivo amyloids of huntingtin-exon 1 show different cytotoxicity. Proc Natl Acad Sci USA 106:9679–9684

Chapter 3

Consequences of Stress in the Secretary Pathway: The ER Stress Response and Its Role in the Metabolic Syndrome

Martin Schröder and Louise Sutcliffe

Abstract

The unfolded protein response (UPR) was originally identified as a signaling network coordinating adaptive and apoptotic responses to accumulation of unfolded proteins in the endoplasmic reticulum (ER). More recent work has shown that UPR signaling can be triggered by a multitude of cellular events and that the UPR plays a critical role in the prevention, and also the progression, of a wide variety of diseases. Much attention has been paid to the role of the UPR in neurodegenerative diseases in the past. More recently, important roles for the UPR in diseases associated with the metabolic syndrome have been discovered. Here we review the role of the UPR in these diseases, including type 2 diabetes, atherosclerosis, fatty liver disease, and ischemia.

Key words: Apoptosis, Conformational disease, Diabetes, Endoplasmic reticulum, Inflammation, Neurodegenerative disease, Unfolded protein response

1. Introduction

The mammalian endoplasmic reticulum (ER) is the site for the synthesis, folding, posttranslational modification, and quality control of most secretory and transmembrane proteins, an intracellular Ca^{2+} store involved in Ca^{2+} signaling, the site for synthesis of phospholipids and other lipids such as cholesterol, and the site for detoxification of xenobiotic substances. Recently, a role for the ER in cellular signaling leading to apoptosis, inflammation, and activation of immune responses has been identified. Perturbation of any of these functions causes ER stress which is characterized by an accumulation of unfolded proteins in the ER (1). The different functions of the ER are tightly interconnected. Perturbation of one function affects another (Fig. 1). In obese

Peter Bross and Niels Gregersen (eds.), *Protein Misfolding and Cellular Stress in Disease and Aging: Concepts and Protocols*, Methods in Molecular Biology, vol. 648, DOI 10.1007/978-1-60761-756-3_3, © Springer Science+Business Media, LLC 2010

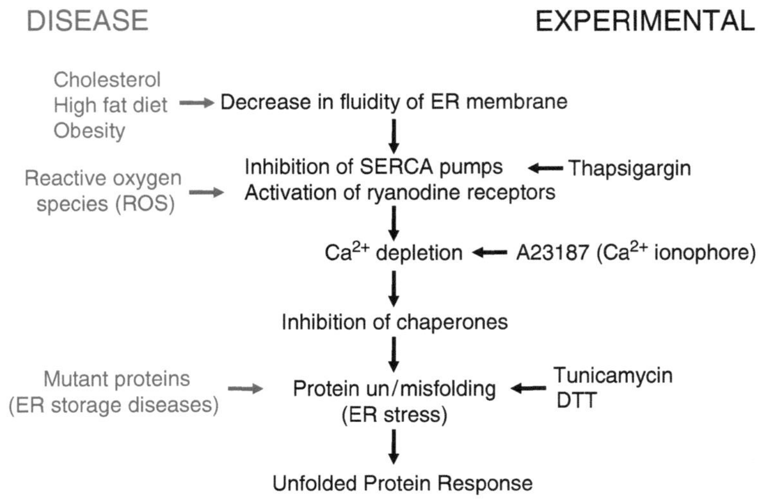

Fig. 1. Events causing ER stress and UPR activation.

individuals increased cholesterol and saturated fatty acid levels decrease the fluidity of the ER membrane, leading to inhibition of SERCA Ca^{2+}-ATPases, depletion of ER luminal Ca^{2+} stores, inhibition of ER-resident molecular chaperones, and the accumulation of unfolded proteins in the ER. Cytokine action on pancreatic β-cells generates nitric oxide, a reactive nitrogen species, which depletes ER luminal Ca^{2+} stores by irreversibly nitrosylating thiol groups in ryanodine Ca^{2+} release channels in the ER membrane (2), inhibiting protein folding chaperone machineries in the ER, triggering apoptotic UPR signaling, β-cell death, and insulin-dependent (type 1) diabetes. These two examples illustrate the importance of the ER and the complexity of positive feedback loops involved in ER stress signaling, progression, and prevention of disease.

2. The UPR

Central to ER homeostasis is a signaling network activated by unfolded proteins in the ER, the unfolded protein response (UPR). The UPR maintains homeostasis of the ER by coordinating the folding demand imposed on the ER-resident protein folding machinery by unfolded proteins with the capacity of this machinery by increasing expression of molecular chaperones and foldases of the ER, stimulating phospholipid synthesis, ER-associated protein degradation (ERAD) and autophagy, selectively degrading mRNAs encoding secretory proteins, activation of an antioxidant response, and attenuating general

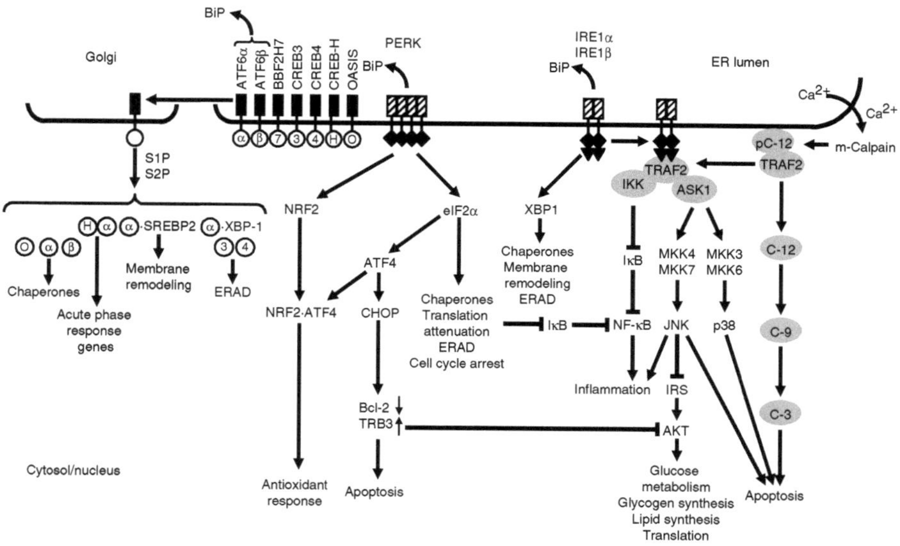

Fig. 2. Principal signal transduction pathways in the mammalian UPR. Reprinted in modified form from Fig. 6 published in Schröder (1), copyright 2007, Birkhäuser Basel, with kind permission of Springer Science and Business.

translation and transcription of genes encoding secretory proteins (Fig. 2). In addition, the UPR activates inflammatory and apoptotic signaling pathways (Fig. 3) and signaling pathways involved in immune responses.

Folding stress signals are transduced across the ER membrane by several transmembrane proteins, type 2 transmembrane transcription factors such as ATF6α and ATF6β, the protein kinase PERK, and the protein kinases–endoribonucleases IRE1α and IRE1β. Two competing models for how these proteins sense ER stress are discussed in the literature. In the competition model, the ER stress sensors are kept in an inactive form through interaction with ER luminal chaperones, especially the HSP70 class chaperone BiP/GRP78 (3). Accumulation of unfolded proteins in the ER sequesters BiP away from the ER luminal domains of the ER stress sensors, leading to their activation. In IRE1 and PERK BiP release unmasks dimerization motifs, in ATF6α sequences mediating transit of ATF6 to the Golgi complex (4, 5). While widely accepted, this model cannot explain several experimental observations. Most importantly, yeast Ire1p deleted for all BiP binding sites was still activated by ER stress, and remained inactive in the absence of ER stress (6). Solution of the crystal structure of the core region of the ER luminal domain of Ire1p revealed an MHC-like peptide-binding pocket in an Ire1p dimer (7). Consequently, it was proposed that direct interaction of the ER luminal domain with unfolded proteins induces oligomerization of Ire1p. In in vitro aggregation assays, this core region displayed chaperone activity,

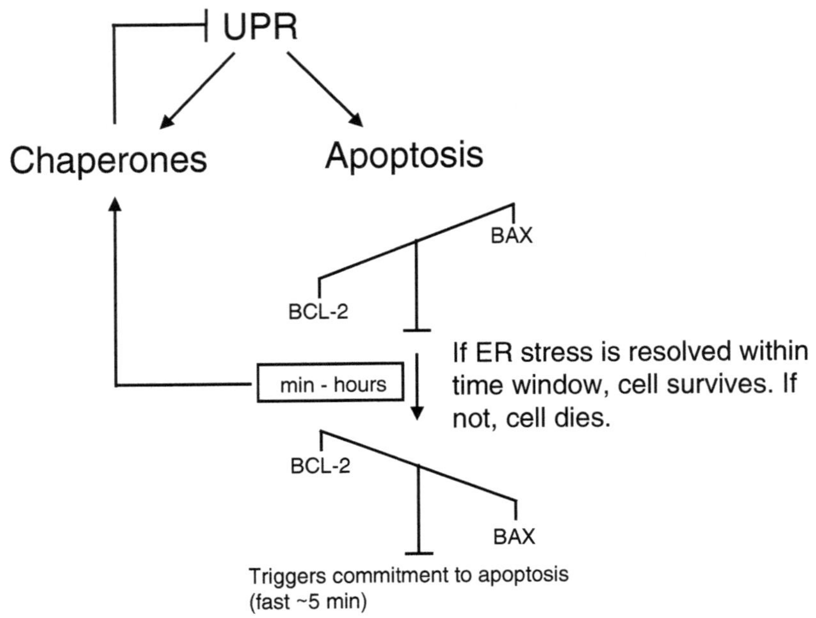

Fig. 3. Model for the decision-making process between cell survival and death by the UPR. ER stress sensors, such as IRE1, ATF6, and PERK, activate chaperone expression (protective) and apoptotic signaling pathways simultaneously. Over time, apoptotic UPR signaling tilts the balance of anti- and proapoptotic BCL-2 proteins toward an excess of proapoptotic BCL-2 proteins, committing the cell to apoptosis. If in this time window activation of chaperone and ERAD systems by the UPR has remedied the cause for ER stress, and as a consequence turned off the UPR, the cell survives. Apoptosis ensues if the UPR cannot relieve ER stress in the time required to tilt the BCL-2 protein balance in favor of pro-apoptotic BCL-2 proteins.

which could be destroyed by introducing ER stress activation-impaired mutations (8). These data suggest that Ire1p is activated by directly interacting with unfolded proteins. However, this peptide-binding pocket is oriented toward the ER membrane. In mammalian IRE1α, this pocket is too narrow for peptide binding and access to this pocket is blocked by an α-helix (9), arguing against a direct role for this pocket in unfolded protein binding.

Once activated, ATF6 translocates to the Golgi complex where its cytosolic bZIP transcription factor domain is proteolytically released by sequential cleavage by site-specific proteases 1 and 2 (S1P and S2P) (10). The cytosolic domain of ATF6 including its bZIP and transcriptional activation domains translocates to the nucleus where it activates transcription of ER chaperone genes, genes involved in ERAD, phospholipid biosynthetic genes, and acute phase response genes. Other type 2 transmembrane bZIP transcription factors that contribute to the ER stress response are BBF2H7, CREB3, CREB4, CREB-H, and OASIS. An ATF6·CREB-H heterodimer activates acute phase and inflammatory genes (11). PERK signals via phosphorylation of at least two proteins, the translation initiation factor eIF2α (12, 13) and the bZIP transcription factor NRF2 (14, 15). Phosphorylation of eIF2α attenuates general translation and serves to decrease the influx of nascent, unfolded polypeptide chains into the ER. Translational attenuation also clears short-lived proteins from the cell, including D-type cyclins and the NF-κB inhibitory IκB proteins. Loss of D-type cyclins causes cell cycle arrest in G_1 (16). Depletion of IκB activates the transcription factor NF-κB (17) and subsequently innate immune response, inflammatory, and antioxidant genes (18, 19). eIF2α phosphorylation promotes translation of mRNAs containing several short upstream open reading frames in their 5′ untranslated region, for example, the mRNA for the transcription factor ATF4 (20). ATF4, in concert with NRF2, activates an antioxidant response, induces the inhibitor of mRNA 5′-cap-binding protein 4E-BP1 (21), and the proapoptotic transcription factor CHOP (22). Induction of 4E-BP1 contributes to translational arrest. CHOP induction is countered by induction of antioxidant response genes, such as glutathione *S*-transferase and heme-oxygenase 1 (23), and stimulation of translation of cIAP1/hIAP2 mRNA (24, 25), encoding an inhibitor of apoptosis. Translational attenuation during ER stress is transient. eIF2α phosphorylation is countered by induction of *GADD34*, encoding a regulatory subunit of protein phosphatase 1 (PP1) that directs PP1 toward phosphorylated eIF2α (26). *gadd34*$^{-/-}$ cells are protected from ER stress-induced cell death, suggesting that early recovery from translational arrest contributes to apoptotic cell death (27), possibly via the activation of apoptotic signaling in response to elevated ER stress.

IRE1α initiates non-spliceosomal splicing of the mRNA for the bZIP transcription factor XBP-1 (28–31). XBP-1 contributes to full induction of many chaperone and protein foldase genes such as *BiP*, *GRP94*, *p58*^{IPK}, *ERdj4*, *ERO1-Lα*, -β, and *ERP72* (32, 33). XBP-1 increases the activity of phosphocholine cytidyltransferase (34), the rate-limiting enzyme for phosphatidylcholine synthesis. An XBP-1·ATF6 heterodimer induces genes involved in ERAD (35, 36). IRE1 also triggers inflammatory and apoptotic signaling via activation of MAP kinase modules, leading to the activation of the MAP kinases JNK and p38. Activation of these MAP kinases by IRE1 requires its interaction with the adaptor protein TRAF2 (37). TRAF2 is a member of the TRAF protein family, which encompasses six different proteins in humans. TRAF proteins contain a *C*-terminal TRAF domain which mediates their interactions with transmembrane receptors. The *N*-terminal coiled-coil portion of the TRAF domain is required for homo- and heterotrimer formation. The *N*-terminal effector domains of TRAF2, a RING domain which may possess ubiquitin ligase activity, and five Zn^{2+}-fingers are required for the activation of JNK and NF-κB by TRAF2 (38). In the UPR, TRAF2 activates the IκB kinase IKK (39) and the MAP kinase kinase kinase ASK1 (40), which activates p38 kinase and JNK via the MAP kinase kinases MKK3/6 and MKK4/7, respectively. Potentiation of activity of the transcription factor c-Jun by JNK phosphorylation induces proapoptotic genes such as *BIM* and *FasL* (41). JNK activates the proapoptotic BCL-2 proteins BIM and BMF by phosphorylation (42, 43). Activation of the JNK pathway contributes to the development of insulin resistance in ER stressed cells (cf. below). In mice, clustering and activation of caspase-12 is triggered by sequestration of TRAF2 by IRE1 (44). Cleavage of procaspase-9 by caspase-12 activates a caspase cascade resulting in activation of the executioner caspase caspase-3 and apoptotic cell death (45). In humans, caspase-4 substitutes for caspase-12 as humans lack a functional *CASPASE-12* gene (46, 47).

3. The UPR in the Metabolic Syndrome

The metabolic syndrome is caused by the combination of a sedentary life style and a calorie-rich diet. Hyperlipidemia, glucose intolerance, hypertension, visceral adiposity, and obesity are early manifestations of the metabolic syndrome, which, in combination, are a major risk factor for type 2 diabetes and cardiovascular disease (Fig. 4). Leptin and insulin resistance are two key intermediary stages in the metabolic syndrome. Persistent insulin resistance causes β-cell failure giving rise to type 2 diabetes. Nonalcoholic fatty liver disease (NAFLD), which develops from benign fatty liver (hepatosteatosis) to inflammation, fibrosis, and hepatocyte injury (steatohepatitis) and liver cirrhosis, liver failure, and hepatocellular carcinoma, is another manifestation of the metabolic syndrome. Inherited hyperhomocyst(e)inemia or hyperhomocysteinemia caused by chronic alcohol abuse or as a consequence of obesity (48) is a risk factor for fatty liver disease. Alcoholic fatty liver disease (AFLD) closely resembles NAFLD. High plasma lipid, glucose, and homocysteine levels are risk factors for atherosclerosis. Thrombosis, a consequence of atherosclerosis, causes loss of blood supply to tissues (ischemia) and ultimately stroke and myocardial infarction. In the following sections, we will review evidence pointing toward ER stress being a cellular event at the heart of the metabolic syndrome and its clinical manifestations.

3.1. Leptin Resistance

The protein hormone leptin signals satiety and energy expenditure by action on two neuronal populations in the actuate nuclei of the hypothalamus (reviewed in (49, 50)). After binding to the alternative splice variant b of the leptin receptor (LRb) leptin stimulates synthesis of the hormone pro-opiomelanocortin

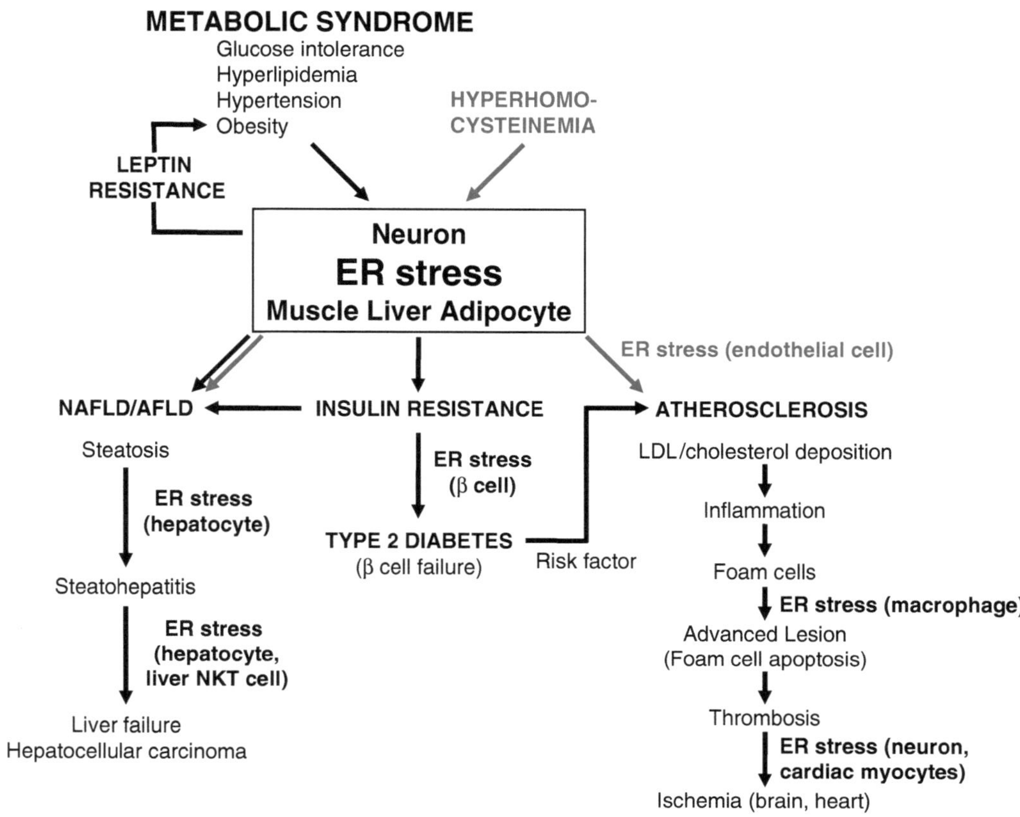

Fig. 4. Roles of ER stress in progression of the metabolic syndrome and of hyperhomocysteinemia to diseases. ER stress plays a central role in the onset of non-alcoholic fatty liver disease (NAFLD), alcoholic fatty liver disease (AFLD), type 2 diabetes, and atherosclerosis in hyperhomocysteinemia. Legend: *Black arrows* – Progression of the metabolic syndrome to disease, *gray arrows* – progression of hyperhomocysteinemia to disease.

(POMC), which is processed to the anorexic (appetite-decreasing) α-melanocyte-stimulating hormone (αMSH) in LRb/POMC neurons. At the same time, leptin, by acting through LRb, inhibits synthesis of the orexigenic (appetite-stimulating) peptide hormones neuropeptide Y (NPY) and agouti-related peptide (AgRP) by NPY/AgRP neurons. At the molecular level, the unliganded, dimeric LRb is associated with the protein tyrosine kinase JAK2. Conformational changes in the LRb upon leptin binding trigger *trans*-autophosphorylation and tyrosine phosphorylation of LRb by JAK2. As a consequence SRC homology 2 (SH2) domain containing proteins such as the transcription factor STAT3 and the tyrosine phosphatase, SHP-2 bind to tyrosine phosphorylated LRb. Tyrosine phosphorylation of STAT3 triggers nuclear translocation and subsequent transcriptional activation of STAT3 target genes. Tyrosine phosphorylation of insulin receptor (IR) substrate (IRS) proteins activates phosphoinositide (PI) 3-kinase (PI3K) signaling (cf. below) promoting growth, cell division, and energy expenditure. Protein tyrosine phosphatase 1B (PTP1B) and suppressor of cytokine signaling (SOCS)-3 are negative regulators of leptin signaling. SOCS3 binds to tyrosine 985 of LRb and JAK2 and inhibits LRb signaling via STAT3.

ER stress caused by increased circulating fatty acid and homocysteine levels may interfere with leptin signaling at several levels. Leptin is produced in adipose tissue. To exert its effects on hypothalamic neurons, leptin crosses the blood brain barrier using a saturable transport mechanism which may involve soluble splice variants of the LR. Secretion of these soluble LR splice variants may be inhibited by ER stress. Activation of NF-κB by ER stressed hypothalamic neurons induces expression of SOCS3, inhibiting LRb signaling (51). ER stress also activates PTP1B which inhibits leptin signaling (52). Serine phosphorylation of IRS proteins by JNK may inhibit their tyrosine phosphorylation by JAK2, again attenuating LRb signaling.

3.2. Insulin Resistance

Insulin resistance is characterized by the inhibition of glucose uptake by muscle cells for glycogen synthesis (53), hepatic glucose secretion even when blood glucose levels are already high (54), and inhibition of negative control of the hormone-sensitive lipase by insulin in adipocytes (55). At the molecular level, insulin resistance interferes with signaling by the IR (reviewed in (56–58)). A conformational change in the IR triggered by binding of insulin to the IR activates its protein tyrosine kinase activity. The activated IR autophosphorylates itself and tyrosine phosphorylates IRS1-4 proteins, and several SH-2 domain containing (SHC) proteins. Several proteins containing SH-2 domains including PI3K are activated by recruitment to tyrosine phosphorylated IRS and SHC proteins. Activated PI3K converts PI-3-phosphate to PI-3,4-bisphosphate and PI-3,4,5-trisphosphate which recruit protein kinase B (PKB/AKT) isoforms and phosphoinositide-dependent kinases (PDK) 1 and 2 to the plasma membrane. There PDKs activate PKB1, -2, and -3 by phosphorylation. Activated PKB controls several cellular events, such as protein and glycogen synthesis, glucose transport, and cell survival and proliferation.

ER stress interferes with insulin signaling through at least three mechanisms (Fig. 4). Activation of JNK by IRE1α results in serine phosphorylation of IRS proteins, which inhibits IRS tyrosine phosphorylation by the IR (59), recruitment of PI3K to IRS proteins (60), and stimulates degradation of IRS1 (61). Induction of BiP, elevated eIF2α, and PERK phosphorylation in mouse models of obesity (59) show that ER stress is an early molecular hallmark of the metabolic syndrome. In cultured cells, saturated fatty acids induce *XBP-1* splicing (62, 63), suggesting that elevated circulating saturated fatty acid levels in obese individuals (64) cause ER stress. Elevated plasma homocysteine in obese patients is the second cause for ER stress in obesity (48).

ER stress also elevates ROS levels in cells (65, 66), which inhibit JNK phosphatases by oxidizing catalytic cysteines (67), leading to the accumulation of phosphorylated, activated JNK. TRB3, whose expression is induced by CHOP (68, 69), binds to and inhibits AKT (70). XBP-1 positively controls expression of several lipogenic genes such as diacylglycerol-*O*-acyltransferase (*DGAT2*), acetyl CoA carboxylase β (*ACC2*), and stearoyl-CoA desaturase 1 (*SCD1*) and is required for hepatic triglyceride secretion (71). In enterocytes, IRE1β mediates degradation of the mRNA for microsomal triglyceride transfer protein (MTP) (72), which is required for assembly of chylomicrons in the ER. *ire1β*$^{-/-}$ mice display hyperlipidemia (72), suggesting that IRE1β contributes to lipid homeostasis and plays a positive role in prevention of the metabolic syndrome. However, *IRE1β* mRNA levels are decreased by a cholesterol-rich or high-fat diet (72). Thus, stimulation of lipid synthesis by an activated UPR may lead to self-perpetuating ER stress and UPR activation and chronic, ever-worsening insulin resistance.

3.3. Type 2 Diabetes

Type 2 diabetes is characterized by a ~60% loss in β cells in islets, accumulation of islet amyloid, and increased β cell apoptosis (73). ER stress contributes to β cell failure and apoptosis in type 2 diabetes (Fig. 4). β cells respond to insulin resistance by increasing their insulin secretion. As insulin resistance becomes more severe and ever more insulin is required to stimulate IR receptor signaling, β cells cross a threshold of insulin synthesis that triggers ER stress, ultimately leading to β cell apoptosis. Deletion of CHOP protects β cells from apoptosis, but also exacerbates diet or genetically induced obesity (74, 75). Mouse models with mutations in the proinsulin gene preventing proper folding of proinsulin are associated with ER stress, apoptotic loss of β cells, and development of diabetes (76–81). A second major secretory client protein of β cells is islet amyloid precursor protein (IAPP). Accumulation of IAPP oligomers in β cells contributes to β cell failure (82, 83). Human, but not rodent, IAPP forms nonselective ion channels and disrupts membrane function (83, 84). ER stress, through depletion of chaperones, may contribute to aggregation of IAPP in β cells. IAPP oligomers may act in similar ways as neurodegenerative amyloids, that is, amyloid β, to cause ER stress (85) and β cell death.

3.4. Non-alcoholic Fatty Liver Disease

The early stage of NAFLD is hepatosteatosis. Elevated plasma free fatty acid and homocysteine levels cause ER stress, leading to insulin resistance. Insulin resistance increases lipid secretion by adipocytes, exacerbating ER stress (Fig. 1). Hepatosteatosis is associated with markers of ER stress such as *XBP-1* splicing and *BiP* induction (63), suggesting that a decrease in the fluidity of the ER membrane interferes with function of critical ER membrane proteins, such as the SERCA Ca^{2+} pumps, causing ER stress. *XBP-1* splicing and eIF2α phosphorylation have been linked to the regulation of lipid metabolism in the liver. XBP-1 stimulates lipid and triglyceride synthesis in the liver (71). *gcn2*$^{-/-}$ mice develop liver steatosis caused by decreased lipid mobilization and elevated expression of lipogenic genes such as SREBP1-c and fatty acid synthase (FAS) upon leucine deprivation (86). Enforced expression of GADD34 in liver decreased eIF2α phosphorylation and liver glycogen pools, improved glucose tolerance, and diminished steatosis in mice fed a high-fat diet. Decreased expression of the lipogenic gene peroxisome proliferator-activated receptor γ and its activators, the genes encoding the bZIP transcription factors *C/EBPα* and *C/EBPβ* in the livers of transgenic mice expressing the C-terminal fragment of GADD34 from the liver-specific albumin promoter are the likely cause for decreased steatosis in these animals (87, 88). Whether initial steatosis leading to ER stress or whether ER stress leading to steatosis, followed in each case by self-enforcing loops, is the initiating event in NAFLD is not easy to distinguish. Tunicamycin injection into lean, healthy mice caused hepatosteatosis, suggesting that ER stress is the priming event in NAFLD (Fig. 4) (89). In steatosis, ER stress decreases the number of natural killer T (NKT) cells in the liver (90) by inhibiting trafficking of CD1d to the plasma membrane (Fig. 4) (89). Surface displayed CD1d is involved in the activation of liver NKTs by fatty hepatocytes. A reduced liver NKT population impairs clearance of tumors and of microbial agents, giving rise to hepatocellular carcinoma and inflammation, which further elevates steatosis (90).

3.5. Hyperhomocysteinemia and Alcoholic Fatty Liver Disease

Hyperhomocysteinemia is characterized by elevated plasma homocysteine levels. Increased homocysteine plasma levels are a risk factor for NAFLD, AFLD (91), and atherosclerosis (Fig. 4) (92). Loss-of-function mutations in genes whose products are involved in methionine metabolism, such as the genes encoding cystathionine β-synthase, 5-methyltetrahydrofolate-homocysteine methyltransferase or 5,10-methylenetetrahydrofolate reductase cause an inherited, autosomal recessive form of this metabolic disease (92). Chronic alcohol abuse decreases expression of several genes involved in homocysteine metabolism, such as methionine adenosyltransferase, glycine methyltransferase, methionine synthase, betaine homocysteine methyltransferase, and cystathionine β-synthase (93) leading to elevated plasma homocysteine levels. Several mechanisms through which homocysteine impairs protein folding are known (91, 94). Nitrosylated homocysteine escapes the proofreading activity of methionine tRNA synthase and can be incorporated into protein. Conversion of homocysteine to a highly reactive thiolactone by the proofreading activity of aminoacyl-tRNA synthetases can modify the ε-NH_2 groups of lysine.

Homocysteine interferes with proper disulfide bond formation and its oxidation to homocystine is a source for ROS, which damage proteins by carbonylation of the peptide backbone, and also reversibly or irreversibly activate ryanodine Ca^{2+} release channels in the ER membrane. Through a combination of these mechanisms homocysteine prevents transport of thrombomodulin (95) and von Willebrand factor to the cell surface (96), and induces UPR markers such as BiP, GRP94, CHOP, and HERP (97), damaging the endothelial cell layer and contributing to initiation of atherosclerosis (Fig. 4). Homocysteine also activates lipogenic signaling via the SREBPs leading to cholesterol accumulation (98), which contributes to the development of ER stress in hyperhomocysteinemia and fatty liver disease. *chop*$^{-/-}$ mice were protected from alcohol-induced hepatic apoptosis, but not ER stress, fatty liver, or hyperhomocysteinemia (99), suggesting that apoptotic cell death in response to alcohol abuse or hyperhomocysteinemia is triggered by ER stress signaling pathways.

3.6. Atherosclerosis

Atherosclerosis is a major cause for myocardial infarction, stroke, and heart disease. Damage of the vascular endothelial cell layer allows formation of lipid-, especially cholesterol-rich deposits of low density lipoprotein (LDL) in the subendothelial intima (Fig. 4) (100). LDL oxidation induces endothelial cells to produce inflammatory cytokines such as monocyte chemoattractant protein-1 (MCP-1) leading to invasion of the intima by monocytes, which then differentiate into macrophages. These macrophages endocytose highly oxidized LDL, which is produced by ROS released by endothelial cells and macrophages. Cholesterol esterification transforms macrophages into foam cells. Macrophages secrete apolipoprotein E (apoE), which promotes cholesterol efflux to high density lipoproteins (HDL) (100). Release of lipids, mostly cholesterol and cholesterol esters by dying foam cells contributes to formation of fibrous plaques, which may initiate thrombosis (100). ER stress plays important roles in two stages of atherosclerosis (Fig. 4). ER stress is involved in initial damage of the vascular endothelial layer in hyperhomocysteinemia (101–104) by inducing endothelial cell apoptosis via the IRE1α-JNK pathway (103). Atherosclerotic lesion-resident macrophages display markers of UPR activation (104, 105), including PERK phosphorylation, and expression of CHOP, ATF4, and spliced XBP-1 (106). *perk*$^{-/-}$ macrophages are sensitized to, whereas *chop*$^{-/-}$ macrophages are protected from cholesterol-induced apoptosis (106). Cholesterol induces ER stress by decreasing the fluidity of the ER membrane, resulting in inhibition of SERCA Ca^{2+} pumps and Ca^{2+} depletion of the ER (106, 107). Thus, ER stress may be responsible for foam cell death and formation of the necrotic core of fibrous plaques. Cholesterol activation of the UPR in macrophages also contributes to production of inflammatory cytokines such as TNF-α and IL-6 (108) and thus contributes to the inflamed state of atherosclerotic lesions. ER stress in macrophages may decrease apoE secretion and cholesterol efflux from atherosclerotic lesions.

3.7. Ischemia and Hypoxia

Ischemia, the restriction of blood, oxygen, and nutrient supply to tissues and organs, such as the brain (stroke) or the heart (myocardial infarction or heart attack), activates the UPR (109) as evidenced by *XBP-1* splicing (110) and increased eIF2α and PERK (111) phosphorylation after cerebral ischemia and induction of various genes encoding ER chaperones and foldases, such as ERO1-L, ERP72, BiP, GRP94, ORP150, SERP1, and stanniocalcin 2, genes involved in ERAD such as HRD1, and GADD34 (Fig. 4) (109). Cardiac ischemia activates the UPR (112). Overexpression of chaperones, that is, ORP150 and GRP94, protects from ischemia-induced neuronal apoptosis. *chop*$^{-/-}$ cells are protected from neuronal apoptosis induced by hypoxia (113). Cardiac ischemia or exposure of cultured cardiac myocytes to hypoxia activated *XBP-1* splicing, JNK, and expression of BiP and PDI, show that the UPR is activated in ischemic cardiocytes (113).

Whether activation of the UPR in cerebral or cardiac ischemia is protective or injuring is not resolved. Cerebral ischemia activates CHOP expression (114) and caspase-12 (115, 116). These potential apoptotic ER stress signals are balanced by protection against ischemia mediated by induction of ER-resident molecular chaperones. Several studies have shown that overexpression of ER-resident chaperones such as ORP150, BiP, GRP94, or PDI or activation of the UPR with tunicamycin (112, 113, 117) protects neurons and cardiac myocytes from ischemia. Transgenic mice expressing ATF6, which predominantly activates expression of ER chaperone genes and genes involved in ERAD, in the heart showed better recovery from ex vivo ischemia/reperfusion, and less necrotic and apoptotic cell death (118). The outcome of UPR activation on cell fate, survival or apoptosis, most likely depends on the severity and duration of ischemia and, as a consequence, the magnitude and duration of the UPR with a low level, transient UPR being protective and a strong, long lasting UPR being apoptotic.

Hypoxia is often encountered in solid tumors, because tumor growth is often faster than angiogenesis in the tumor region. Hypoxia activates PERK (119), ATF6, and XBP-1 (120). *perk*$^{-/-}$ and *xbp1*$^{-/-}$ MEFs or cells expressing a dominant-negative form of PERK are more sensitive to hypoxia than WT cells (121, 122). Several ER chaperones, such as BiP, GRP94, and calreticulin, and HERP protect cancer cells from apoptosis in culture (123). UPR activation is also involved in angiogenesis in the vicinity of the tumor mass. ER stress promotes secretion of proangiogenic VEGF-A by increasing *VEGF-A* mRNA levels via activation of

ATF4 by PERK (121, 124) and induction of ORP150, which is required for VEGF-A secretion. Furthermore, IRE1α is also required for induction of VEGF-A in tumors (125). Tumor growth and vascularization was decreased in transgenic mice expressing a dominant-negative form of IRE1α compared to WT mice (125). These findings have let to the proposal to exploit inhibition of the UPR and of ER chaperones as a new chemotherapeutic strategy.

What activates the UPR in hypoxia and ischemia? Glucose depletion and ROS are two candidates. Glucose depletion induces ER chaperones (126, 127) through depletion of cellular ATP stores, stalling of chaperone ATPase cycles and inhibition of *N*-linked glycosylation. Ischemia generates superoxide and NO radicals, which, as discussed above, oxidatively damage proteins and may irreversibly deplete ER Ca^{2+}-stores by activating ryanodine receptors and inhibiting SERCA Ca^{2+} pumps (128).

4. Conclusions

Research into the UPR has revealed a surprisingly direct link between protein metabolism in the ER and energy metabolism pointing toward a central role for ER stress in energy homeostasis in mammals. UPR signaling is also linked to nitrogen metabolism in the unicellular eukaryote *Saccharomyces cerevisiae* (109, 110), indicating that control of metabolism by the UPR has evolved in lower eukaryotes. Future studies using mouse models will continue to unravel the contribution of the UPR to energy homeostasis in mammals. Considering the apparent conservation of the role of the UPR in metabolism, additional studies in other eukaryotic model organisms should be encouraged, which, for example, in yeast may contribute to unraveling functions of the *IRE1* branch of the UPR in the absence of insulin, but not necessarily glucose, signaling. Despite the exciting finding of the importance of ER stress in the metabolic syndrome, open questions remaining in the UPR proper will need to be addressed in future work. Important future research directions include the identification of the 'ER stress' sensing mechanism employed by IRE1, PERK, and ATF6, the identification of how UPR signaling controls cell fate, that is, to answer the question why UPR signaling promotes cell survival in one situation and apoptosis in another, and last, but not least, the identification of negative regulatory mechanisms turning off the UPR once ER stress has been resolved. Addressing the first and last questions will benefit from studies in eukaryotic model organisms such as yeast, as evolutionary grafting of additional layers of regulation onto basic regulatory mechanisms is likely to complicate studies in mouse or human cells.

Acknowledgements

We apologize to all whose work could not be cited because of space limitations. This work was supported by the Biotechnology and Biological Sciences Research Council [BB/C513418/1, BB/D01588X/1, BB/E006035/1]; the European Commission [HEALTH-F7-2007-201608]; and the Wellcome Trust [079821] to M.S. and the Biotechnology and Biological Sciences Research Council [BB/D526188/1] to L.S. and M.S. The research leading to these results has received funding from the European Community's seventh Framework Program (FP7/2007-2013) under grant agreement n° 201608.

References

1. Schröder M (2008) Endoplasmic reticulum stress responses. Cell Mol Life Sci 65:862–894
2. Hidalgo C (2005) Cross talk between Ca^{2+} and redox signalling cascades in muscle and neurons through the combined activation of ryanodine receptors/Ca^{2+} release channels. Philos Trans R Soc Lond B Biol Sci 360:2237–2246
3. Bertolotti A, Zhang Y, Hendershot LM, Harding HP, Ron D (2000) Dynamic interaction of BiP and ER stress transducers in the unfolded-protein response. Nat Cell Biol 2:326–332
4. Chen X, Shen J, Prywes R (2002) The luminal domain of ATF6 senses endoplasmic reticulum (ER) stress and causes translocation of ATF6 from the ER to the Golgi. J Biol Chem 277:13045–13052
5. Shen J, Chen X, Hendershot L, Prywes R (2002) ER stress regulation of ATF6 localization by dissociation of BiP/GRP78 binding and unmasking of Golgi localization signals. Dev Cell 3:99–111
6. Kimata Y, Oikawa D, Shimizu Y, Ishiwata-Kimata Y, Kohno K (2004) A role for BiP as an adjustor for the endoplasmic reticulum stress-sensing protein Ire1. J Cell Biol 167:445–456
7. Credle JJ, Finer-Moore JS, Papa FR, Stroud RM, Walter P (2005) Inaugural article: on the mechanism of sensing unfolded protein in the endoplasmic reticulum. Proc Natl Acad Sci USA 102:18773–18784
8. Kimata Y, Ishiwata-Kimata Y, Ito T, Hirata A, Suzuki T, Oikawa D, Takeuchi M, Kohno K (2007) Two regulatory steps of ER-stress sensor Ire1 involving its cluster formation and interaction with unfolded proteins. J Cell Biol 179:75–86
9. Zhou J, Liu CY, Back SH, Clark RL, Peisach D, Xu Z, Kaufman RJ (2006) The crystal structure of human IRE1 luminal domain reveals a conserved dimerization interface required for activation of the unfolded protein response. Proc Natl Acad Sci USA 103:14343–14348
10. Ye J, Rawson RB, Komuro R, Chen X, Dave UP, Prywes R, Brown MS, Goldstein JL (2000) ER stress induces cleavage of membrane-bound ATF6 by the same proteases that process SREBPs. Mol Cell 6: 1355–1364
11. Zhang K, Shen X, Wu J, Sakaki K, Saunders T, Rutkowski DT, Back SH, Kaufman RJ (2006) Endoplasmic reticulum stress activates cleavage of CREBH to induce a systemic inflammatory response. Cell 124: 587–599
12. Harding HP, Zhang Y, Ron D (1999) Protein translation and folding are coupled by an endoplasmic-reticulum-resident kinase. Nature 397:271–274
13. Shi Y, An J, Liang J, Hayes SE, Sandusky GE, Stramm LE, Yang NN (1999) Characterization of a mutant pancreatic eIF-2α kinase, PEK, and co-localization with somatostatin in islet delta cells. J Biol Chem 274:5723–5730
14. Cullinan SB, Diehl JA (2004) PERK-dependent activation of Nrf2 contributes to redox homeostasis and cell survival following endoplasmic reticulum stress. J Biol Chem 279:20108–20117

15. Cullinan SB, Zhang D, Hannink M, Arvisais E, Kaufman RJ, Diehl JA (2003) Nrf2 is a direct PERK substrate and effector of PERK-dependent cell survival. Mol Cell Biol 23:7198–7209
16. Brewer JW, Diehl JA (2000) PERK mediates cell-cycle exit during the mammalian unfolded protein response. Proc Natl Acad Sci USA 97:12625–12630
17. Deng J, Lu PD, Zhang Y, Scheuner D, Kaufman RJ, Sonenberg N, Harding HP, Ron D (2004) Translational repression mediates activation of nuclear factor kappa B by phosphorylated translation initiation factor 2. Mol Cell Biol 24:10161–10168
18. Li Q, Verma IM (2002) NF-κB regulation in the immune system. Nat Rev Immunol 2:725–734
19. Pham CG, Bubici C, Zazzeroni F, Papa S, Jones J, Alvarez K, Jayawardena S, De Smaele E, Cong R, Beaumont C, Torti FM, Torti SV, Franzoso G (2004) Ferritin heavy chain upregulation by NF-κB inhibits TNFα-induced apoptosis by suppressing reactive oxygen species. Cell 119:529–542
20. Harding HP, Novoa I, Zhang Y, Zeng H, Wek R, Schapira M, Ron D (2000) Regulated translation initiation controls stress-induced gene expression in mammalian cells. Mol Cell 6:1099–1108
21. Yamaguchi S, Ishihara H, Yamada T, Tamura A, Usui M, Tominaga R, Munakata Y, Satake C, Katagiri H, Tashiro F, Aburatani H, Tsukiyama-Kohara K, Miyazaki J-i, Sonenberg N, Oka Y (2008) ATF4-mediated induction of 4E-BP1 contributes to pancreatic β cell survival under endoplasmic reticulum stress. Cell Metab 7:269–276
22. Ma Y, Brewer JW, Diehl JA, Hendershot LM (2002) Two distinct stress signaling pathways converge upon the CHOP promoter during the mammalian unfolded protein response. J Mol Biol 318:1351–1365
23. Nguyen T, Sherratt PJ, Pickett CB (2003) Regulatory mechanisms controlling gene expression mediated by the antioxidant response element. Annu Rev Pharmacol Toxicol 43:233–260
24. Warnakulasuriyarachchi D, Cerquozzi S, Cheung HH, Holcik M (2004) Translational induction of the inhibitor of apoptosis protein HIAP2 during endoplasmic reticulum stress attenuates cell death and is mediated via an inducible internal ribosome entry site element. J Biol Chem 279:17148–17157
25. Yoshimura FK, Luo X, Zhao X, Gerard HC, Hudson AP (2008) Up-regulation of a cellular protein at the translational level by a retrovirus. Proc Natl Acad Sci USA 105:5543–5548
26. Novoa I, Zeng H, Harding HP, Ron D (2001) Feedback inhibition of the unfolded protein response by *GADD34*-mediated dephosphorylation of eIF2α. J Cell Biol 153:1011–1022
27. Marciniak SJ, Yun CY, Oyadomari S, Novoa I, Zhang Y, Jungreis R, Nagata K, Harding HP, Ron D (2004) CHOP induces death by promoting protein synthesis and oxidation in the stressed endoplasmic reticulum. Genes Dev 18:3066–3077
28. Calfon M, Zeng H, Urano F, Till JH, Hubbard SR, Harding HP, Clark SG, Ron D (2002) IRE1 couples endoplasmic reticulum load to secretory capacity by processing the *XBP-1* mRNA. Nature 415:92–96
29. Yoshida H, Matsui T, Yamamoto A, Okada T, Mori K (2001) XBP1 mRNA is induced by ATF6 and spliced by IRE1 in response to ER stress to produce a highly active transcription factor. Cell 107:881–891
30. Shen X, Ellis RE, Lee K, Liu C-Y, Yang K, Solomon A, Yoshida H, Morimoto R, Kurnit DM, Mori K, Kaufman RJ (2001) Complementary signaling pathways regulate the unfolded protein response and are required for *C. elegans* development. Cell 107:893–903
31. Lee K, Tirasophon W, Shen X, Michalak M, Prywes R, Okada T, Yoshida H, Mori K, Kaufman RJ (2002) IRE1-mediated unconventional mRNA splicing and S2P-mediated ATF6 cleavage merge to regulate XBP1 in signaling the unfolded protein response. Genes Dev 16:452–466
32. Lee AH, Iwakoshi NN, Glimcher LH (2003) XBP-1 regulates a subset of endoplasmic reticulum resident chaperone genes in the unfolded protein response. Mol Cell Biol 23:7448–7459
33. Sriburi R, Bommiasamy H, Buldak GL, Robbins GR, Frank M, Jackowski S, Brewer JW (2007) Coordinate regulation of phospholipid biosynthesis and secretory pathway gene expression in XBP-1(S)-induced endoplasmic reticulum biogenesis. J Biol Chem 282:7024–7034
34. Sriburi R, Jackowski S, Mori K, Brewer JW (2004) XBP1: a link between the unfolded protein response, lipid biosynthesis, and biogenesis of the endoplasmic reticulum. J Cell Biol 167:35–41
35. Wu J, Rutkowski DT, Dubois M, Swathirajan J, Saunders T, Wang J, Song B, Yau GD-Y,

Kaufman RJ (2007) ATF6α optimizes long-term endoplasmic reticulum function to protect cells from chronic stress. Dev Cell 13:351–364
36. Yamamoto K, Sato T, Matsui T, Sato M, Okada T, Yoshida H, Harada A, Mori K (2007) Transcriptional induction of mammalian ER quality control proteins is mediated by single or combined action of ATF6α and XBP1. Dev Cell 13:365–376
37. Urano F, Wang X, Bertolotti A, Zhang Y, Chung P, Harding HP, Ron D (2000) Coupling of stress in the ER to activation of JNK protein kinases by transmembrane protein kinase IRE1. Science 287:664–666
38. Dempsey PW, Doyle SE, He JQ, Cheng G (2003) The signaling adaptors and pathways activated by TNF superfamily. Cytokine Growth Factor Rev 14:193–209
39. Hu P, Han Z, Couvillon AD, Kaufman RJ, Exton JH (2006) Autocrine tumor necrosis factor alpha links endoplasmic reticulum stress to the membrane death receptor pathway through IRE1α-mediated NF-κB activation and down-regulation of TRAF2 expression. Mol Cell Biol 26:3071–3084
40. Nishitoh H, Matsuzawa A, Tobiume K, Saegusa K, Takeda K, Inoue K, Hori S, Kakizuka A, Ichijo H (2002) ASK1 is essential for endoplasmic reticulum stress-induced neuronal cell death triggered by expanded polyglutamine repeats. Genes Dev 16:1345–1355
41. Dunn C, Wiltshire C, MacLaren A, Gillespie DA (2002) Molecular mechanism and biological functions of c-Jun N-terminal kinase signalling via the c-Jun transcription factor. Cell Signal 14:585–593
42. Lei K, Davis RJ (2003) JNK phosphorylation of Bim-related members of the Bcl2 family induces Bax-dependent apoptosis. Proc Natl Acad Sci USA 100:2432–2437
43. Putcha GV, Le S, Frank S, Besirli CG, Clark K, Chu B, Alix S, Youle RJ, LaMarche A, Maroney AC, Johnson EM Jr (2003) JNK-mediated BIM phosphorylation potentiates BAX-dependent apoptosis. Neuron 38:899–914
44. Yoneda T, Imaizumi K, Oono K, Yui D, Gomi F, Katayama T, Tohyama M (2001) Activation of caspase-12, an endoplastic reticulum (ER) resident caspase, through tumor necrosis factor receptor-associated factor 2-dependent mechanism in response to the ER stress. J Biol Chem 276:13935–13940
45. Nakagawa T, Zhu H, Morishima N, Li E, Xu J, Yankner BA, Yuan J (2000) Caspase-12 mediates endoplasmic-reticulum-specific apoptosis and cytotoxicity by amyloid-β. Nature 403:98–103
46. Fischer H, Koenig U, Eckhart L, Tschachler E (2002) Human caspase 12 has acquired deleterious mutations. Biochem Biophys Res Commun 293:722–726
47. Hitomi J, Katayama T, Eguchi Y, Kudo T, Taniguchi M, Koyama Y, Manabe T, Yamagishi S, Bando Y, Imaizumi K, Tsujimoto Y, Tohyama M (2004) Involvement of caspase-4 in endoplasmic reticulum stress-induced apoptosis and Aβ-induced cell death. J Cell Biol 165:347–356
48. Narin F, Atabek ME, Karakukcu M, Narin N, Kurtoglu S, Gumus H, Coksevim B, Erez R (2005) The association of plasma homocysteine levels with serum leptin and apolipoprotein B levels in childhood obesity. Ann Saudi Med 25:209–214
49. Münzberg H, Björnholm M, Bates SH, Myers MG Jr (2005) Leptin receptor action and mechanisms of leptin resistance. Cell Mol Life Sci 62:642–652
50. Myers MG, Cowley MA, Münzberg H (2008) Mechanisms of leptin action and leptin resistance. Annu Rev Physiol 70: 537–556
51. Zhang X, Zhang G, Zhang H, Karin M, Bai H, Cai D (2008) Hypothalamic IKKβ/NF-κB and ER stress link overnutrition to energy imbalance and obesity. Cell 135:61–73
52. Hosoi T, Sasaki M, Miyahara T, Hashimoto C, Matsuo S, Yoshii M, Ozawa K (2008) Endoplasmic reticulum stress induces leptin resistance. Mol Pharmacol 74:1610–1619
53. Shulman GI (2000) Cellular mechanisms of insulin resistance. J Clin Invest 106: 171–176
54. Groop LC (1999) Insulin resistance: the fundamental trigger of type 2 diabetes. Diab Obes Metab 1(Suppl 1):S1–S7
55. Langin D (2006) Adipose tissue lipolysis as a metabolic pathway to define pharmacological strategies against obesity and the metabolic syndrome. Pharmacol Res 53:482–491
56. Shulman GI (1999) Cellular mechanisms of insulin resistance in humans. Am J Cardiol 84:3J–10J
57. Saltiel AR, Kahn CR (2001) Insulin signalling and the regulation of glucose and lipid metabolism. Nature 414:799–806
58. Draznin B (2006) Molecular mechanisms of insulin resistance: serine phosphorylation of insulin receptor substrate-1 and increased

expression of p85alpha: the two sides of a coin. Diabetes 55:2392–2397

59. Özcan U, Cao Q, Yilmaz E, Lee A-H, Iwakoshi NN, Ozdelen E, Tuncman G, Görgün C, Glimcher LH, Hotamisligil GS (2004) Endoplasmic reticulum stress links obesity, insulin action, and type 2 diabetes. Science 306:457–461
60. White MF (2003) Insulin signaling in health and disease. Science 302:1710–1711
61. Shah OJ, Wang Z, Hunter T (2004) Inappropriate activation of the TSC/Rheb/mTOR/S6K cassette induces IRS1/2 depletion, insulin resistance, and cell survival deficiencies. Curr Biol 14:1650–1656
62. Wei Y, Wang D, Topczewski F, Pagliassotti MJ (2006) Saturated fatty acids induce endoplasmic reticulum stress and apoptosis independently of ceramide in liver cells. Am J Physiol Endocrinol Metab 291:E275–E281
63. Wang D, Wei YR, Pagliassotti MJ (2006) Saturated fatty acids promote endoplasmic reticulum stress and liver injury in rats with hepatic steatosis. Endocrinology 147:943–951
64. Bergman RN, Ader M (2000) Free fatty acids and pathogenesis of type 2 diabetes mellitus. Trends Endocrinol Metab 11:351–356
65. Haynes CM, Titus EA, Cooper AA (2004) Degradation of misfolded proteins prevents ER-derived oxidative stress and cell death. Mol Cell 15:767–776
66. Harding HP, Zhang Y, Zeng H, Novoa I, Lu PD, Calfon M, Sadri N, Yun C, Popko B, Paules R, Stojdl DF, Bell JC, Hettmann T, Leiden JM, Ron D (2003) An integrated stress response regulates amino acid metabolism and resistance to oxidative stress. Mol Cell 11:619–633
67. Kamata H, Honda S, Maeda S, Chang L, Hirata H, Karin M (2005) Reactive oxygen species promote TNFα-induced death and sustained JNK activation by inhibiting MAP kinase phosphatases. Cell 120:649–661
68. Örd D, Örd T (2005) Characterization of human NIPK (*TRB3*, *SKIP3*) gene activation in stressful conditions. Biochem Biophys Res Commun 330:210–218
69. Ohoka N, Yoshii S, Hattori T, Onozaki K, Hayashi H (2005) TRB3, a novel ER stress-inducible gene, is induced via ATF4-CHOP pathway and is involved in cell death. EMBO J 24:1243–1255
70. Du K, Herzig S, Kulkarni RN, Montminy M (2003) TRB3: a *tribbles* homolog that inhibits Akt/PKB activation by insulin in liver. Science 300:1574–1577
71. Lee AH, Scapa EF, Cohen DE, Glimcher LH (2008) Regulation of hepatic lipogenesis by the transcription factor XBP1. Science 320:1492–1496
72. Iqbal J, Dai K, Seimon T, Jungreis R, Oyadomari M, Kuriakose G, Ron D, Tabas I, Hussain MM (2008) IRE1β inhibits chylomicron production by selectively degrading MTP mRNA. Cell Metab 7:445–455
73. Butler AE, Janson J, Bonner-Weir S, Ritzel R, Rizza RA, Butler PC (2003) β-cell deficit and increased β-cell apoptosis in humans with type 2 diabetes. Diabetes 52:102–110
74. Song B, Scheuner D, Ron D, Pennathur S, Kaufman RJ (2008) *Chop* deletion reduces oxidative stress, improves β cell function, and promotes cell survival in multiple mouse models of diabetes. J Clin Invest 118: 3378–3389
75. Ariyama Y, Shimizu H, Satoh T, Tsuchiya T, Okada S, Oyadomari S, Mori M (2007) *Chop*-deficient mice showed increased adiposity but no glucose intolerance. Obesity (Silver Spring) 15:1647–1656
76. Nozaki J, Kubota H, Yoshida H, Naitoh M, Goji J, Yoshinaga T, Mori K, Koizumi A, Nagata K (2004) The endoplasmic reticulum stress response is stimulated through the continuous activation of transcription factors ATF6 and XBP1 in *Ins2*$^{+/Akita}$ pancreatic β cells. Genes Cells 9:261–270
77. Zuber C, Fan JY, Guhl B, Roth J (2004) Misfolded proinsulin accumulates in expanded pre-Golgi intermediates and endoplasmic reticulum subdomains in pancreatic beta cells of Akita mice. FASEB J 18:U341–U360
78. Wang J, Takeuchi T, Tanaka S, Kubo SK, Kayo T, Lu D, Takata K, Koizumi A, Izumi T (1999) A mutation in the insulin 2 gene induces diabetes with severe pancreatic β-cell dysfunction in the *Mody* mouse. J Clin Invest 103:27–37
79. Herbach N, Rathkolb B, Kemter E, Pichl L, Klaften M, de Angelis MH, Halban PA, Wolf E, Aigner B, Wanke R (2007) Dominant-negative effects of a novel mutated *Ins2* allele causes early-onset diabetes and severe beta-cell loss in Munich *Ins2*^{C95S} mutant mice. Diabetes 56:1268–1276
80. Oyadomari S, Koizumi A, Takeda K, Gotoh T, Akira S, Araki E, Mori M (2002) Targeted disruption of the *Chop* gene delays endoplasmic reticulum stress-mediated diabetes. J Clin Invest 109:525–532
81. Colombo C, Porzio O, Liu M, Massa O, Vasta M, Salardi S, Beccaria L, Monciotti C,

Toni S, Pedersen O, Hansen T, Federici L, Pesavento R, Cadario F, Federici G, Ghirri P, Arvan P, Iafusco D, Barbetti F (2008) Seven mutations in the human insulin gene linked to permanent neonatal/infancy-onset diabetes mellitus. J Clin Invest 118: 2148–2156

82. Zhang S, Liu J, Saafi EL, Cooper GJ (1999) Induction of apoptosis by human amylin in RINm5F islet β-cells is associated with enhanced expression of p53 and p21WAF1/CIP1. FEBS Lett 455:315–320
83. Janson J, Ashley RH, Harrison D, McIntyre S, Butler PC (1999) The mechanism of islet amyloid polypeptide toxicity is membrane disruption by intermediate-sized toxic amyloid particles. Diabetes 48:491–498
84. Mirzabekov TA, Lin MC, Kagan BL (1996) Pore formation by the cytotoxic islet amyloid peptide amylin. J Biol Chem 271: 1988–1992
85. Huang C-j, Lin C-y, Haataja L, Gurlo T, Butler AE, Rizza RA, Butler PC (2007) High expression rates of human islet amyloid polypeptide induce endoplasmic reticulum stress-mediated β cell apoptosis, a characteristic of humans with type 2 but not type 1 diabetes. Diabetes 56:2016–2027
86. Guo F, Cavener DR (2007) The GCN2 eIF2α kinase regulates fatty-acid homeostasis in the liver during deprivation of an essential amino acid. Cell Metab 5:103–114
87. Oyadomari S, Harding HP, Zhang Y, Oyadomari M, Ron D (2008) Dephosphorylation of translation initiation factor 2α enhances glucose tolerance and attenuates hepatosteatosis in mice. Cell Metab 7:520–532
88. Rutkowski DT, Wu J, Back S-H, Callaghan MU, Ferris SP, Iqbal J, Clark R, Miao H, Hassler JR, Fornek J, Katze MG, Hussain MM, Song B, Swathirajan J, Wang J, Yau GD, Kaufman RJ (2008) UPR pathways combine to prevent hepatic steatosis caused by ER stress-mediated suppression of transcriptional master regulators. Dev Cell 15:829–840
89. Yang L, Jhaveri R, Huang J, Qi Y, Diehl AM (2007) Endoplasmic reticulum stress, hepatocyte CD1d and NKT cell abnormalities in murine fatty livers. Lab Invest 87: 927–937
90. Guebre-Xabier M, Yang S, Lin HZ, Schwenk R, Krzych U, Diehl AM (2000) Altered hepatic lymphocyte subpopulations in obesity-related murine fatty livers: potential mechanism for sensitization to liver damage. Hepatology 31:633–640
91. Ji C (2008) Dissection of endoplasmic reticulum stress signaling in alcoholic and non-alcoholic liver injury. J Gastroenterol Hepatol 23(Suppl 1):S16–S24
92. Loscalzo J (1996) The oxidant stress of hyperhomocyst(e)inemia. J Clin Invest 98:5–7
93. Avila MA, Berasain C, Torres L, Martin-Duce A, Corrales FJ, Yang H, Prieto J, Lu SC, Caballeria J, Rodes J, Mato JM (2000) Reduced mRNA abundance of the main enzymes involved in methionine metabolism in human liver cirrhosis and hepatocellular carcinoma. J Hepatol 33:907–914
94. Malhotra JD, Kaufman RJ (2007) The endoplasmic reticulum and the unfolded protein response. Semin Cell Dev Biol 18:716–731
95. Lentz SR, Sadler JE (1991) Inhibition of thrombomodulin surface expression and protein C activation by the thrombogenic agent homocysteine. J Clin Invest 88:1906–1914
96. Lentz SR, Sadler JE (1993) Homocysteine inhibits von Willebrand factor processing and secretion by preventing transport from the endoplasmic reticulum. Blood 81:683–689
97. Austin RC, Lentz SR, Werstuck GH (2004) Role of hyperhomocysteinemia in endothelial dysfunction and atherothrombotic disease. Cell Death Differ 11(Suppl 1):S56–S64
98. Werstuck GH, Lentz SR, Dayal S, Hossain GS, Sood SK, Shi YY, Zhou J, Maeda N, Krisans SK, Malinow MR, Austin RC (2001) Homocysteine-induced endoplasmic reticulum stress causes dysregulation of the cholesterol and triglyceride biosynthetic pathways. J Clin Invest 107:1263–1273
99. Ji C, Mehrian-Shai R, Chan C, Hsu YH, Kaplowitz N (2005) Role of CHOP in hepatic apoptosis in the murine model of intragastric ethanol feeding. Alcohol Clin Exp Res 29:1496–1503
100. Lusis AJ (2000) Atherosclerosis. Nature 407:233–241
101. Kokame K, Kato H, Miyata T (1996) Homocysteine-respondent genes in vascular endothelial cells identified by differential display analysis. GRP78/BiP and novel genes. J Biol Chem 271:29659–29665
102. Roybal CN, Yang SJ, Sun CW, Hurtado D, Vander Jagt DL, Townes TM, Abcouwer SF (2004) Homocysteine increases the expression of vascular endothelial growth factor by a mechanism involving endoplasmic reticulum stress and transcription factor ATF4. J Biol Chem 279:14844–14852

103. Zhang C, Cai Y, Adachi MT, Oshiro S, Aso T, Kaufman RJ, Kitajima S (2001) Homocysteine induces programmed cell death in human vascular endothelial cells through activation of the unfolded protein response. J Biol Chem 276:35867–35874
104. Zhou J, Lhotak S, Hilditch BA, Austin RC (2005) Activation of the unfolded protein response occurs at all stages of atherosclerotic lesion development in apolipoprotein E-deficient mice. Circulation 111: 1814–1821
105. Zhou J, Werstuck GH, Lhoták Š, de Koning AB, Sood SK, Hossain GS, Møller J, Ritskes-Hoitinga M, Falk E, Dayal S, Lentz SR, Austin RC (2004) Association of multiple cellular stress pathways with accelerated atherosclerosis in hyperhomocysteinemic apolipoprotein E-deficient mice. Circulation 110:207–213
106. Feng B, Yao PM, Li Y, Devlin CM, Zhang D, Harding HP, Sweeney M, Rong JX, Kuriakose G, Fisher EA, Marks AR, Ron D, Tabas I (2003) The endoplasmic reticulum is the site of cholesterol-induced cytotoxicity in macrophages. Nat Cell Biol 5:781–792
107. Li Y, Ge M, Ciani L, Kuriakose G, Westover EJ, Dura M, Covey DF, Freed JH, Maxfield FR, Lytton J, Tabas I (2004) Enrichment of endoplasmic reticulum with cholesterol inhibits sarcoplasmic-endoplasmic reticulum calcium ATPase-2b activity in parallel with increased order of membrane lipids: implications for depletion of endoplasmic reticulum calcium stores and apoptosis in cholesterol-loaded macrophages. J Biol Chem 279:37030–37039
108. Li Y, Schwabe RF, Devries-Seimon T, Yao PM, Gerbod-Giannone MC, Tall AR, Davis RJ, Flavell R, Brenner DA, Tabas I (2005) Free cholesterol-loaded macrophages are an abundant source of TNF-α and IL-6. Model of NF-κB- and MAP kinase-dependent inflammation in advanced atherosclerosis. J Biol Chem 280:21763–21772
109. Paschen W, Mengesdorf T (2005) Cellular abnormalities linked to endoplasmic reticulum dysfunction in cerebrovascular disease – therapeutic potential. Pharmacol Ther 108:362–375
110. Paschen W, Aufenberg C, Hotop S, Mengesdorf T (2003) Transient cerebral ischemia activates processing of xbp1 messenger RNA indicative of endoplasmic reticulum stress. J Cereb Blood Flow Metab 23:449–461
111. Kumar R, Azam S, Sullivan JM, Owen C, Cavener DR, Zhang PC, Ron D, Harding HP, Chen JJ, Han AP, White BC, Krause GS, DeGracia DJ (2001) Brain ischemia and reperfusion activates the eukaryotic initiation factor 2α kinase, PERK. J Neurochem 77:1418–1421
112. Azfer A, Niu J, Rogers LM, Adamski FM, Kolattukudy PE (2006) Activation of endoplasmic reticulum stress response during the development of ischemic heart disease. Am J Physiol Heart Circ Physiol 291:H1411–H1420
113. Glembotski CC (2008) The role of the unfolded protein response in the heart. J Mol Cell Cardiol 44:453–459
114. Paschen W, Gissel C, Linden T, Althausen S, Doutheil J (1998) Activation of gadd153 expression through transient cerebral ischemia: evidence that ischemia causes endoplasmic reticulum dysfunction. Brain Res Mol Brain Res 60:115–122
115. Shibata M, Hattori H, Sasaki T, Gotoh J, Hamada J, Fukuuchi Y (2003) Activation of caspase-12 by endoplasmic reticulum stress induced by transient middle cerebral artery occlusion in mice. Neuroscience 118: 491–499
116. Mouw G, Zechel JL, Gamboa J, Lust WD, Selman WR, Ratcheson RA (2003) Activation of caspase-12, an endoplasmic reticulum resident caspase, after permanent focal ischemia in rat. NeuroReport 14:183–186
117. Zhang PL, Lun M, Teng J, Huang J, Blasick TM, Yin L, Herrera GA, Cheung JY (2004) Preinduced molecular chaperones in the endoplasmic reticulum protect cardiomyocytes from lethal injury. Ann Clin Lab Sci 34:449–457
118. Martindale JJ, Fernandez R, Thuerauf D, Whittaker R, Gude N, Sussman MA, Glembotski CC (2006) Endoplasmic reticulum stress gene induction and protection from ischemia/reperfusion Injury in the hearts of transgenic mice with a tamoxifen-regulated form of ATF6. Circ Res 98:1186–1193
119. Koumenis C, Naczki C, Koritzinsky M, Rastani S, Diehl A, Sonenberg N, Koromilas A, Wouters BG (2002) Regulation of protein synthesis by hypoxia via activation of the endoplasmic reticulum kinase PERK and phosphorylation of the translation initiation factor eIF2α. Mol Cell Biol 22: 7405–7416
120. Moenner M, Pluquet O, Bouchecareilh M, Chevet E (2007) Integrated endoplasmic reticulum stress responses in cancer. Cancer Res 67:10631–10634

121. Bi MX, Naczki C, Koritzinsky M, Fels D, Blais J, Hu NP, Harding H, Novoa I, Varia M, Raleigh J, Scheuner D, Kaufman RJ, Bell J, Ron D, Wouters BG, Koumenis C (2005) ER stress-regulated translation increases tolerance to extreme hypoxia and promotes tumor growth. EMBO J 24:3470–3481
122. Romero-Ramirez L, Cao H, Nelson D, Hammond E, Lee A-H, Yoshida H, Mori K, Glimcher LH, Denko NC, Giaccia AJ, Le Q-T, Koong AC (2004) XBP1 is essential for survival under hypoxic conditions and is required for tumor growth. Cancer Res 64:5943–5947
123. Lee AS (2007) GRP78 induction in cancer: therapeutic and prognostic implications. Cancer Res 67:3496–3499
124. Abcouwer SF, Marjon PL, Loper RK, Vander Jagt DL (2002) Response of VEGF expression to amino acid deprivation and inducers of endoplasmic reticulum stress. Invest Ophthalmol Vis Sci 43:2791–2798
125. Drogat B, Auguste P, Nguyen DT, Bouchecareilh M, Pineau R, Nalbantoglu J, Kaufman RJ, Chevet E, Bikfalvi A, Moenner M (2007) IRE1 signaling is essential for ischemia-induced vascular endothelial growth factor-A expression and contributes to angiogenesis and tumor growth in vivo. Cancer Res 67:6700–6707
126. Pouysségur J, Shiu RP, Pastan I (1977) Induction of two transformation-sensitive membrane polypeptides in normal fibroblasts by a block in glycoprotein synthesis or glucose deprivation. Cell 11:941–947
127. Shiu RP, Pouyssegur J, Pastan I (1977) Glucose depletion accounts for the induction of two transformation-sensitive membrane proteins in Rous sarcoma virus-transformed chick embryo fibroblasts. Proc Natl Acad Sci USA 74:3840–3844
128. Doutheil J, Althausen S, Treiman M, Paschen W (2000) Effect of nitric oxide on endoplasmic reticulum calcium homeostasis, protein synthesis and energy metabolism. Cell Calcium 27:107–115

Chapter 4

What Role Does Mitochondrial Stress Play in Neurodegenerative Diseases?

Alicia Mae Pickrell and Carlos Torres Moraes

Abstract

The essential need for mitochondrial function has been extensively shown to relate to neuronal health. Neurodegeneration and neurodegeneration-related diseases have been associated with multiple mitochondrial dysfunctions. This review highlights key findings related to commonly studied mitochondrial dysfunctions: imbalance of mitochondrial dynamics, mutations in the mitochondrial genome, excessive reactive oxygen species, and misfolded protein associations/interactions with the mitochondria. Future research in mitochondrial function will help elucidate complex neurodegenerative events while impacting both individual and societal health.

Key words: Mitochondrial stress, Neurodegeneration, Mitochondrial dysfunction, Reactive oxygen species, Mitochondrial genome, Misfolded proteins

1. Introduction

Mitochondrial ATP production is vital for cellular function, signaling pathways, and overall cell viability. This is true for all cells; however, the reliance on proper mitochondrial function is particularly high for neurons due to their postmitotic status, unique electrophysiological properties, and high ATP demand. Neurons cannot easily be replaced when dysfunctions of the mitochondria can no longer be compensated for. Neurodegenerative diseases encompass an age-related regional and selective loss of specific neural cell populations that causes behavioral and mental decline depending on the extent and location of the cells that were lost. Perturbations associated with mitochondria are thought to correlate with or directly cause neurodegenerative events.

Peter Bross and Niels Gregersen (eds.), *Protein Misfolding and Cellular Stress in Disease and Aging: Concepts and Protocols*, Methods in Molecular Biology, vol. 648, DOI 10.1007/978-1-60761-756-3_4, © Springer Science+Business Media, LLC 2010

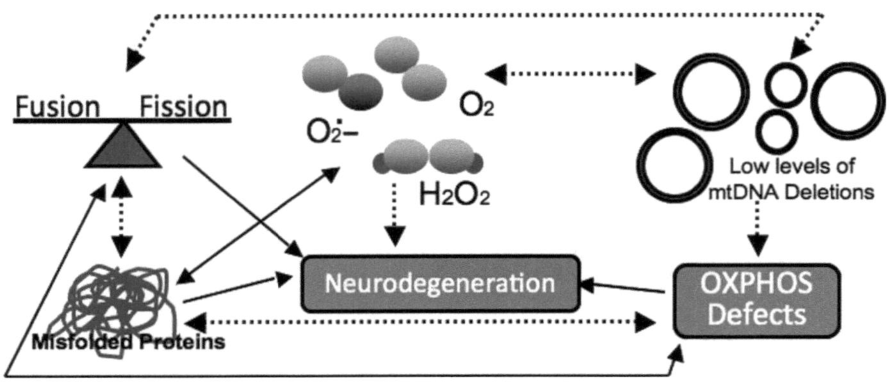

Fig. 1. Known and suggested interactions between mitochondrial dynamics, ROS, misfolded proteins, and low levels of mtDNA deletions with neurodegenerative processes. *Solid black arrows* indicate known interactions whereas *dashed arrows* indicate suggested ones.

Controversy and interest lies in which types of disruptions are causal or possibly secondary to this eventual undesirable cell death (Fig. 1).

This introductory chapter will highlight key findings on four different types of mitochondrial malfunctions and their relation to neurodegenerative disease. Defects in mitochondrial dynamics, mutations in the mitochondrial genome, the creation and presence of reactive oxygen species, and protein aggregate-associated dysfunctions of mitochondria exemplify how different stressors to this important organelle can contribute to neurodegenerative disease.

2. Mitochondrial Dynamics and Importance to Neuronal Function and Physiology

Mitochondria are not static organelles suspended in isolation inside of the cell; they are dynamic. There are two aspects related to mitochondrial dynamics in neurons. (1) Mitochondrial movement along axons and dendritic processes. (2) Mitochondrial fusion and fission. Both the compartmentalization inside of the neuron and dynamics of mitochondrial fission and fusion are extremely important to their neuronal functioning.

As a highly polarized cell, the neuron has different physiological properties and demands for bioenergetic needs regionalized in certain locations. Anterograde mitochondrial transport to mobilize mitochondria into presynaptic terminals is necessary for normal neuronal function, health, and electrophysiological properties needed to reverse ion fluxes and alter Ca^{+2} signaling demanded by synaptic transmission (1). Also, the presence of the mitochondria at this locale allows it to provide ATP that is used to mobilize the vesicle reserve pool during high frequency stimulation (2). Not only are mitochondria involved in the presynaptic side of the cell, but also at the postsynapse. The postsynaptic dendritic morphology, plasticity, and quantity rely on mitochondrial mobilization to postsynaptic densities regulated in an activity dependent manner (3). This type of plasticity alters different postsynaptic targets mediating different mitochondrial demands on the presynaptic side of the cell causing differences in Ca^{+2} buffering mechanisms and dynamics (4).

Specialized proteins have been discovered for the specific movement of mitochondria for axonal transport. The *Drosophila* protein, Milton, discovered through genetic mutational screens, was the first neuronal-specific mitochondrial transport protein identified for synapse and terminal mitochondrial localization (5). Miro and kinesin heavy chain (KHC) build a complex with Milton to facilitate anterograde movement down microtubules dependent on the absence or presence of Ca^{+2} in the synapse (6). Miro seems to act as the calcium sensor in this complex to determine where mitochondria should be released and positioned around postsynaptic sides of the synapse where glutaminergic neuronal activity induces large Ca^{+2} fluxes (7).

Besides trafficking, mitochondria have an intrinsic need to fuse and segregate from each other to maintain normal operations. Fis1 and Drp1 are responsible for fission; while mitofusins (Mfn1, Mfn2) and OPA 1 mediate fusion events (reviewed in (8)). Although the mechanisms responsible for what benefits fusion and fission provide for mitochondria are unknown, alterations in their balance lead to cell death. Disturbances in either fusion or fission as opposed to improper trafficking have been found to be consistently associated with neurodegenerative diseases.

2.1. Familial Mutations of Genes Associated with Mitochondrial Dynamics

Mutations found in genes ubiquitously expressed in the body that control mitochondrial fusion and fission events have been found to cause exclusive neurodegenerative events (reviewed in (9, 10)). By examining pedigrees of familial cases of dominant optic atrophy (DOA) and Charcot Marie Tooth 2A (CMT2A) patients, frameshift and missense mutations were found in GTPase mitochondrial fusion proteins OPA1 and Mitofusin2, respectively (11, 12). DOA selectively causes the loss of retinal ganglion cells, and CMT2A affects axonal periphery sensorimotor neuropathy. Studies examining the molecular basis behind these mutations reveal the importance behind mitochondrial morphology, dynamics, and its relation to these degenerative observations. In the case of OPA1, the dysfunction of this protein leads to a fragmented mitochondrial network, causes a loss of mitochondrial membrane potential ($\Delta\psi_m$), and upregulates the release of cytochrome c into the cytoplasm increasing apoptosis (13). More work is needed to dissect the molecular mechanisms behind defects of Mfn2 seen in CMT2A. However, defects in Mfn2 created by a region specific-targeted knockout in the cerebellum caused a shortening of the Purkinje cells' dendritic length eventually leading to death, which seems to be caused by defects in oxidative phosphorylation (OXPHOS) activity as a result of an abnormal segregating mitochondrial DNA (mtDNA) pool leaving some cells without a genome in some mitochondria (14). This need for mitochondrial fusion does not seem restricted to developmental requirements, but rather to an intrinsic need to maintain mitochondrial function in different tissues (reviewed in (15)).

2.2. Mitochondrial Dynamics Involvement in Common Neurodegenerative Diseases

Mitochondrial dynamics recently have been found to be affected in the four most common neurodegenerative diseases: Parkinson's disease (PD), Huntington's disease (HD), Amyotrophic lateral sclerosis (ALS) and Alzheimer's disease (AD). Each of these neurodegenerative diseases shows an age-related loss of dopaminergic substantia nigra *pars compacta* neurons, striatal spiny neurons, and cortical/hippocampal neurons, respectively. PTEN-induced kinase 1 (PINK1) is a gene, mutated in forms of recessive familial cases of PD, where deficiency leads to PD with decreases in catecholamine release, quanta release, and diminishing long-term potentiation synaptic plasticity (16). PINK1 has been found to regulate mitochondrial fission through Drp1 and Fis1, and mutations in these genes cause defects in mitochondrial morphology (17). However, this may not be a sole regulatory role for dynamics. Knocking out PINK1 has also been shown to affect other mitochondrial functions causing decreases in cell respiration, membrane potential, and calcium-handling capacities (18). Amyloid beta (Aβ), known to accumulate as different oligomers in AD, also plays a role in affecting mitochondrial dynamics as well and seems to be responsible for a fragmented morphology leading to an abnormal cellular distribution (19). An even more direct connection showed that Aβ at physiological levels caused increases in S-nitrosylation of Drp1, thus increasing GTPase activity leading to mitochondrial fission events, increased cell death, and decreased the quantity of dendrites (20). The N′ terminus of mutant huntingtin protein in HD also has been shown to associate with mitochondria and inhibit its trafficking ability in striatal cultures (21). The CAG repeat length on huntingtin increases the propensity for mitochondria to fragment, which is abrogated by overexpression of Mfn2 or a dominant negative Drp1K38A (22). Most recently, familial mutations in the ALS associated -Cu, Zn, superoxide dismutase (SOD1) protein have been shown to induce mitochondria fragmentation and impair trafficking (23). Defects in mitochondrial dynamics are beginning to appear as phenotypes in more common neurodegenerative diseases as opposed to isolated neurodegenerative cases where mutations are found directly affecting mitochondrial fission/fusion proteins. Although these changes are likely secondary, they may play an important role in the pathophysiology of these diseases.

2.3. Future Directions on Mitochondrial Dynamics

The importance of mitochondrial dynamics seems to be much more crucial to neuronal cell types than other cells as mutations of fission/fusion proteins found in these patients are ubiquitous in nature, but their disease states are only characterized by a selective neuronal loss of a certain population. The most common suggestion to answer why this is may be that the long distance mitochondria have to travel in axons would make neurons more susceptible to defects in mitochondrial dynamics. Definite proof of this model is lacking. However, new studies have now begun to show associations between Mfn2 and Miro/Milton complexes tying together mitochondrial dynamics and trafficking in neurons (24). Another interesting question is whether the differences in which neurons become susceptible is related to specific energy demands different neuronal types may have. The mechanisms associated with neuronal death and defects in mitochondrial dynamics also remain to be elucidated. With increasing evidence that mitochondrial dynamics contributes to neurodegenerative disease, it will be important to understand whether these changes are causal or secondary characteristics of neuronal cell death.

3. Mitochondrial DNA Mutations and Their Occurrences in Neurons

Besides the nuclei, mitochondria are the only animal organelles that contain multiple copies of their own genome, which code for 37 genes: 13 polypeptides, 22 transfer RNAs, and 2 (small and large) ribosomal RNAs. Mitochondrial DNA (mtDNA) in humans is 16,569 base pairs-long and it contributes to oxidative phosphorylation (OXPHOS) components needed for ATP production. In an assembly line fashion, the electron chain transports electrons through Complexes I–IV to drive a protein gradient in the inner membrane space for Complex V, an ATPase synthase pump to produce adenosine triphosphate (ATP). Mitochondrial DNA, which is transcribed and translated in the mitochondria, contributes subunits to all of these complexes except Complex II (25).

A connection between senescence, mtDNA deletions, and neuronal dysfunction has been reported; however, the details to date are still unclear. Aging postmortem brain tissues of healthy individuals and those suffering from neurodegenerative diseases show accumulation of partially deleted mtDNAs (26, 27, 28). The most commonly reported deletion in humans, named the "common deletion," of 4,977 base pairs in length lies between nucleotides 8,470–8,483 to 13,447–13,459 and is thought to arise by intramolecular recombination, possibly because of the stalling of the mtDNA polymerase during replication of the genome (reviewed in (29)). Replication of the genome is not the only way to introduce large-scale deletions found in mtDNA. Pauses in replication can also cause double-strand breaks (DSB) that are recombinogenic. Accordingly, DSB in mtDNA, elicited by restriction enzymes targeted to mitochondria in vivo, undergo homologous and nonhomologous repair to create a multiple array of diverse deletions. Using a mouse model with a mitochondria-targeted restriction endonuclease, it was shown that partially deleted mtDNAs had a replicative advantage over the full-length wild-type molecule in both postmitotic neurons and muscle fibers (30, 31). Natural mechanisms of double-strand breaking could occur during replication pausing, ultraviolent radiation, or oxidative stress with repair of these damages as a source for deletions (reviewed in (32)). In postmortem tissues, these partially deleted mtDNAs were found at different levels in regionally different areas within the human brain (27). Because their overall levels are relatively low, the functional significance of these mutated mtDNA species is unknown. However, the accumulation of mtDNA deletions at relatively high levels (40–60%) has been reported in the aging substantia nigra, and these high deletion levels were associated with a cytochrome *c* oxidase (COX) deficiency in individual neurons (26, 33, 34).

These large missing portions of DNA spanning the mitochondrial genome are not the only alterations found over time in aging organisms. Somatic point mutations in the mtDNA can lead to both silent, missense, and nonsense mutations varying in the degrees of functional change depending on their location. In human AD patients, there is an increase in the amount of point mutations accumulating in the control region of the genome positively correlating with age (35). However, this finding may not extend to the rest of the genome as an increased burden of mutations was also found in age-matched patients whether they were AD positive or nondiseased (36). In a *Drosophilia melanogaster* model where mtDNA point mutations were introduced in the germline, animals harboring different base pair substitutions changing the amino acid code in the cytochrome *c* oxidase subunit I gene showed age-related neurodegenerative phenyotypes and myopathies (37).

3.1. Models of mtDNA Deletions/Mutations Mimicking Neurodegeneration

Transgenic mouse models affecting the repair or replicative abilities of the mtDNA in neuronal specific populations have been reported. A targeted dopaminergic mitochondrial transcription factor A (TFAM) knockout mouse developed parkinsonian-like symptoms after severe mtDNA depletion causing a specific loss of this neuronal population (38). Another TFAM knockout mouse targetting forebrain neurons showed similar results in this different population of neurons leading to COX-deficient cells, neuronal loss, astrogliosis, and increased suspectibility to kaniac acid insults (39). Trans-neuronal degeneration has also been reported in chimeric mitochondrial late-onset neurodegeneration (MILON) mice. Defect in specific neurons caused the death of both neurons with mtDNA depletion (caused by TFAM deletion) and normal OXPHOS functioning neurons nearby (40). Another transgenic model reported to accumulate mtDNA deletions and point mutations by mutating the catalytic proofreading subunit of polymerase gamma (polγ) reported high mutations in the brain with coinciding respiratory defects (41). However, two other independent groups with a systematic and neuronal-specific mutant polγ model did not report neurodegeneration in the brain (42, 43). New reported evidence indicates that previously reported deletions from this model may really be replication intermediates due to breakage at the origins of the heavy and light strand (44).

3.2. Future Directions of Understanding mtDNA Mutations

Severity and time of induction for these alterations in the mtDNA are limiting factors in the current in vivo models in teasing out this relationship between age-related neurodegeneration and the state of mitochondrial genome. The accumulation of mtDNA mutations or depletion levels cannot be controlled in these models allowing them to accrue at folds higher than reported in postmortem studies. This leaves us with the question of whether the physiological levels of mtDNA mutations seen during natural aging contributed to functional impairment. The induction of a defect during development or in a mature organism also matters and can change the possible phenotypic outcomes observed. With the same mtDNA mutation load in the heart, the polγ mouse accumulates mutations beginning at 7 days postconception and the cardiac-specific version of this mouse with mutations generated after birth, show distinctly different cardiac phenotypes (reviewed in (45)). Finally, most models manipulate genes with products that control mitochondrial genome replication or repair (reviewed in (46)). These genetic manipulations leave the mitochondrial genome in a state of defenselessness exacerbating deficits that would normally be repaired or compensated for, even in postmitotic neurons. Better models are needed to better define the role of mtDNA and its ability to cause neurodegenerative events.

New models are being developed to study mtDNA deletions in vivo and in culture specifically to study the mutations impact on neurons. Fukui and Moraes created a mouse model that can inducibly express a mitochondria-targeted restriction endonuclease (*Pst*I). They showed that expression of mito-Pst1 leads to the formation of large-scale deletions (30). Another new exciting model has created neurons derived from embryonic stem cell (ES) cybrids harboring mtDNA point mutations. These models have already shown how point mutations in complexes I and IV affect electrophysiological properties and neurodegeneration in a neuronal cell phenotype (47, 48).

In the future, these models may better reflect normal aging and neurodegenerative processes. With the biggest risk factor for neurodegenerative diseases being aging, accumulation of mtDNA mutations in postmitotic cells could have a major impact on senescence (reviewed in (49)).

4. Reactive Oxygen Species Role in Neurodegeneration

Due to the high-energy demand of brain tissues, impairments in OXPHOS can have particularly severe consequences in the CNS. ATP synthesis is highly correlated with brain activity levels (50). However, mitochondrial oxidative functions may cause unwanted damage to the system over time. The electron transport chain (ETC) is not perfect and mitochondria produce reactive oxygen species (ROS) from an estimated 0.2–2% of the oxygen consumed. Most of this production is observed at complexes I and III where large changes in the energy potential of electrons relative to oxygen levels occur (reviewed in (51)). ROS are free stable oxygen-based molecules with an unpaired electron with the ability to be donated or accepted by DNA, proteins, fatty acids, or other biological molecules causing unwarranted modifications. Superoxide molecules (O_2^-) can also react with hydrogen to become hydrogen peroxide (H_2O_2). Although ROS can have functional roles, such as promote cell growth (52), they can also signal apoptosis and global changes in gene expression (reviewed in (53)). The mitochondria has evolved multiple defense mechanisms against ROS to neutralize hydrogen and oxygen-based insults (reviewed in (54)). Superoxide dismutase 1, 2 (SOD1, SOD2), glutathione and catalase are antioxidant defense enzymes used to scavenge these molecules to safely deduce them to hydrogen peroxide or H_2O and O_2, respectively. ROS is a recognized source of damage the cell needs to protect itself against. Although it is generally accepted that oxidative damage is part of the aging process (reviewed in (55)), the exact role of ROS in normal aging and neurodegeneration is far from understood.

4.1. Aging, ROS, and Neurodegeneration

The free radical theory of aging states that a vicious cycle of free radical damage harming mtDNA and OXPHOS machinery leads to the feedforward production of more ROS (56). However, a number of recent reports provided evidence against this hypothesis. In the polγ model, which was discussed earlier, even though the causation of why these mice have defective OXPHOS functioning is unknown, they do not show increased ROS levels in the brain (41, 42). And, other genetic defects decreasing OXPHOS function do not seem to lead to increases in ROS or oxidative stress (reviewed in (57)). Also, common mouse mtDNA variants display different phenotypes with changes in ROS levels and ROS sensitive enzymes all seemingly compensate to have the same overall cellular respiration between different haplotypes (58).

There is conflicting evidence that ROS plays a role in age-related degeneration. It seems that superoxide formation has a strong inverse relation to maximum lifespan potential (LSP); and the rate of neurodegeneration is also inversely related to LSP (59). Whether this is actually a causal relationship is unknown. Nuclear DNA undergoes ROS damage in a regional specific manner in the brain increasing with age (60). These different regional damages seen in protein carbonyl modifications by ROS also seem to have correlated age-associated decline behaviorally in cognition and motor coordination (61). But definitive evidence demonstrating that differences in brain regions or neuronal populations have different amounts of ROS production or energy consumption demands has yet to be shown.

4.2. Nonmitochondrial Sources of ROS and Neurodegeneration

Seemingly, although nonmitochondrial ROS production from lipid metabolism, cytosolic enzymes, and NADPH oxidases are thought to contribute only a tenth of what mitochondrial sources of ROS do, they are often associated with neurodegenerative diseases. This has been most extensively shown with AD, but not exclusively. AD brains with abnormal ceramide metabolism and free cholesterol levels had increased 4-hydroxynoneal (HNE) adducts that are indicative of oxidative modifications, but from a nonmitochondrial source of damage (62). HNE adducts are found to be highly reactive and when decreasing aldehyde dehydrogenase 2 activity in transgenic mice, there are increases in ROS adjunct proteins and neurodegeneration with decreases in cognition (63). In addition, mitochondrial ROS may be a secondary contributor after an initial activation of NADPH oxidase by Aβ in AD causing the dysfunction of supporting astrocytes unable to maintain neurons (64). Recently, the involvement of NADPH oxidase as the main contributor to superoxide formation, as opposed to having a mitochondrial origin, was associated with excitotoxic NMDA receptor-mediated death in primary culture and in vivo (65). These studies reinforce the idea that mitochondria are not the sole generators of ROS.

4.3. Future Directions on the Mitochondrial ROS and Neurodegeneration Connection

How much ROS damage produced by mitochondrial dysfunction leads to neuronal cell death in diseased states? It is not clear whether the source of this damage is in any way mitochondrial based during each neurodegenerative insult. The effects of oxidative stress are frequently reported; however, the measurements of the quantity and source are not easy to perform or interpret. Conclusions cannot be drawn from isolated mitochondria preparations held with reduction/oxidization conditions at unnatural states. The relationship between mitochondrial function, ROS damage, and neurodegeneration is not clear-cut, as these three can possibly be very separate and independent phenomena.

5. Protein Aggregations Associated with Mitochondria

In most neurodegenerative diseases, two events appear to coincide: a region-specific degeneration of neuronal processes and a misfolding/aggregation of abnormal proteins (66). The composition and identity of these mutant proteins differ depending on the type of neurodegenerative disease; however, despite their differences, at some stage of their aggregation, they all seem to negatively impact on neuronal health. Some of the most common examples of these aggregations found in neurodegenerative diseases include: Aβ and hyperphosphorylated tau in Alzheimer's disease, α-synuclein in Parkinson's disease, and polyglutamine (CAG) repeats on protein in Huntington's disease (67). This correlative relationship between different forms of protein aggregates and the rate of neurodegeneration is not under debate; however, the causal relationship is still questionable. Evidence for abnormal protein aggregations causing a molecular cascade of deteriorating events has been shown, and it is the increased proper clearance of intermediate oligomers (as opposed to endpoint plaques or building monomers), that improves symptoms of neurodegeneration (66, 68, 69). These aggregated or mutant proteins in these pathologies have been shown to affect mitochondria, leading to dysfunction as an additional source of organelle stress.

5.1. Common Aggregates and Mitochondrial Associations

Aβ has been found to interact with mitochondria and other proteins thus leading to neuronal dysfunctions. Cyclophilin D, a suspected mitochondrial permeability transition pore protein, has an antagonistic association with Aβ leading to decreases in calcium buffering capacity, respiration, complex IV activity, and subsequently causes dysfunction at the synaptic and neuronal level (70). Mice overexpressing Amyloid Precursor Protein (APP) show abnormal increases in Aβ that cause decreases in complex IV activity, ATP synthesis, and loss of membrane potential (71). Aβ interacting with Aβ alcohol dehydrogenase (ABAD) also has been shown to cause mitochondrial dysfunction, increased ROS, and increased cell death (72).

There is also evidence of a mitochondria/aggregate interaction in Parkinson's disease (PD). The accumulations of overexpressed α-synuclein protein specifically inhibits Complex I activity in PD (73). Aged yeast models of PD that overexpress wild-type and mutant A52T α-synuclein demonstrate that functional mitochondria with OXPHOS capacity are required and necessary for this protein's toxicity (74). Fluorescence resonance energy transfer (FRET) and biochemical fractionation gradient assays further confirmed a α-synuclein protein interaction with native brain mitochondrial membranes (75). This association may even be extended to PD-related proteins besides α-synuclein. As mentioned above, PINK1 has been shown to both influence mitochondrial morphology and function. Parkin, associated with familial forms of PD, is recruited selectively to dysfunctional mitochondria with low membrane potential for mitophagy elimination. Failure to remove these dysfunctional organelles contributes to the pathogenesis of PD in these families (76, 77).

A mitochondrial involvement with huntingtin aggregates has been described for Huntington's disease (HD). The mutant huntingtin protein, which aggregates once the CAG polyglutamine repeat reaches a certain critical threshold, has been shown to directly affect energy metabolism (ATP/ADP ratio), $\Delta\psi_m$, and intracellular Ca^{+2} levels (78). Mutant huntingtin has also been showed to cause a transcriptional repression of PGC-1α, a transcription factor regulating the control of mitochondrial biogenesis in the cell, leading to another source of dysfunction indirect from immediate associations with the mitochondria (79).

Likewise, impairment in mitochondrial function has also been reported in mouse models of ALS that involve mutations in the Cu, Zn superoxide dismutase (SOD1). Both wild-type and mutant G93A-SOD1 have been found to reside inside the mitochondrial matrix with mutant SOD1 forming high molecular aggregations here in an age and symptomatic-dependent manner (80). Overexpression of the copper chaperone responsible for the maturation of SOD1 in the G93A-SOD1 transgenic mouse background leads to an accelerated neurodegenerative phenotype, mitochondrial vacuole formation, and increasing OXPHOS deficiencies (81).

Mitochondrial perturbations in astrocytes and other glia-derived cell types can potentiate death in neurons. ALS-associated G93A-SOD1 mutation in astrocytes displays severe impairment of oxygen consumption and ADP respiratory control causing a decline in motor neuron survival (82). Microglia activates neuroinflammatory responses where it synergistically acts with α-synuclein overexpression models to contribute to the demise of dopaminergic neuron (83).

5.2. Future Directions on Aggregates and Mitochondrial Dysfunctions

Although there has been direct evidence that abnormal or overabundant accumulations of aggregation-prone proteins affect mitochondria, the causal mechanism is still a topic of debate. Complex IV dysfunction in AD was suggested to precede and perhaps initiate AD-protein aggregation; however, recently this was disproven as complex IV impairment has been shown to decrease Aβ accumulations in a mouse model (84). A more likely scenario is one in which the abnormal Aβ accumulation affects mitochondrial function and synaptic function (reviewed in (85)).

6. Final Remarks and Conclusions

What role does mitochondrial stress play in neurodegenerative diseases? The massive amount of evidence supporting the importance of mitochondrial integrity in neurodegenerative events cannot be ignored. However, more scientific research is needed in all four of the areas discussed in this review to understand the complex cellular and molecular mechanisms of neurodegeneration (Fig. 1). Hopefully, this knowledge will lead to more preventative and curative measures for neurodegenerative disorders.

References

1. Guo X, Macleod GT, Wellington A, Hu F, Panchumarthi S, Schoenfield M, Marin L, Charlton MP, Atwood HL, Zinsmaier KE (2005) The GTPase dMiro is required for axonal transport of mitochondria to Drosophila synapses. Neuron 47:379–393
2. Verstreken P, Ly CV, Venken KJ, Koh TW, Zhou Y, Bellen HJ (2005) Synaptic mitochondria are critical for mobilization of reserve pool vesicles at Drosophila neuromuscular junctions. Neuron 47:365–378
3. Li Z, Okamoto K, Hayashi Y, Sheng M (2004) The importance of dendritic mitochondria in the morphogenesis and plasticity of spines and synapses. Cell 119:873–887
4. Lee D, Lee KH, Ho WK, Lee SH (2007) Target cell-specific involvement of presynaptic mitochondria in post-tetanic potentiation at hippocampal mossy fiber synapses. J Neurosci 27:13603–13613
5. Stowers RS, Megeath LJ, Gorska-Andrzejak J, Meinertzhagen IA, Schwarz TL (2002) Axonal transport of mitochondria to synapses depends on milton, a novel Drosophila protein. Neuron 36:1063–1077
6. Wang X, Schwarz TL (2009) The mechanism of Ca2+ -dependent regulation of kinesin-mediated mitochondrial motility. Cell 136: 163–174
7. Macaskill AF, Rinholm JE, Twelvetrees AE, Arancibia-Carcamo IL, Muir J, Fransson A, Aspenstrom P, Attwell D, Kittler JT (2009) Miro1 is a calcium sensor for glutamate receptor-dependent localization of mitochondria at synapses. Neuron 61:541–555
8. Chen H, Chan DC (2005) Emerging functions of mammalian mitochondrial fusion and fission. Hum Mol Genet 14(Spec No. 2):R283–R289
9. Knott AB, Perkins G, Schwarzenbacher R, Bossy-Wetzel E (2008) Mitochondrial fragmentation in neurodegeneration. Nat Rev Neurosci 9:505–518
10. Chen H, Chan DC (2006) Critical dependence of neurons on mitochondrial dynamics. Curr Opin Cell Biol 18:453–459
11. Delettre C, Lenaers G, Griffoin JM, Gigarel N, Lorenzo C, Belenguer P, Pelloquin L, Grosgeorge J, Turc-Carel C, Perret E, Astarie-Dequeker C, Lasquellec L, Arnaud B, Ducommun B, Kaplan J, Hamel CP (2000) Nuclear gene OPA1, encoding a mitochondrial dynamin-related protein, is mutated in dominant optic atrophy. Nat Genet 26:207–210
12. Zuchner S, Mersiyanova IV, Muglia M, Bissar-Tadmouri N, Rochelle J, Dadali EL, Zappia M, Nelis E, Patitucci A, Senderek J, Parman Y, Evgrafov O, Jonghe PD, Takahashi Y, Tsuji S, Pericak-Vance MA, Quattrone A, Battaloglu E, Polyakov AV, Timmerman V, Schroder JM, Vance JM (2004) Mutations in the mitochondrial GTPase mitofusin 2 cause Charcot-Marie-Tooth neuropathy type 2A. Nat Genet 36:449–451
13. Olichon A, Baricault L, Gas N, Guillou E, Valette A, Belenguer P, Lenaers G (2003) Loss of OPA1 perturbates the mitochondrial inner membrane structure and integrity, leading to cytochrome c release and apoptosis. J Biol Chem 278:7743–7746
14. Chen H, McCaffery JM, Chan DC (2007) Mitochondrial fusion protects against neurodegeneration in the cerebellum Cell 130: 548–562
15. Manfredi G, Beal MF (2007) Merging mitochondria for neuronal survival. Nat Med 13:1140–1141
16. Kitada T, Pisani A, Porter DR, Yamaguchi H, Tscherter A, Martella G, Bonsi P, Zhang C, Pothos EN, Shen J (2007) Impaired dopamine release and synaptic plasticity in the

striatum of PINK1-deficient mice. Proc Natl Acad Sci U S A 104:11441–11446
17. Poole AC, Thomas RE, Andrews LA, McBride HM, Whitworth AJ, Pallanck LJ (2008) The PINK1/Parkin pathway regulates mitochondrial morphology. Proc Natl Acad Sci U S A 105:1638–1643
18. Gandhi S, Wood-Kaczmar A, Yao Z, Plun-Favreau H, Deas E, Klupsch K, Downward J, Latchman DS, Tabrizi SJ, Wood NW, Duchen MR, Abramov AY (2009) PINK1-associated Parkinson's disease is caused by neuronal vulnerability to calcium-induced cell death. Mol Cell 33:627–638
19. Wang X, Su B, Siedlak SL, Moreira PI, Fujioka H, Wang Y, Casadesus G, Zhu X (2008) Amyloid-beta overproduction causes abnormal mitochondrial dynamics via differential modulation of mitochondrial fission/fusion proteins. Proc Natl Acad Sci U S A 105:19318–19323
20. Cho DH, Nakamura T, Fang J, Cieplak P, Godzik A, Gu Z, Lipton SA (2009) S-nitrosylation of Drp1 mediates beta-amyloid-related mitochondrial fission and neuronal injury. Science 324:102–105
21. Orr AL, Li S, Wang CE, Li H, Wang J, Rong J, Xu X, Mastroberardino PG, Greenamyre JT, Li XJ (2008) N-terminal mutant huntingtin associates with mitochondria and impairs mitochondrial trafficking. J Neurosci 28:2783–2792
22. Wang H, Lim PJ, Karbowski M, Monteiro MJ (2009) Effects of overexpression of huntingtin proteins on mitochondrial integrity. Hum Mol Genet 18:737–752
23. Jordi Magrané, Isabel Hervias , Matthew S. Henning, Maria Damiano, Hibiki Kawamata and Giovanni Manfredi (2009) Mutant SOD1 in neuronal mitochondria causes toxicity and mitochondrial dynamics abnormalities. Hum Mol Genetics 18(23):4552–4564
24. Misko A, Jiang S, Wegorzewska I, Milbrandt J, Baloh RH. (2010) Mitofusin 2 is necessary for transport of axonal mitochondria and interacts with the Miro/Milton complex." J Neurosci 30(12):4232–40
25. Anderson S, Bankier AT, Barrell BG, de Bruijn MH, Coulson AR, Drouin J, Eperon IC, Nierlich DP, Roe BA, Sanger F, Schreier PH, Smith AJ, Staden R, Young IG (1981) Sequence and organization of the human mitochondrial genome. Nature 290:457–465
26. Bender A, Krishnan KJ, Morris CM, Taylor GA, Reeve AK, Perry RH, Jaros E, Hersheson JS, Betts J, Klopstock T, Taylor RW, Turnbull DM (2006) High levels of mitochondrial DNA deletions in substantia nigra neurons in aging and Parkinson disease. Nat Genet 38:515–517
27. Corral-Debrinski M, Horton T, Lott MT, Shoffner JM, Beal MF, Wallace DC (1992) Mitochondrial DNA deletions in human brain: regional variability and increase with advanced age. Nat Genet 2:324–329
28. Soong NW, Hinton DR, Cortopassi G, Arnheim N (1992) Mosaicism for a specific somatic mitochondrial DNA mutation in adult human brain. Nat Genet 2:318–323
29. Krishnan KJ, Greaves LC, Reeve AK, Turnbull DM (2007) Mitochondrial DNA mutations and aging. Ann N Y Acad Sci 1100:227–240
30. Fukui H, Moraes CT (2009) Mechanisms of formation and accumulation of mitochondrial DNA deletions in aging neurons. Hum Mol Genet 18:1028–1036
31. Srivastava S, Moraes CT (2005) Double-strand breaks of mouse muscle mtDNA promote large deletions similar to multiple mtDNA deletions in humans. Hum Mol Genet 14:893–902
32. Krishnan KJ, Reeve AK, Samuels DC, Chinnery PF, Blackwood JK, Taylor RW, Wanrooij S, Spelbrink JN, Lightowlers RN, Turnbull DM (2008) What causes mitochondrial DNA deletions in human cells? Nat Genet 40:275–279
33. Kraytsberg Y, Kudryavtseva E, McKee AC, Geula C, Kowall NW, Khrapko K (2006) Mitochondrial DNA deletions are abundant and cause functional impairment in aged human substantia nigra neurons. Nat Genet 38:518–520
34. Reeve AK, Krishnan KJ, Elson JL, Morris CM, Bender A, Lightowlers RN, Turnbull DM (2008) Nature of mitochondrial DNA deletions in substantia nigra neurons. Am J Hum Genet 82:228–235
35. Coskun PE, Beal MF, Wallace DC (2004) Alzheimer's brains harbor somatic mtDNA control-region mutations that suppress mitochondrial transcription and replication. Proc Natl Acad Sci U S A 101:10726–10731
36. Lin MT, Simon DK, Ahn CH, Kim LM, Beal MF (2002) High aggregate burden of somatic mtDNA point mutations in aging and Alzheimer's disease brain. Hum Mol Genet 11:133–145
37. Xu H, DeLuca SZ, O'Farrell PH (2008) Manipulating the metazoan mitochondrial genome with targeted restriction enzymes. Science 321:575–577
38. Ekstrand MI, Terzioglu M, Galter D, Zhu S, Hofstetter C, Lindqvist E, Thams S, Bergstrand A, Hansson FS, Trifunovic A,

Hoffer B, Cullheim S, Mohammed AH, Olson L, Larsson NG (2007) Progressive parkinsonism in mice with respiratory-chain-deficient dopamine neurons. Proc Natl Acad Sci U S A 104:1325–1330
39. Sorensen L, Ekstrand M, Silva JP, Lindqvist E, Xu B, Rustin P, Olson L, Larsson NG (2001) Late-onset corticohippocampal neurodepletion attributable to catastrophic failure of oxidative phosphorylation in MILON mice. J Neurosci 21:8082–8090
40. Dufour E, Terzioglu M, Sterky FH, Sorensen L, Galter D, Olson L, Wilbertz J, Larsson NG (2008) Age-associated mosaic respiratory chain deficiency causes trans-neuronal degeneration. Hum Mol Genet 17:1418–1426
41. Trifunovic A, Wredenberg A, Falkenberg M, Spelbrink JN, Rovio AT, Bruder CE, Bohlooly YM, Gidlof S, Oldfors A, Wibom R, Tornell J, Jacobs HT, Larsson NG (2004) Premature ageing in mice expressing defective mitochondrial DNA polymerase. Nature 429:417–423
42. Kujoth GC, Hiona A, Pugh TD, Someya S, Panzer K, Wohlgemuth SE, Hofer T, Seo AY, Sullivan R, Jobling WA, Morrow JD, Van Remmen H, Sedivy JM, Yamasoba T, Tanokura M, Weindruch R, Leeuwenburgh C, Prolla TA (2005) Mitochondrial DNA mutations, oxidative stress, and apoptosis in mammalian aging. Science 309:481–484
43. Kasahara T, Kubota M, Miyauchi T, Noda Y, Mouri A, Nabeshima T, Kato T (2006) Mice with neuron-specific accumulation of like phenotypes. Mol Psychiatry 11(577–593):23
44. Bailey LJ, Cluett TJ, Reyes A, Prolla TA, Poulton J, Leeuwenburgh C, Holt IJ (2009) Mice expressing an error-prone DNA polymerase in mitochondria display elevated replication pausing and chromosomal breakage at fragile sites of mitochondrial DNA. Nucleic Acids Res 37:2327–2335
45. Khrapko K, Vijg J (2007) Mitochondrial DNA mutations and aging: a case closed? Nat Genet 39:445–446
46. Tyynismaa H, Suomalainen A (2009) Mouse models of mitochondrial DNA defects and their relevance for human disease. EMBO Rep 10:137–143
47. Trevelyan AJ, Kirby DM, Smulders-Srinivasan TK, Nooteboom M, Acin-Perez R, Enriquez JA, Whittington MA, Lightowlers RN, Turnbull DM. (2010) Mitochondrial DNA mutations affect calcium handling in differentiated neurons. Brain. 133(Pt 3):787–96
48. Abramov AY, Smulders-Srinivasan TK, Kirby DM, Acin-Perez R, Enriquez JA, Lightowlers RN, Duchen MR, Turnbull DM. (2010) Mechanism of neurodegeneration of neurons with mitochondrial DNA mutations. Brain. 133(Pt 3):797–807
49. Mattson MP, Magnus T (2006) Ageing and neuronal vulnerability. Nat Rev Neurosci 7:278–294
50. Du F, Zhu XH, Zhang Y, Friedman M, Zhang N, Ugurbil K, Chen W (2008) Tightly coupled brain activity and cerebral ATP metabolic rate. Proc Natl Acad Sci U S A 105: 6409–6414
51. Balaban RS, Nemoto S, Finkel T (2005) Mitochondria, oxidants, and aging. Cell 120:483–495
52. Xia C, Meng Q, Liu LZ, Rojanasakul Y, Wang XR, Jiang BH (2007) Reactive oxygen species regulate angiogenesis and tumor growth through vascular endothelial growth factor. Cancer Res 67:10823–10830
53. Giorgio M, Trinei M, Migliaccio E, Pelicci PG (2007) Hydrogen peroxide: a metabolic byproduct or a common mediator of ageing signals? Nat Rev Mol Cell Biol 8:722–728
54. Lin MT, Beal MF (2006) Mitochondrial dysfunction and oxidative stress in neurodegenerative diseases. Nature 443:787–795
55. Finkel T, Holbrook NJ (2000) Oxidants, oxidative stress and the biology of ageing. Nature 408:239–247
56. Harman D (1956) Aging: a theory based on free radical and radiation chemistry. J Gerontol 11:298–300
57. Fukui H, Moraes CT (2008) The mitochondrial impairment, oxidative stress and neurodegeneration connection: reality or just an attractive hypothesis? Trends Neurosci 31:251–256
58. Moreno-Loshuertos R, Acin-Perez R, Fernandez-Silva P, Movilla N, Perez-Martos A, Rodriguez de Cordoba S, Gallardo ME, Enriquez JA (2006) Differences in reactive oxygen species production explain the phenotypes associated with common mouse mitochondrial DNA. Nat Genet 38:1261–1268
59. Wright AF, Jacobson SG, Cideciyan AV, Roman AJ, Shu X, Vlachantoni D, McInnes RR, Riemersma RA (2004) Lifespan and mitochondrial control of neurodegeneration. Nat Genet 36:1153–1158
60. Cardozo-Pelaez F, Brooks PJ, Stedeford T, Song S, Sanchez-Ramos J (2000) DNA damage, repair, and antioxidant systems in brain regions: a correlative study. Free Radic Biol Med 28:779–785
61. Forster MJ, Dubey A, Dawson KM, Stutts WA, Lal H, Sohal RS (1996) Age-related losses of cognitive function and motor skills in mice are associated with oxidative protein

damage in the brain. Proc Natl Acad Sci U S A 93:4765–4769
62. Cutler RG, Kelly J, Storie K, Pedersen WA, Tammara A, Hatanpaa K, Troncoso JC, Mattson MP (2004) Involvement of oxidative stress-induced abnormalities in ceramide and cholesterol metabolism in brain aging and Alzheimer's disease. Proc Natl Acad Sci U S A 101:2070–2075
63. Ohsawa I, K, Murakami Y, Suzuki Y, Ishikawa M, Ohta S (2008) Age-dependent neurodegeneration accompanying memory loss in transgenic mice defective in mitochondrial aldehyde dehydrogenase 2 activity. J Neurosci 28:6239–6249
64. Abramov AY, Canevari L, Duchen MR (2004) Beta-amyloid peptides induce mitochondrial dysfunction and oxidative stress in astrocytes and death of neurons through activation of NADPH oxidase. J Neurosci 24:565–575
65. Brennan AM, Suh SW, Won SJ, Narasimhan P, Kauppinen TM, Lee H, Edling Y, Chan PH, Swanson RA (2009) NADPH oxidase is the primary source of superoxide induced by NMDA receptor activation. Nat Neurosci 12:857–863
66. Lansbury PT, Lashuel HA (2006) A century-old debate on protein aggregation and neurodegeneration enters the clinic. Nature 443:774–779
67. Rubinsztein DC (2006) The roles of intracellular protein-degradation pathways in neurodegeneration. Nature 443:780–786
68. Oddo S, Billings L, Kesslak JP, Cribbs DH, LaFerla FM (2004) Abeta immunotherapy leads to clearance of early, but not late, hyperphosphorylated tau aggregates via the proteasome. Neuron 43:321–332
69. Morgan D, Diamond DM, Gottschall PE, Ugen KE, Dickey C, Hardy J, Duff K, Jantzen P, DiCarlo G, Wilcock D, Connor K, Hatcher J, Hope C, Gordon M, Arendash GW (2000) A beta peptide vaccination prevents memory loss in an animal model of Alzheimer's disease. Nature 408: 982–985
70. Du H, Guo L, Fang F, Chen D, Sosunov AA, McKhann GM, Yan Y, Wang C, Zhang H, JD, Gunn-Moore FJ, Vonsattel JP, Arancio O, Chen JX, Yan SD (2008) Cyclophilin D deficiency attenuates mitochondrial and neuronal perturbation and ameliorates learning and memory in Alzheimer's disease. Nat Med 14:1097–1105
71. Lin MT, Beal MF (2006) Alzheimer's APP mangles mitochondria. Nat Med 12:1241–1243
72. Lustbader JW, Cirilli M, Lin C, Xu HW, Takuma K, Wang N, Caspersen C, Chen X, Pollak S, Chaney M, Trinchese F, Liu S, Gunn-Moore F, Lue LF, Walker DG, Kuppusamy P, Zewier ZL, Arancio O, Stern D, Yan SS, Wu H (2004) ABAD directly links Abeta to mitochondrial toxicity in Alzheimer's disease. Science 304:448–452
73. Devi L, Raghavendran V, Prabhu BM, Avadhani NG, Anandatheerthavarada HK (2008) Mitochondrial import and accumulation of alpha-synuclein impair complex I in human dopaminergic neuronal cultures and Parkinson disease brain. J Biol Chem 283: 9089–9100
74. Buttner S, Bitto A, Ring J, Augsten M, Zabrocki P, Eisenberg T, Jungwirth H, Hutter S, Carmona-Gutierrez D, Kroemer G, Winderickx J, Madeo F (2008) Functional mitochondria are required for alpha-synuclein toxicity in aging yeast. J Biol Chem 283:7554–7560
75. Nakamura K, Nemani VM, Wallender EK, Kaehlcke K, Ott M, Edwards RH (2008) Optical reporters for the conformation of alpha-synuclein reveal a specific interaction with mitochondria. J Neurosci 28: 12305–12317
76. Narendra D, Tanaka A, Suen DF, Youle RJ (2008) Parkin is recruited selectively to impaired mitochondria and promotes their autophagy. J Cell Biol 183:795–803
77. Narendra D, Tanaka A, Suen DF, Youle RJ (2009) Parkin-induced mitophagy in the pathogenesis of Parkinson disease. Autophagy 5:706–708
78. Seong IS, Ivanova E, Lee JM, Choo YS, Fossale E, Anderson M, Gusella JF, Laramie JM, Myers RH, Lesort M, MacDonald ME (2005) HD CAG repeat implicates a dominant property of huntingtin in mitochondrial energy metabolism. Hum Mol Genet 14:2871–2880
79. Cui L, Jeong H, Borovecki F, Parkhurst CN, Tanese N, Krainc D (2006) Transcriptional repression of PGC-1alpha by mutant huntingtin leads to mitochondrial dysfunction and neurodegeneration. Cell 127:59–69
80. Vijayvergiya C, Beal MF, Buck J, Manfredi G (2005) Mutant superoxide dismutase 1 forms aggregates in the brain mitochondrial matrix of amyotrophic lateral sclerosis mice. J Neurosci 25:2463–2470
81. Son M, Puttaparthi K, Kawamata H, Rajendran B, Boyer PJ, Manfredi G, Elliott JL (2007) Overexpression of CCS in G93A-SOD1 mice leads to accelerated neurological deficits with severe mito-

chondrial pathology. Proc Natl Acad Sci U S A 104:6072–6077

82. Cassina P, Cassina A, Pehar M, Castellanos R, Gandelman M, de Leon A, Robinson KM, Mason RP, Beckman JS, Barbeito L, Radi R (2008) Mitochondrial dysfunction in SOD1G93A-bearing astrocytes promotes motor neuron degeneration: prevention by mitochondrial-targeted antioxidants. J Neurosci 28:4115–4122
83. Gao HM, Kotzbauer PT, Uryu K, Leight S, Trojanowski JQ, Lee VM (2008) Neuroinflammation and oxidation/nitration of alpha-synuclein linked to dopaminergic neurodegeneration. J Neurosci 28:7687–7698
84. Fukui H, Diaz F, Garcia S, Moraes CT (2007) Cytochrome c oxidase deficiency in neurons decreases both oxidative stress and amyloid formation in a mouse model of Alzheimer's disease. Proc Natl Acad Sci U S A 104:14163–14168
85. Reddy PH, Beal MF (2008) Amyloid beta, mitochondrial dysfunction and synaptic damage: implications for cognitive decline in aging and Alzheimer's disease. Trends Mol Med 14:45–53

Chapter 5

Autophagy in Disease

Dalibor Mijaljica, Mark Prescott, and Rodney J. Devenish

Abstract

Autophagy is a cellular quality control process by which cytoplasmic constituents including proteins, protein aggregates, organelles, and invading pathogens can be delivered to lysosomes for degradation. Autophagy is activated in response to changes in the internal status of the cell and/or changes in the extracellular environment. It is therefore essential for the maintenance of cellular homeostasis and for an efficient response to cellular stresses. As such autophagy has been implicated either in the pathogenesis, or response to a wide variety of diseases, bacterial, and viral infections, and ageing.

Key words: Autophagy, Degradation, Disease, Pathophysiology, Stress

1. Introduction

Autophagy is a highly conserved process of quality control occurring inside cells by which cytoplasmic constituents including long-lived proteins, protein aggregates, organelles, and invading pathogens can be delivered to lysosomes for degradation and subsequent recycling of the degradation products. Autophagy is activated in response to changes in the internal status of the cell and/or changes in the extracellular milieu. It is therefore essential for the maintenance of cellular homeostasis and for an efficient response to cellular stresses. Autophagy has been implicated either in the pathogenesis or response to a wide variety of diseases, including cancer, neurodegeneration, myopathies, liver, and heart disease, diabetes, as well as bacterial and viral infections, and ageing.

In this chapter, we will first highlight the basic mechanisms and regulation of autophagic processes and then secondly discuss autophagy in human disease. Many aspects that we touch on have

Peter Bross and Niels Gregersen (eds.), *Protein Misfolding and Cellular Stress in Disease and Aging: Concepts and Protocols*, Methods in Molecular Biology, vol. 648, DOI 10.1007/978-1-60761-756-3_5, © Springer Science+Business Media, LLC 2010

been the subject of recent comprehensive reviews and we direct readers interested in the full details of specific aspects of autophagy to the relevant sources (1–14).

2. Autophagy: Basic Principles

2.1. Major Types of Autophagy in Mammalian Cells

In mammalian cells, three forms of autophagic process have been described: macroautophagy, microautophagy, and chaperone-mediated autophagy (CMA). These differ morphologically and mechanistically, and presumably crosstalk between them can occur depending on the particular circumstances, although the details remain to be elucidated. When reports refer to autophagy, unless specified otherwise, it is macroautophagy that is being considered.

2.1.1. Macroautophagy

Macroautophagy involves the sequestration of cytoplasmic constituents including macromolecular structures (e.g., ribosomes), long-lived proteins, protein aggregates, organelles (e.g., mitochondria, peroxisomes, endoplasmic reticulum (ER)), and even pathogens into double-membrane vesicles called autophagosomes (APs). These APs fuse with lysosomal membranes liberating single-membrane vesicles, into the lumen of the lysosome where they are degraded by the resident hydrolases (11, 14) (Fig. 1). The origin of AP membranes remains a matter of debate, with some support for the suggestion that they originate from the ribosome-free ER (11, 15). The mechanistic details of AP formation and degradation have been best studied in the yeast, *Saccharomyces cerevisiae*, facilitated by its genetic malleability and the isolation of mutants defective in autophagy and related processes (16); 31 *ATG* genes (AuTophaGy) have been described. The steps in AP formation and degradation are: (1) induction, (2) AP nucleation, (3) cargo recognition, (4) AP completion, (5) Atg protein cycling, (6) AP fusion with lysosomes, (7) breakdown, and (8) recycling (17). The process of macroautophagy in higher eukaryotes is essentially the same as that in yeast. To date, human orthologs of some 17 yeast *ATG* genes have been identified (4, 18) (Fig. 1, as indicated in brackets). For a more detailed description of the molecular mechanism of macroautophagy, the reader is referred to specialised reviews (4, 17–20).

Extracellular (nutrient starvation, hormone, or pharmacologic treatment) as well as intracellular stimuli (accumulation of misfolded proteins and protein aggregates, invasion by pathogens) are able to modulate the autophagic response. As autophagy occurs at a basal, constitutive level under normal conditions, there must be mechanisms by which extracellular and/or intracellular signals are transmitted to the regulatory factors to promote or

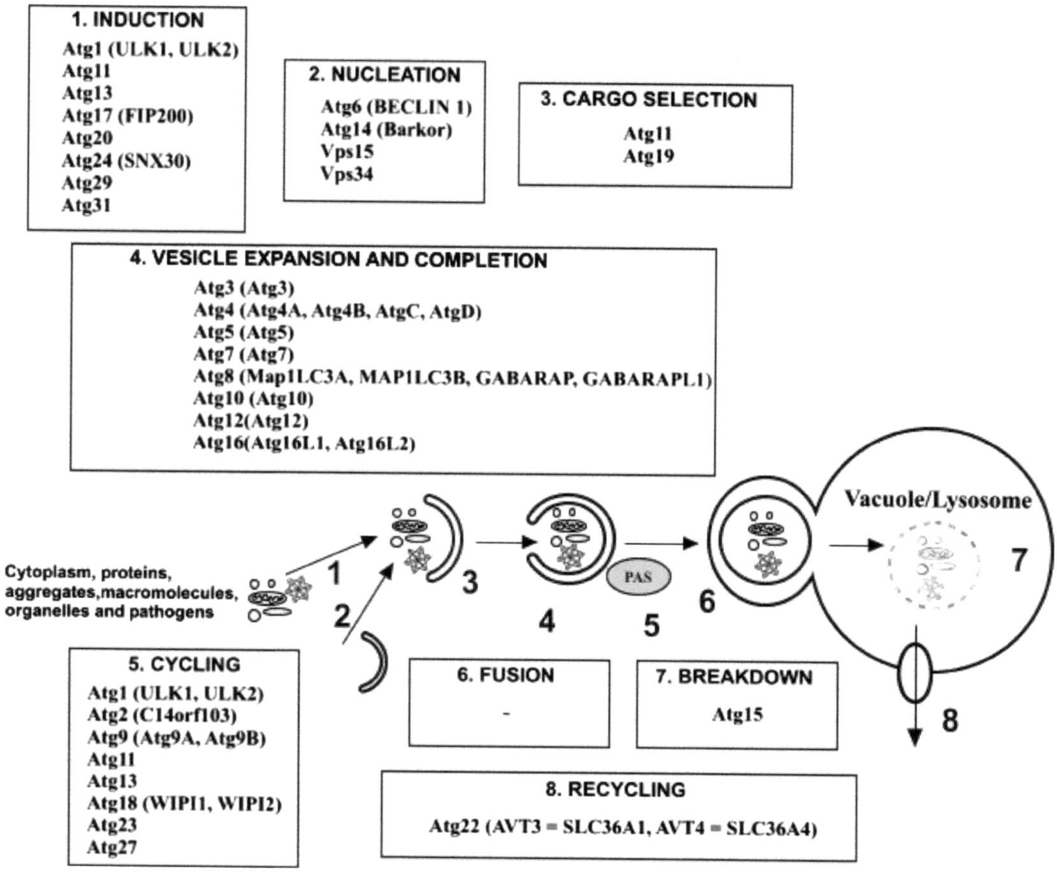

Fig. 1. Schematic representation of the protein machinery involved in macroautophagy. The pathway can be dissected into several steps including (1) induction, (2) vesicle (AP) nucleation, (3) cargo selection, (4) vesicle (AP) expansion and completion, (5) cycling of Atg proteins, (6) fusion, (7) breakdown, and (8) recycling. The Atg proteins (in yeast) and their human homologs/orthologs (as indicated in brackets) involved in each step are shown. *Atg* autophagy related protein, *PAS* pre-autophagosomal structure.

inhibit autophagy when needed (2, 21). Autophagy signalling (relating largely to macroautophagy) is being revealed as increasingly intricate and remains far from fully elucidated. In brief, the central player is a serine/threonine kinase, TOR (*t*arget *o*f *r*apamycin) which has other known roles in regulating cell growth, proliferation, motility, and survival, as well as transcription and protein synthesis (22). Importantly, mammalian TOR (mTOR) is known to be comprised of two distinct complexes: TORC1 and TORC2 (22); the former being a nutrient-sensitive complex which is inhibited by rapamycin, a commonly used pharmacological inducer of autophagy. Despite mTOR being considered a major negative regulator of autophagy, how it controls autophagy is still far from fully elucidated. Most other regulatory molecules (e.g., hormones such as insulin and glucagon, glucose, amino acids (in particular L-glutamine flux; (23)), adenosine triphosphate

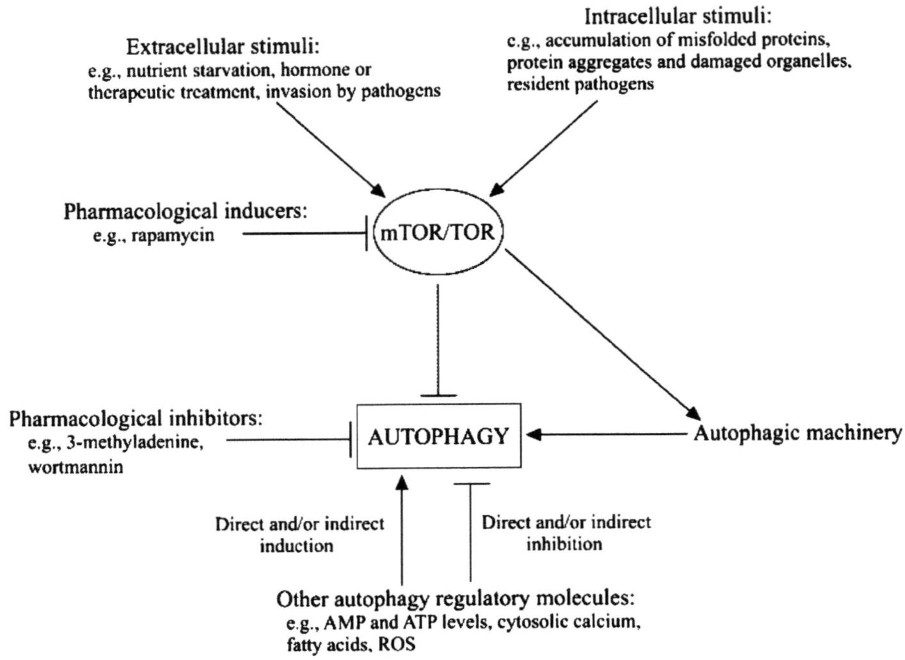

Fig. 2. A simplified model of signalling pathways and major players that regulate macroautophagy. See text for more details. *AMP* adenosine monophosphate, *ATP* adenosine-5′-triphosphate, *(m)TOR* (mammalian) target of rapamycin, *ROS* reactive oxygen species.

(ATP), reactive oxygen species (ROS), calcium, ceramide) either directly or indirectly inhibit or stimulate autophagy through TOR. The cell surface transporters that facilitate amino acid entry into cells for activation of autophagy remain poorly characterised. A simplified model of the signalling which regulates macroautophagy is presented in Fig. 2. Even though a complete and fully integrated picture of autophagy regulation is not currently available, some aspects have been covered in focussed reviews (21, 24, 25).

2.1.2. Microautophagy

During microautophagy, the lysosomal membrane invaginates so as to sequester and internalise cytosolic components in single membrane vesicles (derived from the lysosomal membrane). These vesicles are then degraded in the lysosomal lumen (14). In mammalian cells, microautophagy seems to be unresponsive to classical macroautophagic stimuli and the dissection of the mechanism, together with the identification of proteins involved is still in its infancy. Relatively little is known about the regulation of microautophagy. Indeed, it is rarely considered by investigators when reporting on autophagy. Microautophagic processes are better described in yeast where some components of the characterised macroautophagic machinery have been implicated in the selective microautophagy of peroxisomes (26), the nucleus (27), and mitochondria (28). In yeast, in addition to TOR, complexes involved in the regulation of microautophagy are EGO (a vacuolar membrane-associated protein complex required for the activation of microautophagy during exit from rapamycin-induced growth arrest) (29) and the vacuolar transport chaperone (VTC) complex (30).

2.1.3. Chaperone-Mediated Autophagy

CMA differs from the other two forms of autophagy in that vesicular traffic is not involved in transfer of the "cargo" to the lysosomes. Cytosolic proteins containing an RNAse A, pentapeptide sequence motif, KFERQ, are target substrates for CMA. This signal peptide is recognised by cytosolic HSC70 chaperones, which unfold substrates and promote, with the help of co-chaperones, their docking with a lysosomal membrane receptor, lysosome-associated membrane protein type 2A (LAMP-2A). An intralysosomal HSC70 is required for substrate entry into the lysosomal lumen. CMA is responsible for the degradation of 30% of cytosolic proteins under conditions of prolonged nutrient deprivation. LAMP-2A levels in the lysosomal membrane can be increased by reduced degradation and/or redistribution from the lysosomal lumen to the lysosomal membrane. CMA is also activated by oxidative stress as a result of transcriptional regulation. CMA activity can be reduced by inhibitors of glucose-6-phosphate dehydrogenase, and of the 90 kDa heat shock protein. Furthermore, the levels of cytosolic and lysosomal HSC70 and LAMP-2A vary depending on cell type and result in tissue-specific variations in CMA activity. For example, maximal activation of CMA occurs in organs such as the liver, kidney, heart, and spleen under conditions of stress (e.g., starvation). In other tissues, such as skeletal muscle, CMA is not up-regulated under similar conditions (31).

3. The Functions of Autophagy in Health and Disease

In yeast, autophagy mainly acts as a survival mechanism induced under starvation conditions. However, in mammals (and other multicellular organisms), the roles of autophagy are more diverse. Thus, beyond the classical function of autophagy during starvation, it is involved in programmed cell death and in different tissue-specific functions, and thereby contributes to the pathogenesis of several human diseases, including, cancer, neurodegeneration, muscle disorders (myopathies), liver disease, heart disease, and diabetes (1–14) (Table 1).

3.1. Cancer

The first indication that a defect in autophagy may contribute to tumourigenesis came from the characterisation of the gene encoding Beclin 1 (*BECN1/ATG6*) as a tumour suppressor gene. *BECN1* is mono-allelically deleted in a large proportion of human breast and ovarian cancers (32). Impaired autophagy can contribute to

Table 1
The positive (benefits) and/or negative (disadvantages) contributions of autophagy with respect to the pathogenesis of various human diseases

	Autophagy in human disease	
Condition/disease	**Positive contributions (benefits)**	**Negative contributions (disadvantages)**
Cancer	Removal of damaged organelles Reduces chromosome instability	Resistance of cancer cells to anticancer treatment damage Survival in conditions of low nutrient supply
Neurodegeneration	Prevents accumulation of intracellular proteins and aggregates to toxic levels	Defective autophagy results in accumulation of toxic aggregates and proteins
Myopathies	Prevents accumulation of aggregate prone proteins and/or autophagic vacuoles	Muscle wasting and accumulation of non-degraded APs
Heart disease	Protects during ischemia and pressure overload	Can be deleterious in excess
Liver disease	Degrades portions of the ER (including misfolded proteins), damaged mitochondria, and peroxisomes	Can be deleterious in excess, leading to inflammation, cirrhosis, and cell death
Infection and immunity	Removal of pathogens from host cell Antigen processing for MHC II presentation	Some pathogens can subvert, escape, and exploit autophagy to replicate and survive
Diabetes	Maintenance of the architecture and function of pancreatic β-cells	Reduced ability to produce insulin arising from defective β-cell mitochondria and ER
Ageing and longevity	Removal of damaged organelles Limits production of ROS	Increased ROS production leads to organelle dysfunction (e.g., mitochondria)

APs autophagosome(s), *ER* endoplasmic reticulum, *MHC* major histocompatibility complex, *ROS* reactive oxygen species tumour formation via impaired regulation of cell growth, and/or decreased cell death. Furthermore, it has been shown that failure to sustain metabolism via autophagy (due to allelic loss of *BECN1*, or deficiency in *ATG5*) results in increased DNA damage, gene amplification, aneuploidy, and enhanced chromosomal instability, which in turn enhance tumour progression (32, 33). Another source of genetic instability may arise from autophagy defects that lead to the accumulation of dysfunctional mitochondria, with consequent increased oxidative stress leading to damage to membrane, proteins, and DNA (34) (see also concept Chapter 6 in this volume).

Autophagy can contribute to a form of cell death, usually referred to as Type II programmed cell death or autophagic cell death (ACD) which has different features to apoptosis. ACD is characterised by a large-scale sequestration of portions of the cytoplasm in APs, giving the cell a characteristic vacuolated appearance. ACD also occurs in a caspase-independent manner, without key features of apoptotic morphology, such as chromatin condensation and nuclear fragmentation (35, 36).

The outcome of induction of autophagy can be variable in terms of cell survival and will depend, as indicated above, not only on the genotype of the cell, but also on the environment (for example, metabolic stress induced by the treatment regime). Many of the signals that enable unrestricted cell proliferation also inhibit autophagy, which is normally induced to sustain cells during nutrient limitation (4, 9, 34). Thus autophagy can promote survival of non-cancerous cells in the setting of metabolic stress in the tumour microenvironment. On the other hand, in some (mostly advanced) cancers, autophagy could benefit the progression of the tumour, because it can enable cells to resist damage induced by cytotoxic anti-cancer treatments and enhance their survival in conditions of nutrient limitation and decreased energy production (3, 9, 34) (Table 1).

3.2. Neurodegeneration

Autophagic activity is essential for normal function of the nervous system. In mice, the tissue-specific knockout of autophagy genes in neurons causes a massive accumulation of ubiquitin-positive protein aggregates and neurodegeneration (37, 38), indicating that autophagy is necessary for the constitutive clearance of aggregate-prone proteins (39, 40). Many other studies indicate that autophagy can contribute to constitutive and/or enhanced clearance of some aggregate-prone proteins, but not others (2). However, even in these cases, at some point autophagy fails to cope with increased levels of aggregation leading to decreased capability of cells to clear aggregates which in turn contributes to disease progression. The role of autophagy has been demonstrated in Huntington's disease, (accumulation of Huntingtin), familial Parkinson disease (accumulation of α-synuclein; CMA can degrade normal α-synuclein, but not variant forms) (3, 11), and in Alzheimer's disease (deposits of β-amyloid precursors that are too large to be removed by the proteasomal degradation system) (41). Evidence in support of a role for autophagy in promoting neuronal cell death, particularly in neurons with acute injury has also been presented (42). Possible differences in the regulation and adaptations to local physiology in the axons, of autophagy have recently been reviewed (43).

Overall, the positive role of autophagy in neurodegeneration can be considered as prevention of the accumulation of intracellular proteins and aggregates to toxic levels. However, defective autophagy can result in the accumulation of these proteins and aggregates to lethal levels (9) (Table 1). In some instances, enhanced autophagy, by pharmacological induction using small molecules in a tissue-specific manner may prove to be of benefit in treatment of disease, as demonstrated in animal models where enhanced autophagy improved clearance of the aggregated proteins and reduced the symptoms of neurodegeneration (44).

3.3. Muscle Disorders (Various Myopathies) and Heart Disease

Muscular disorders (also known as vacuolar myopathies), are associated with the accumulation of autophagic and/or lysosomal vacuoles in the skeletal muscles and the heart (2, 45), as well as in several other tissues including liver, pancreas, spleen, kidney, and lymph nodes (2). Such a pathological situation can be observed in patients suffering from Danon's disease which has clinical characteristics including cardiomyopathy, myopathy, and variable mental retardation (45), resulting from mutation of the gene encoding LAMP-2A (46). Other myopathies associated with vacuole accumulation are X-linked myopathy with excessive autophagy, inclusion body myositis, and Marinesco-Sjögren syndrome. Defective autophagy in these myopathies may contribute to muscle wasting (9) (Table 1).

The importance of autophagy in the heart is emphasised by the observation that in adult mice, cardiac-specific deficiency of *ATG5* leads to cardiac hypertrophy, left-ventricular dilatation, and contractile dysfunction. The heart tissue from such animals also exhibited an increased level of ubiquitination and mitochondrial aggregation (47, 48). A basal level of autophagy under normal conditions presumably is cytoprotective and seemingly autophagy is required to ensure the availability of sufficient energy substrates in normal heart tissues and to control cardiomyocyte size and overall cardiac structure and function (2). By contrast, as indicated above, defects in autophagy may contribute to cardiac dysfunction and heart failure. The evidence suggests that autophagy is rapidly up-regulated under environmental stress conditions, but it is not clear what benefit might be derived from the augmentation of autophagy in the failing heart. One outcome could be autophagy-mediated survival without activating death pathways. However, in the case of pressure overload stress in cardiomyocytes, enhanced levels of autophagy produce adverse consequences. Many questions concerning the role of autophagy in heart disease remain to be fully resolved, including the connections between autophagy and atherosclerosis (49).

3.4. Liver Disease

It has been suggested that autophagy may play significant roles in three areas of liver physiology: (1) the balance of nutrients and energy resulting from neonatal adaptation, fasting, and metabolic disturbance, (2) the removal of misfolded proteins resulting from genetic mutations, pathological stimulation or ER stress, and (3) turnover of organelles such as mitochondria, peroxisomes, and ER under normal or pathophysiological conditions resulting from a range of chemical treatments, hypoxia, ischemia, and oxidative stress (12). The disturbance of autophagy function in the liver could lead to metabolic imbalance, inflammation, cell death, cirrhosis, and hepatocellular carcinoma (12) (Table 1).

3.5. Infection, Immunity, and Crohn's Disease

Autophagy has emerged as a central component of antimicrobial host defense against diverse bacterial (both gram-positive and gram-negative), parasitic (protozoa and fungi), and viral (including both RNA and DNA viruses) infections. In addition to pathogen degradation (sometimes referred to as xenophagy), autophagy has other functions during infection such as innate and adaptive immune activation (antigen presentation, immune cell development and homeostasis, and intestinal homeostasis) (50) and cell survival (nutrient generation, degradation of damaged proteins, and organelles and possibly degradation of pro-death and cytolytic molecules). As autophagy is an important host defense pathway, microbes, and viruses have also evolved mechanisms to evade, subvert, or exploit autophagic pathways. However, some pathogens are able to modulate cellular components of autophagy in a unique manner in order to establish an environmental niche and promote pathogenesis (51–53) (Table 1).

Recently autophagy has attracted attention as a contributor to Crohn's disease (CD), an inflammatory bowel disease affecting the small intestine. On the basis of genetic linkage data, polymorphisms in two genes, one (*IRGM1*) encoding an autophagy-inducing signalling molecule and the other (*ATG16L1*) a component of the ubiquitin-like conjugation system involved in AP membrane expansion, have been reported as being strongly associated with the development of CD (54–57). The pathogenic mechanisms of CD may involve a dysregulated immune response to commensal bacteria, altered mucosal barrier function and/or defects in bacterial clearance. It is plausible that defects in autophagy could contribute to one or more of these potential pathogenic mechanisms, however, the precise contribution of autophagy to CD remains to be elucidated.

3.6. Diabetes

Autophagy is inhibited by insulin and amino acid signalling to the mTOR complex, and can be activated by amino acid depletion or the pharmacological inducer rapamycin. Insulin inhibits autophagy in two ways (58): first, by activating mTOR in synergy with amino acids, which results in the phosphorylation and inhibition of the protein kinase Atg1/Ulk1 (in mammals) (59); second, by protein kinase B-mediated phosphorylation, and inhibition of the transcription factor FoxO3, which contributes to the regulation of *ATG* gene expression. Somewhat paradoxically, it has been suggested that moderately defective insulin signalling can actually enhance life span (58–60), suggesting any deficiency of autophagy in such circumstances can be compensated for.

To date, little information has been available concerning autophagy in insulin-resistant states. (58). Autophagy is known to act as a protection mechanism against oxidative damage arising from mitochondrial ROS production (see concept Chapter 6 in this volume). In insulin-sensitive cells, including β-cells, the consequences of such stress include ubiquitination and the storage of proteins and protein aggregates, which can be cleared by autophagy (61). Two studies have reported islet degeneration, decreased glucose tolerance, and reduced insulin secretion in β-cell-specific *ATG7* knockout mice (whose principal phenotypic features are hypoinsulinemia and hyperglycemia) (62, 63). In the study by Lee and colleagues (62), autophagy was apparently severely impaired due to accumulation of protein aggregates. The presence of aberrant, malformed, functionally defective and swollen mitochondria, and distended ER in β-cells presumably contributed to the reduced ability to produce insulin. However, the mechanism by which mitochondrial dysfunction and/or ER stress leads to insulin resistance is not clear (62).

In summary, these studies suggest that autophagy is necessary to maintain the structure, mass and function of pancreatic β-cells under conditions such as oxidative stress, and that reduction in autophagic activity may contribute to a reduced ability to produce insulin (Table 1).

3.7. Ageing and Longevity

Proper functioning of autophagy has been related to ageing and longevity (3, 9, 64, 65). The efficiency of macroautophagy and CMA are known to decrease with age (66). Alterations in autophagy signalling by amino acids and hormones may contribute to up- or down-regulation of macroautophagy (64). Defective CMA arises as a consequence of diminished levels of the lysosomal receptor that mediates uptake of substrate proteins (66). The cellular consequences of diminished autophagic activity include inefficient removal of damaged intracellular structures, alterations in cellular homeostasis, inability to adapt to extracellular changes, and reduced defensive response against damaging agents (64). Clearly the cumulative effects of incomplete autophagic clearance over prolonged periods and decline in autophagy function could contribute to the deterioration of cellular function that occurs in ageing (3).

Limited dietary intake decreases the incidence of age-related diseases and increases life span in numerous experimental models from yeast and invertebrates to mammals including primates (66). Such treatment seems to improve autophagy induction, possibly owing to lower levels of insulin (9, 65). However, the relationship between ageing and autophagy is far from clear in higher organisms. If the role of autophagy in age-associated decline of cellular function is clarified, not only will the role autophagy in age-associated diseases be better understood, but it may be possible to partially arrest ageing through modulation of autophagic activity.

4. Conclusions and Outlook

Although starvation or stress adaptation seems to be an evolutionary conserved function of autophagy under physiological conditions, the degradation of intracellular components may be a more important function considering the contribution of autophagy (benefits vs. disadvantage) in a range of human diseases. As a possible contribution to therapeutic intervention in disease it will be necessary to devise the means to specifically stimulate, or inhibit, autophagy and often it would be desirable to achieve this in a tissue-specific manner. The rational design of small molecules able to modulate autophagy has already commenced (67). To further this goal, and to achieve optimal outcomes from therapeutic intervention, we need a better understanding of the following: (1) the molecular actions of the component proteins of autophagic systems, (2) the signalling control and regulation of all autophagic processes, (3) the interplay between the various forms of autophagy under a variety of conditions, and (4) the complex (and sometimes seemingly paradoxical roles) of autophagy in various tissues and organs which appear often to be context specific depending on the cell type and the nature and extent of the damage sustained.

Acknowledgments

We apologise to investigators whose important original contributions have not been cited; where possible we have chosen to cite recent reviews in which details of such contributions can be found.

References

1. Huang J, Klionsky DJ (2007) Autophagy and human disease. Cell Cycle 6:1837–1849
2. Eskelinen EL, Saftig P (2009) Autophagy: A lysosomal degradation pathway with a central role in health and disease. Biochim Biophys Acta 1793:664–673
3. Levine B, Kroemer G (2008) Autophagy in the pathogenesis of disease. Cell 132:27–42
4. Kundu M, Thompson CB (2008) Autophagy: Basic principles and relevance to disease. Annu Rev Pathol 3:427–455
5. Colombo MI (2007) Autophagy: A pathogen driven process. IUBMB Life 59:238–242
6. Mizushima N, Klionsky DJ (2007) Protein turnover via autophagy: Implications for metabolism. Annu Rev Nutr 27:19–40
7. Cecconi F, Levine B (2008) The role of autophagy in mammalian development: Cell makeover rather than cell death. Dev Cell 15:344–357
8. Hussey S, Terebiznik MR, Jones NL (2008) Autophagy: Healthy eating and self digestion for gastroenterologists. J Pediatr Gastroenterol Nutr 46:496–506
9. Mizushima N, Levine B, Cuervo AM, Klionsky DJ (2008) Autophagy fights disease through cellular self-digestion. Nature 451:1069–1075
10. Uchiyama Y, Shibata M, Koike M, Yoshimura K, Sasaki M (2008) Autophagy-physiology and pathophysiology. Histochem Cell Biol 129:407–420
11. van der Vaart A, Mari M, Reggiori F (2008) A picky eater: Exploring the mechanisms of selective autophagy in human pathologies. Traffic 9:281–289
12. Yin XM, Ding WX, Gao W (2008) Autophagy in the liver. Hepatology 47:1773–1785
13. Yu L, Strandberg L, Leonardo MJ (2008) The selectivity of autophagy and its role in cell death and survival. Autophagy 4:567–573
14. Todde V, Veenhuis M, van der Klei IJ (2009) Autophagy: Principles and significance in health and disease. Biochim Biophys Acta 1792:3–13
15. Axe EL, Walker SA, Manifava M, Chandra P, Roderick HL, Habermann A, Griffiths G, Ktistakis NT (2008) Autophagosome formation from membrane compartments enriched in phosphatidylinositol 3-phosphate and dynamically connected to the endoplasmic reticulum. J Cell Biol 182: 685–701
16. Klionsky DJ (2007) Autophagy: From phenomenology to molecular understanding in less than a decade. Nat Rev Mol Cell Biol 8:931–937
17. Xie Z, Klionsky DJ (2007) Autophagosome formation: Core machinery and adaptations. Nat Cell Biol 9:1102–1109
18. Meijer WH, van der Klei IJ, Veenhuis M, Kiel JA (2007) *ATG* genes involved in non-selective autophagy are conserved from yeast to man, but the selective Cvt and pexophagy pathways also require organism-specific genes. Autophagy 3:106–116
19. Yorimitsu T, Klionsky DJ (2005) Autophagy: Molecular machinery for self-eating. Cell Death Differ 12(Suppl 2):1542–1552
20. Suzuki K, Ohsumi Y (2007) Molecular machinery of autophagosome formation in yeast, *Saccharomyces cerevisiae*. FEBS Lett 581:2156–2161
21. Yang YP, Liang ZQ, Gu ZL, Qin ZH (2005) Molecular mechanism and regulation of autophagy. Acta Pharmacol Sin 26: 1421–1434
22. Wullschleger S, Loewith R, Hall MN (2006) TOR signaling in growth and metabolism. Cell 124:471–484
23. Nicklin P, Bergman P, Zhang B, Triantafellow E, Wang H, Nyfeler B, Yang H, Hild M, Kung C, Wilson C, Myer VE, MacKeigan JP, Porter JA, Wang YK, Cantley LC, Finan PM, Murphy LO (2009) Bidirectional transport of amino acids regulates mTOR and autophagy. Cell 136:521–534
24. Kadowaki M, Karim MR, Carpi A, Miotto G (2006) Nutrient control of macroautophagy in mammalian cells. Mol Aspects Med 27:426–443
25. Meijer AJ, Codogno P (2006) Signalling and autophagy regulation in health, aging and disease. Mol Aspects Med 27:411–425
26. Dunn WA Jr, Cregg JM, Kiel JA, van der Klei IJ, Oku M, Sakai Y, Sibirny AA, Stasyk OV, Veenhuis M (2005) Pexophagy: The selective autophagy of peroxisomes. Autophagy 1:75–83
27. Krick R, Muehe Y, Prick T, Bremer S, Schlotterhose P, Eskelinen EL, Millen J, Goldfarb DS, Thumm M (2008) Piecemeal microautophagy of the nucleus requires the core macroautophagy genes. Mol Biol Cell 19:4492–4505
28. Kišsová I, Salin B, Schaeffer J, Bhatia S, Manon S, Camougrand N (2007) Selective and non-selective autophagic degradation of mitochondria in yeast. Autophagy 3:329–336

29. Dubouloz F, Deloche O, Wanke V, Cameroni E, De Virgilio C (2005) The TOR and EGO protein complexes orchestrate microautophagy in yeast. Mol Cell 19:15–26
30. Uttenweiler A, Schwarz H, Neumann H, Mayer A (2007) The vacuolar transporter chaperone (VTC) complex is required for microautophagy. Mol Biol Cell 18:166–175
31. Kaushik S, Cuervo AM (2008) Chaperone-mediated autophagy. Meth Mol Biol 445:227–244
32. Liang XH, Jackson S, Seaman M, Brown K, Kempkes B, Hibshoosh H, Levine B (1999) Induction of autophagy and inhibition of tumorigenesis by beclin 1. Nature 402: 672–676
33. Jin S (2006) Autophagy, mitochondrial quality control, and oncogenesis. Autophagy 2:80–84
34. Mathew R, Karantza-Wadsworth V, White E (2007) Role of autophagy in cancer. Nat Rev Cancer 7:961–967
35. Kroemer G, Levine B (2008) Autophagic cell death: The story of a misnomer. Nat Rev Mol Cell Biol 9:1004–1010
36. Bialik S, Kimchi A (2008) Autophagy and tumor suppression: Recent advances in understanding the link between autophagic cell death pathways and tumor development. Adv Exp Med Biol 615:177–200
37. Komatsu M, Waguri S, Chiba T, Murata S, Iwata J, Tanida I, Ueno T, Koike M, Uchiyama Y, Kominami E, Tanaka K (2006) Loss of autophagy in the central nervous system causes neurodegeneration in mice. Nature 441:880–884
38. Hara T, Nakamura K, Matsui M, Yamamoto A, Nakahara Y, Suzuki Migishima R, Yokoyama M, Mishima K, Saito I, Okano H, Mizushima N (2006) Suppression of basal autophagy in neural cells causes neurodegenerative disease in mice. Nature 441:885–889
39. Ventruti A, Cuervo AM (2007) Autophagy and neurodegeneration. Curr Neurol Neurosci Rep 7:443–451
40. Winslow AR, Rubinsztein DC (2008) Autophagy in neurodegeneration and development. Biochim Biophys Acta 1782: 723–729
41. Yu WH, Cuervo AM, Kumar A, Peterhoff CM, Schmidt SD, Lee JH, Mohan PS, Mercken M, Farmery MR, Tjernberg LO, Jiang Y, Duff K, Uchiyama Y, Näslund J, Mathews PM, Cataldo AM, Nixon RA (2005) Macroautophagy-a novel Beta-amyloid peptide-generating pathway activated in Alzheimer's disease. J Cell Biol 171:87–98
42. Koike M, Shibata M, Tadakoshi M, Gotoh K, Komatsu M, Waguri S, Kawahara N, Kuida K, Nagata S, Kominami E, Tanaka K, Uchiyama Y (2008) Inhibition of autophagy prevents hippocampal pyramidal neuron death after hypoxic-ischemic injury. Am J Pathol 172:454–469
43. Yue Z, Friedman L, Komatsu M, Tanaka K (2009) The cellular pathways of neuronal autophagy and their implication in neurodegenerative diseases. Biochim Biophys Acta 1793:1496–1507
44. Williams A, Jahreiss L, Sarkar S, Saiki S, Menzies FM, Ravikumar B, Rubinsztein DC (2006) Aggregate-prone proteins are cleared from the cytosol by autophagy: Therapeutic implications. Curr Top Dev Biol 76:89–101
45. Nishino I (2003) Autophagic vacuolar myopathies. Curr Neurol Neurosci Rep 3:64–69
46. Nishino I, Fu J, Tanji K, Yamada T, Shimojo S, Koori T, Mora M, Riggs JE, Oh SJ, Koga Y, Sue CM, Yamamoto A, Murakami N, Shanske S, Byrne E, Bonilla E, Nonaka I, DiMauro S, Hirano M (2000) Primary LAMP-2 deficiency causes X-linked vacuolar cardiomyopathy and myopathy (Danon disease). Nature 406:906–910
47. Martinet W, Knaapen MW, Kockx MM, De Meyer GR (2007) Autophagy in cardiovascular disease. Trends Mol Med 13:482–491
48. Nakai A, Yamaguchi O, Takeda T, Higuchi Y, Hikoso S, Taniike M, Omiya S, Mizote I, Matsumura Y, Asahi M, Nishida K, Hori M, Mizushima N, Otsu K (2007) The role of autophagy in cardiomyocytes in the basal state and in response to hemodynamic stress. Nat Med 13:619–624
49. Martinet W, De Meyer GR (2009) Autophagy in atherosclerosis: A cell survival and death phenomenon with therapeutic potential. Circ Res 104:304–317
50. Gannagé M, Münz C (2009) Macroautophagy in immunity and tolerance. Traffic 10:616–620
51. Münz C (2009) Enhancing immunity through autophagy. Annu Rev Immunol 27:423–449
52. Levine B, Deretic V (2007) Unveiling the roles of autophagy in innate and adaptive immunity. Nat Rev Immunol 7:767–777
53. Orvedahl A, Levine B (2009) Eating the enemy within: Autophagy in infectious diseases. Cell Death Differ 16:57–69
54. Hampe J, Franke A, Rosenstiel P, Till A, Teuber M, Huse K, Albrecht M, Mayr G, De La Vega FM, Briggs J, Günther S, Prescott NJ, Onnie CM, Häsler R, Sipos B, Fölsch UR, Lengauer T, Platzer M, Mathew CG, Krawczak

M, Schreiber S (2007) A genome-wide association scan of nonsynonymous SNPs identifies a susceptibility variant for Crohn disease in ATG16L1. Nat Genet 39:207–211
55. Massey DC, Parkes M (2007) Genome-wide association scanning highlights two autophagy genes, ATG16L1 and IRGM, as being significantly associated with Crohn's disease. Autophagy 3:649–651
56. Kuballa P, Huett A, Rioux JD, Daly MJ, Xavier RJ (2008) Impaired autophagy of an intracellular pathogen induced by a Crohn's disease associated ATG16L1 variant. PLoS ONE 3:e3391
57. Zhang H, Massey D, Tremelling M, Parkes M (2008) Genetics of inflammatory bowel disease: Clues to pathogenesis. Br Med Bull 87:17–30
58. Meijer AJ, Codogno P (2008) Autophagy: A sweet process in diabetes. Cell Metab 8:275–276
59. Codogno P, Meijer AJ (2005) Autophagy and signaling: Their role in cell survival and cell death. Cell Death Differ 12(Suppl 2):1509–1518
60. Russell SJ, Kahn CR (2007) Endocrine regulation of ageing. Nat Rev Mol Cell Biol 8:681–91
61. Kaniuk NA, Kiraly M, Bates H, Vranic M, Volchuk A, Brumell JH (2007) Ubiquitinated-protein aggregates form in pancreatic beta-cells during diabetes-induced oxidative stress and are regulated by autophagy. Diabetes 56:930–939
62. Jung HS, Chung KW, Won Kim J, Kim J, Komatsu M, Tanaka K, Nguyen YH, Kang TM, Yoon KH, Kim JW, Jeong YT, Han MS, Lee MK, Kim KW, Shin J, Lee MS (2008) Loss of autophagy diminishes pancreatic beta cell mass and function with resultant hyperglycemia. Cell Metab 8:318–324
63. Ebato C, Uchida T, Arakawa M, Komatsu M, Ueno T, Komiya K, Azuma K, Hirose T, Tanaka K, Kominami E, Kawamori R, Fujitani Y, Watada H (2008) Autophagy is important in islet homeostasis and compensatory increase of beta cell mass in response to high-fat diet. Cell Metab 8:325–332
64. Cuervo AM (2004) Autophagy: In sickness and in health. Trends Cell Biol 14:70–77
65. Yen WL, Klionsky DJ (2008) How to live long and prosper: Autophagy, mitochondria, and aging. Physiology (Bethesda) 23: 248–262
66. Cuervo AM (2008) Autophagy and aging: Keeping that old broom working. Trends Genet 24:604–612
67. Rubinsztein DC, Gestwicki JE, Murphy LO, Klionsky DJ (2007) Potential therapeutic applications of autophagy. Nat Rev Drug Discov 6:304–312

Chapter 6

Mitophagy and Mitoptosis in Disease Processes

Dalibor Mijaljica, Mark Prescott, and Rodney J. Devenish

Abstract

Mitochondria play a very important role in cellular function, not only through key metabolic reactions and energy generation, but also by being a major site for production of reactive oxygen species and a key player in cell death. Therefore, mitochondrial dysfunction or damage may have severe consequences. Mitophagy (autophagic degradation of mitochondria) and mitoptosis (programmed destruction of mitochondria) are the processes by which cells can deal with impaired mitochondria. The efficiency of these processes may be a contributing factor to the pathogenesis of various diseases.

Key words: Damage, Degradation, Disease, Mitochondria, Mitophagy, Mitoptosis

1. Introduction

Mitochondria play very important roles in cellular function; therefore malfunction or damage may have severe consequences and can be associated with human disease and ageing (1, 2). In order to appreciate the involvement of mitophagy or mitoptosis in various disease processes, we briefly introduce: (1) mitochondrial structure, function, and morphology, (2) mitochondrial dynamics (fission and fusion) and (3) mitochondrial turnover. We then discuss how mitochondria and cells deal with various degrees of mitochondrial damage as well as the possible relationship between mitochondrial properties and metabolic status, and the pathophysiology of disease and ageing.

Peter Bross and Niels Gregersen (eds.), *Protein Misfolding and Cellular Stress in Disease and Aging: Concepts and Protocols*, Methods in Molecular Biology, vol. 648, DOI 10.1007/978-1-60761-756-3_6, © Springer Science+Business Media, LLC 2010

2. Mitochondrial Structure, Function, and Morphology: Implications for Health and Disease

2.1. Mitochondria Are Compartmentalised Organelles

The double membrane of the mitochondrion defines four compartments: (1) outer mitochondrial membrane (OMM), (2) inner mitochondrial membrane (IMM), (3) intermembrane space (IMS) and (4) mitochondrial matrix (Fig. 1). The OMM is the interface between mitochondria and the cytosol (and other organelles). The IMM is the site of the respiratory chain (Complexes I, II, III, and IV) and the ATP synthase complex. Within the IMM, two distinct regions can be distinguished: the inner boundary membrane (IBM) and the cristae membranes which represent invaginations of the IBM that protrude into the matrix space (3). The IMS, between the OMM and IMM has a relatively low protein content that includes cytochrome *c* important for respiratory chain function, but which is also an effector of apoptosis. The matrix is densely packed and occupied by a diverse protein population that includes Kreb's cycle enzymes, multiple copies of the

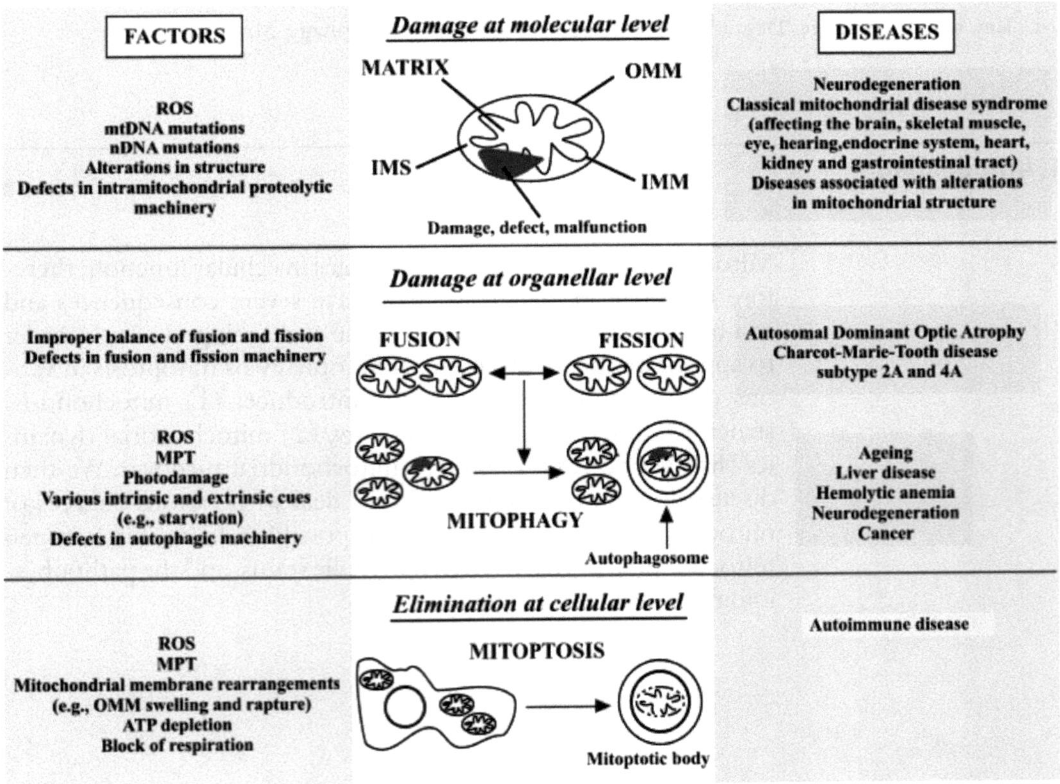

Fig. 1. Schematic model of the mitochondria-related damage (caused by various factors) at molecular, organellar and cellular level and its role in numerous human diseases. See text for more details. Abbreviations ATP, adenosine-5′-triphosphate; IMM, inner mitochondrial membrane; IMS, intermembrene space; MPT, mitochondrial permeability transition; mtDNA, mitochondrial DNA; nDNA, nuclear DNA; OMM, outer mitochondrial membrane; ROS, reactive oxygen species.

circular mitochondrial genome and numerous mitochondrial ribosomes. Mitochondria synthesise only a small number of (mostly respiratory chain) proteins; 90% of the mitochondrial protein mass is encoded by the nuclear genome, synthesised in the cytosol and imported into mitochondria. Thus, mitochondrial biogenesis requires the co-ordination of gene expression from mitochondrial and nuclear genomes (1–4).

Mitochondria have a significant role in determining the bioenergetic status of cells as they are the site of oxidative phosphorylation, whereby the ATP synthase complex generates ATP coupled to electron transfer from respiratory substrates to oxygen by a series of oxidation-reduction reactions and the pumping of protons across the IMM from the matrix space (1, 2). A significant consequence of this activity is the production of reactive oxygen species (ROS); the levels produced can impact negatively not only on mitochondrial function, but more broadly on cellular function (see below).

Mitochondria are also home to key biosynthetic pathways (e.g. heme and lipid synthesis), and participate in the regulation of cellular Ca^{2+} levels. Programmed cell death (apoptosis) can involve mitochondrial events and crucial cell death regulators such as Bcl-2 are associated with mitochondria (1, 2).

2.2. Mitochondrial Dysfunction: Defects in Protein Quality Control Systems, Mitochondrial or Nuclear DNA Mutations, and Changes in Mitochondrial Morphology

A highly conserved, intra-organellar, proteolytic process conducts protein quality control (PQC) within mitochondria (4) by monitoring the folding and assembly of mitochondrial proteins and selectively removing non-assembled, misfolded, damaged and excess proteins from the organelle. Components of this system include a proteasome degradation pathway (located in the OMM), molecular chaperones, and energy-dependent proteases (confined mostly to the IMM and matrix) (4, 5). Once recognised, target proteins are degraded to peptides, and subsequently either exported from mitochondria or degraded further to amino acids by various oligopeptidases (5). As a consequence of their function in PQC, matrix proteases directly impact on the biogenesis of the respiratory chain complexes and ATP synthase. Mutations in genes encoding components of the mitochondrial PQC system are linked to a number of ageing-related neurodegenerative diseases (5, 6) (Fig. 1). For example, mutations in a subunit of the *m*-AAA protease, paraplegin, cause an autosomal recessive form of Hereditary Spastic Paraplegia (HSP). Accordingly, mitochondrial dysfunction and axonal degeneration in the absence of paraplegin may result from the accumulation of non-degraded, misfolded IMM proteins or impaired regulatory steps during mitochondrial biogenesis, or both. Interestingly, mutations affecting expression of the mitochondrial chaperone HSP60, which is localised to the matrix, cause an autosomal dominant form of HSP (5).

Various mitochondrial disease syndromes arise as a consequence of either single or multisystemic mutations or deletions in either mitochondrial DNA (mtDNA) or nuclear DNA (nDNA) leading to the loss of mitochondrial components. These disease syndromes typically have brain and skeletal muscle manifestations, and therefore are often referred to as mitochondrial encephalomyopathies. However, other systems in the body can be affected as well, including the eye (optic atrophy, retinitis pigmentosa, cataracts), hearing (neurosensory deafness), the endocrine system (short stature, diabetes mellitus), the heart (hypertrophic cardiomyopathies), the gastrointestinal tract (exocrine pancreas dysfunction), and the kidney (renal tubular acidosis) (1, 7–10) (Fig. 1).

Certain alterations of mitochondrial structures (different morphological features) seem to be common to many mitochondrial diseases. Onion-shaped giant mitochondria have been observed in a variety of pathologic states, including ATP synthase deficiency, ageing, and hyperoxia, although the exact morphology may not be identical in all instances (11). Mitochondria of lymphoblasts from patients suffering from Barth syndrome appear enlarged with grossly reduced inner membrane surface area and exhibit only a few deranged cristae (11). Again, in Alzheimer's disease, mitochondrial cristae are disrupted, or unusual types of IMM arrangements including concentric and parallel stacks of cristae membranes are observed. Inclusions within mitochondria may also be present (3, 12), (Fig. 1).

3. Mitochondrial Dynamics in Physiology and Pathophysiology

3.1. Fusion and Fission and Their Role in Mitochondrial Network Formation

Mitochondria may vary greatly in length, number, and shape, depending on the type, developmental stage, and physiological state of the cell (13). In many organisms, mitochondria form an interconnected reticulum. For example, in the budding yeast, *Saccharomyces cerevisiae*, mitochondria form a branched tubular network located immediately below the cell cortex, whereas in mammalian fibroblast cells, the interconnected mitochondrial filaments (reticulum) extend throughout the entire cytosol and act as an electrically united system. The mitochondrial reticulum is a highly dynamic structure involving opposing but continuous and (usually) balanced membrane fusion and fission events, mediated by an evolutionarily-conserved protein machinery, that determines overall mitochondrial morphology (14–17) (Fig. 1).

3.2. Fusion and Fission in Cell Life and Death

Mitochondria are propagated by growth and division of pre-existing organelles; as such their inheritance depends on mitochondrial fission during cytokinesis (18). Furthermore, mitochondrial fission also plays an important role during embryonic development in *Caenorhabditis elegans* (19) and the formation of synapses and dendritic spines in rat neurons (20).

As a consequence of their roles in maintenance of the mitochondrial network, fusion and fission can serve to limit the extent of mitochondrial damage generated at the organelle level and play a key role in cell death (17). Fusion serves to mix and unify the mitochondrial compartment, an activity that is thought to constitute a defense mechanism against ageing. As mtDNA is directly located at the site of ROS production, it is particularly vulnerable to ROS-mediated mutations. These mutations accumulate with age until a bioenergetic threshold is breached, resulting in mitochondrial dysfunction that eventually leads to age-associated pathology and death (21). It is proposed that free mixing of mitochondrial genetic components throughout the mitochondrial network protects mammalian mitochondria from direct expression of respiration or other defects caused by accumulated mtDNA mutations (22).

Given the importance of mitochondrial function, it is clear that the rates of fusion and fission must be tightly controlled and balanced, in response to a cohort of intra- and extracellular signals. A disturbance in the balance can lead to a variety of diseases such as Autosomal Dominant Optic Atrophy (a common form of inherited childhood blindness caused by defects in the *OPA1* gene whose product acts to maintain mitochondrial membrane morphology) (23) and Charcot-Marie-Tooth disease subtype 2A (mutations in the *MFN2* gene, whose product is a mediator of mitochondrial fusion, lead to a neurodegenerative disorder clinically characterised by the gradual degeneration of peripheral neurons) (24) (Fig. 1).

Several studies have indicated that mitochondrial morphology changes during apoptosis, resulting in small, rounded, and more numerous organelles. This extensive fission occurs before caspase activation, which causes the ultimate demise of the cell. Inhibition of the mitochondrial fission machinery inhibits cell death, indicating its importance in apoptosis (25). Severe damage of mitochondria or impaired fusion can lead to increased fragmentation of mitochondria, which are then selectively removed by an autophagic process, termed mitophagy (see below) (26), and which prevents the release of pro-apoptotic proteins from damaged mitochondria. Therefore, apoptosis can be suppressed upon induction of autophagy (27).

4. Mitophagy: Mitochondrial Turnover by Autophagy

Mitophagy denotes the (selective) autophagic sequestration, and subsequent degradation, of mitochondria (26, 28, 29). The component amino acids and other basic building blocks are recovered for other uses. Mitochondria could become targets of autophagic degradation during basal, constitutive turnover, during starvation-induced turnover, and turnover of dysfunctional mitochondria (30). How mitochondria are selected and/or recognised by the autophagic machinery is currently under active investigation.

The presence of mitochondria within autophagosomes arising from macroautophagic activity in cells was first observed by electron microscopy (EM) in mammalian cells (31) and later in yeast (32). EM remains a primary technique for detection of mitophagy since it offers the potential to yield information on membrane interactions contributing to mitophagy (see, for example, Chap. 19). Other techniques that can be used to monitor mitophagy in yeast and mammalian cells include: (1) fluorescence microscopy, (2) immunofluorescence, and (3) following protein degradation by western immunoblotting (33); however, they reveal less information concerning the mechanism of the process.

4.1. Factors Involved in Mitophagy

Mitochondrial damage or dysfunction could lead to one or more changes in mitochondrial function, morphology, membrane potential, ATP or ROS production, and Ca^{2+} homeostasis. Presumably such changes are recognised and cells activate autophagy in response. Mitophagy in this context acts as a mechanism for mitochondrial quality control (i.e. maintaining cellular homeostasis) and therefore is an important cytoprotective response (29) (Fig. 1).

4.1.1. The Importance of Fission and Fusion

A recent study tracked individual photolabelled mitochondria through fusion and fission events. It was demonstrated that fission yields some daughter mitochondria having decreased membrane potential and which are less likely to become involved in a subsequent fusion event (34). These observations have led to the proposal of a mechanism by which fusion and fission allow for sequestration of damaged mitochondrial components into daughter mitochondria that are then eliminated by autophagy (35).

4.1.2. ROS and Mitochondrial Permeability Transition

As indicated above, mitochondrial respiration produces ROS, including H_2O_2 and the superoxide anion, especially if respiration is inhibited or otherwise disordered (36). ROS causes oxidative damage to mitochondrial proteins or mtDNA resulting in abnormal mitochondrial protein synthesis, which may then lead to defects in protein folding, aggregate formation. Under conditions of oxidative stress, the opening of non-specific aqueous pores, including mitochondrial permeability transition (MPT) can occur which can provide a signal leading to induction of mitophagy. Pharmacological inhibitors of MPT, such as cyclosporin A (CsA) can block mitophagy (36, 37).

It has been suggested that a graded increase in ROS levels can determine whether the outcome for cells is life (growth and differentiation) or death (apoptotic, autophagic, and necrotic) (38). At low concentrations ROS function as signalling molecules in a number of biological processes, for example, defense against invasion of microorganisms. However, ROS also may cause mild oxidative damage and the MPT. Such changes may induce mitophagy (removal of damaged mitochondria) thereby promoting cell survival (29). If ROS levels increase further, other signalling pathways may be activated that lead to apoptosis. Should ROS levels rise even higher, then cells are likely to die by a sudden necrotic death (28, 29, 38). Note that certain mtDNA mutations give rise to respiratory defects that result in a decrease, rather than an increase, in ROS production. As a consequence, cells harbouring such mtDNA mutations will suffer less oxidative damage making mitochondria less prone to mitophagy (29). Overall it is the severity of mitochondrial damage, and the length of ROS exposure as well as any adverse influence on the autophagic machinery that determines whether mitophagy ultimately leads to cell survival, or contributes to cell death (29). Clearly, the timely elimination of dysfunctional mitochondria is essential to protect cells from the harm of disordered mitochondrial metabolism and the release of pro-apoptotic proteins.

Interestingly, there is evidence for redox regulation of starvation-induced autophagy. The formation of ROS, specifically H_2O_2, was stimulated and the cysteine protease, Atg4, identified as a direct target for oxidation. A cysteine residue located near the catalytic site of this protein was found to be essential for regulation, since its substitution prevented autophagosome formation in cells (39).

4.1.3. Photodamage of Mitochondria

Mitophagy has been shown to occur as a result of irreversible laser-induced photodamage (26). This model provided direct experimental confirmation of a role for autophagy in removing damaged mitochondria. Thus, when cultured hepatocytes were exposed to laser light, the mitochondria of the photodamaged zone dissipated their membrane potential in a light dose-dependent manner. With a longer exposure, membrane potential was irreversibly dissipated, followed by a specific mitophagic response leading to autophagic disposal of photodamaged mitochondria, but not of intact neighbouring organelles (26, 40). Compared with starvation-induced mitophagy, photodamage-induced mitophagy could not be blocked by the use of phosphatidylinositol-3-kinase (PI3K) inhibitors (3-methyladenine and wortmannin), indicating that this specific mitophagic event occurs downstream of PI3K-mediated signalling (40).

4.1.4. Mitophagy and Autophagic Machinery: A Question of Selectivity?

The molecular triggers for mitophagy are still poorly understood. The precise mechanisms also remain to be elucidated, in part due to the absence of a sensitive and convenient method for mitophagy induction and detection. Evidence indicating the mechanistic nature of mitophagy comes primarily from yeast (*S. cerevisiae*). In yeast, two phases of mitophagy, designated non-selective and selective have been described as occurring successively. However, these definitions serve largely to distinguish between morphological features of the two forms rather than to elucidate any mechanistic differences and whether the trigger for each form is different. Which of the two forms will occur depends on nutrient and/or autophagy induction conditions, and the extent of mitochondrial damage (41, 42). Yeast cells subjected to nitrogen starvation after growth on different carbon sources (e.g. lactate, glucose) provided evidence that micromitophagy (e.g. direct engulfment of mitochondria by vacuolar membranes) rather than macromitophagy (i.e. engulfment of mitochondria by autophagosomes), preferentially occurred in cells grown under non-fermentable conditions (i.e. in the presence of lactate) (42). Under a unique set of conditions, mitophagy requires some, but not all, AuTophaGy-related (*ATG*) genes known to be essential for (macro)autophagy (41–44).

In addition to a specific requirement for components of the macroautophagic machinery in the process of mitophagy, it appears that some additional proteins (especially those mitochondrially-localised) may contribute to the selective form of mitophagy (e.g. Uth1p, which is an OMM protein under some conditions) (42). How this protein interacts (if at all) with the characterised (macro)autophagic machinery remains to be elucidated, but it seems possible that it may act as marker labelling mitochondria for mitophagy.

A recent study showed that in mammalian cells, Parkin (a primarily cytosolic, but to some extent mitochondrially localised, ubiquitin E3 ligase), is selectively recruited to dysfunctional mitochondria, in Parkinson's disease (PD). How Parkin can distinguish healthy from damaged mitochondria is not obvious. The signal to recruit may be the initiation of protein aggregation within the OMM, possible second messengers like ROS, or through activation of an unidentified receptor protein. The most obvious candidate for a Parkin recruitment factor would be the OMM kinase Pink1 whose two substrates (the chaperone Trap1/Hsp75 and the serine protease HtrA2/Omi) are also involved in mitochondrial PQC. Parkin recruitment is followed by a complete loss of mitochondria from cells (neurons) in a (macro)autophagy-dependent manner. This study implicates a failure to eliminate dysfunctional mitochondria in the pathogenesis of PD. Interestingly, overexpression of Parkin has been shown to provide some protection against toxin-induced animal models of PD (45, 46).

5. Mitophagy and Disease

The efficiency of mitophagy may play a critical role in ageing and the pathogenesis of various diseases (Fig. 1).

5.1. Mitophagy and Ageing

Mitophagy is proposed as being beneficial in postponing aspects of ageing. As indicated above, in order to maintain healthy cells, ROS-damaged mitochondria must be eliminated. However, as autophagic (and mitophagic) activity declines with age, the efficiency of removal of damaged mitochondria declines and prominent symptoms of ageing become apparent, such as the accumulation of damaged proteins, lipids, and organelles (25, 39, 47, 48).

A restricted food supply (incipient starvation) enhances autophagy in experimental animals, and may partly explain why caloric restriction retards the ageing process. Theoretically, it might be able to offset the natural age-related decline of autophagy and so prolong its essential housekeeping function in cells. Therefore, achieving proper and sufficient function of autophagy makes the manipulation of autophagy a prospect for slowing ageing in mammals and enhancing longevity (4).

5.2. Mitophagy and Liver Disease

Under pathophysiological conditions, a significant change in autophagic removal of mitochondria (and other cellular constituents) can occur. Evidence of enhanced levels of mitophagy are found in liver biopsies of patients with Reye's syndrome (where the disease causes fatty liver with minimal inflammation) (49) and also is observed in a murine disease model using influenza B virus (50, 51). Changes in the autophagic breakdown of mitochondria can contribute to disease pathology (51). For example, *ATG7*-deficient mice accumulate excessive numbers of peroxisomes, deformed mitochondria, and concentric membranous structures, in addition to ubiquitin-positive aggregates, in hepatocytes (52), which can be anticipated to cause liver dysfunction in the long run (51). The precise role of mitophagy in liver injury and potential for its modulation in liver cancer therapy remain to be explored in detail.

5.3. Defective Clearance of Mitochondria in Reticulocytes Results in Hemolytic Anemia

During terminal differentiation, erythroid cells undergo enucleation to become reticulocytes. Subsequently coordinated removal of organelles such as mitochondria occurs leading to the formation of mature red blood cells (RBCs) (53, 54). Nix, a BH3-only member of the Bcl-2 family (also known as Bnip3), is up-regulated in erythroid cells undergoing terminal differentiation and is required for dissipation of mitochondrial membrane potential (54). RBCs from $Nix^{-/-}$ mice show abnormal retention of mitochondria (55).

The loading and unloading of oxygen can induce oxidative stress in RBCs (56). As already discussed, mitochondria are a major site for the generation of ROS, therefore, the retention of mitochondria in RBCs would be expected to be deleterious. Indeed, $Nix^{-/-}$ RBCs are reported to produce more ROS, display increased caspase activation *in vitro* and undergo faster turnover *in vivo* (55). Thus, successful degradation of mitochondria in $Nix^{+/+}$ RBCs will prevent mitochondria-dependent caspase activation and subsequent cell death, reduce ROS production and sustain the survival of erythrocytes. Interestingly, Nix deficiency does not affect the formation of autophagosomes in reticulocytes, instead, mitochondria are clustered outside of autophagosomes (55).

6. Mitoptosis: The Elimination of Mitochondria

6.1. Initial Conception and Characterisation

Mitoptosis can be defined as a type of mitochondrial elimination (57). In 1992, Tedesci and colleagues suggested that mitochondria possess a mechanism of self-elimination (58). This function was ascribed to the MPT, and was not envisaged as requiring any extramitochondrial proteins. Rather, it was conceived as being initiated by a signal originating from a particular mitochondrion, such as ROS production. This is why one can consider the outcome as the programmed destruction of the mitochondrion (mitochondrial suicide). For this event, Skulachev coined the word "mitoptosis," by analogy with apoptosis, programmed death of the cell (59, 60) (Fig. 1).

It has been proposed that mitoptosis represents a mechanism by which mitochondria undergo extensive fragmentation and subsequent caspase-independent elimination during apoptosis (61). On the other hand, apoptotic stimuli target mitochondria for degradation by autophagy, as fragmented mitochondria are found within autophagosomes (62). Recently, in lymphoblastoid T cells, two types of mitoptosis have been described: OMM and IMM. During OMM mitoptosis, swelling, condensation, and the collapse of IMM cristae precede disruption of the OMM, following which the swollen cristae remnants distribute throughout the cell cytoplasm. By contrast, IMM mitoptosis appears to be characterised by the coalescence of mitochondria, loss of density of the matrix and subsequent rupture of IMM cristae, while the OMM remains unchanged (57).

In highly glycolytic human cervical carcinoma (HeLa) cells, a novel and apparently more robust, multi-step (non-autophagic) mitoptotic process has been described. The sequence of events is as follows: (1) fission of mitochondria, (2) clustering of the fragmented mitochondria in the perinuclear region, (3) sequestration

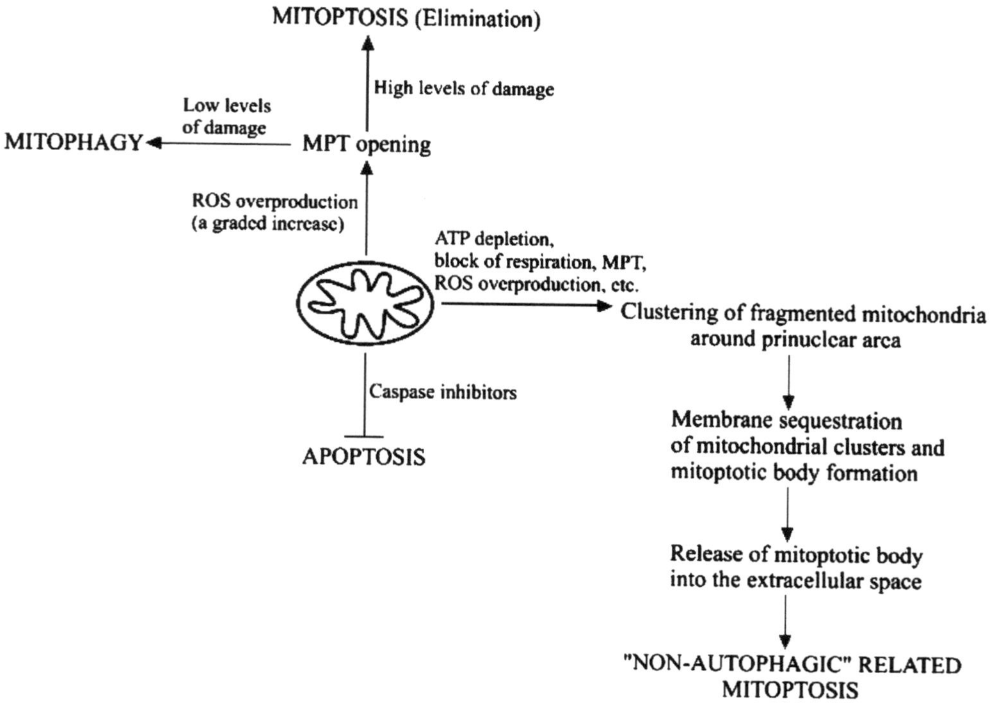

Fig. 2. Scheme illustrating different scenarios of mitochondrial elimination. See text for more details. Abbreviations ATP, adenosine-5′-triphosphate; MPT, mitochondrial permeability transition; ROS, reactive oxygen species.

of mitochondrial clusters by a membrane and formation of a mitoptotic body, (4) decomposition of mitochondria inside the mitoptotic body to small membrane vesicles, (5) protrusion of the mitoptotic body from the cell inside a bleb and its subsequent release into the extracellular space (63) (Fig. 2).

6.2. Mitoptosis in Disease Processes: An Immunogenic Role?

Understanding the role of mitoptosis in pathogenic mechanisms of mitochondria-associated human diseases is far from being fully elucidated (Fig. 1). However, recently, a connection between mitochondria, cell death (perhaps by mitoptosis) and B cell tolerance has been drawn (64). It has been hypothesised that the presence of circulating autoantibodies against mitochondria can be a marker of autoimmune disease as well as a pathogenic determinant (57, 63, 64). Clearly release of mitochondrial material into extracellular compartments arising from mitoptosis could contribute to the production of such autoantibodies. Furthermore, in many cases, death of autoreactive B cells is regulated by the cell intrinsic, or mitochondrial pathway of apoptotic cell death. The pro-apoptotic Bcl-2 family proteins (e.g. Bak, Bax, and Bim) are required, whereas the anti-apoptotic members (e.g. Bcl-2, Bcl-xL) can prevent apoptotic cell death by interfering with the action of Bak and Bax. Notably, Bcl-2 and Bcl-xL have also been shown to regulate autophagic cell death that may also play a role in B cell tolerance (64).

7. Impact and Future Prospects

The presence of healthy mitochondria is essential for vitality and survival. Dysfunction of mitochondria in relation to structure, function, dynamics, and turnover can have severe consequences and is linked to ageing and various diseases in humans including neurodegeneration. Degradation of mitochondria by mitophagy plays an important role in removing of dysfunctional or aged mitochondria. At low levels of damage, mitophagy may be sufficient to remove all dysfunctional organelles, whereas when damage exceeds the capacity of the cell for mitophagy, an alternative could be mitoptosis, or apoptosis (Figs. 1 and 2). Fully understanding the link between mitochondrial dysfunction and subsequent degradation may ultimately offer improvements in human health.

Acknowledgements

We apologise to investigators whose important original contributions have not been cited; where possible we have chosen to cite recent reviews in which details of such contributions can be found.

References

1. Scheffler IE (2001) A century of mitochondrial research: achievements and perspectives. Mitochondrion 1:3–31
2. Duchen MR (2004) Mitochondria in health and disease: perspectives on a new mitochondrial biology. Mol Aspects Med 25:365–451
3. Zick M, Rabl R, Reichert AS (2008) Cristae formation-linking ultrastructure and function of mitochondria. Biochim Biophys Acta 1793:5–19
4. Koppen M, Langer T (2007) Protein degradation within mitochondria: versatile activities of AAA proteases and other peptidases. Crit Rev Biochem Mol Biol 42:221–242
5. Tatsuta T, Langer T (2008) Quality control of mitochondria: protection against neurodegeneration and ageing. EMBO J 27:306–314
6. Lin MT, Beal MF (2006) Mitochondrial dysfunction and oxidative stress in neurodegenerative diseases. Nature 443:787–795
7. DiMauro S (2004) Mitochondrial diseases. Biochim Biophys Acta 1658:80–88
8. DiMauro S (2004) Mitochondrial medicine. Biochim Biophys Acta 1659:107–114
9. Zeviani M, Di Donato S (2004) Mitochondrial disorders. Brain 127(Pt 10):2153–2172
10. Chan DC (2006) Mitochondria: dynamic organelles in disease, aging, and development. Cell 125:1241–1252
11. Acehan D, Xu Y, Stokes DL, Schlame M (2007) Comparison of lymphoblast mitochondria from normal subjects to patients with Barth syndrome using electron microscopy tomography. Lab Invest 87:40–48
12. Baloyannis SJ (2006) Mitochondrial alterations in Alzheimer's disease. J Alzheimers Dis 9:119–126
13. Griparic L, van der Bliek AM (2001) The many shapes of mitochondrial membranes. Traffic 2:235–244
14. Cerveny KL, Tamura Y, Zhang Z, Jensen RE, Sesaki H (2007) Regulation of mitochondrial fusion and division. Trends Cell Biol 17:563–569

15. Detmer SA, Chan DC (2007) Functions and dysfunctions of mitochondrial dynamics. Nat Rev Mol Cell Biol 8:870–879
16. Hoppins S, Lackner L, Nunnari J (2007) The machines that divide and fuse mitochondria. Annu Rev Biochem 76:751–780
17. Westermann B (2008) Molecular machinery of mitochondrial fusion and fission. J Biol Chem 283:13501–13505
18. Warren G, Wickner W (1996) Organelle inheritance. Cell 84:395–400
19. Labrousse AM, Zappaterra MD, Rube DA, van der Bliek AM (1999) *C. elegans* dynamin-related protein DRP-1 controls severing of the mitochondrial outer membrane. Mol Cell 4:815–826
20. Li Z, Okamoto K, Hayashi Y, Sheng M (2004) The importance of dendritic mitochondria in the morphogenesis and plasticity of spines and synapses. Cell 119:873–887
21. Balaban RS, Nemoto S, Finkel T (2005) Mitochondria, oxidants, and aging. Cell 120: 483–495
22. Sato A, Nakada K, Hayashi J (2006) Mitochondrial dynamics and aging: mitochondrial interaction preventing individuals from expression of respiratory deficiency caused by mutant mtDNA. Biochim Biophys Acta 1763:473–481
23. Olichon A, Guillou E, Delettre C, Landes T, Arnauné-Pelloquin L, Emorine LJ, Mils V, Daloyau M, Hamel C, Amati-Bonneau P, Bonneau D, Reynier P, Lenaers G, Belenguer P (2006) Mitochondrial dynamics and disease, OPA1. Biochim Biophys Acta 1763: 500–509
24. Niemann A, Berger P, Suter U (2006) Pathomechanisms of mutant proteins in Charcot-Marie-Toothdisease. Neuromolecular Med 8:217–242
25. Youle RJ, Karbowski M (2005) Mitochondrial fission in apoptosis. Nat Rev Mol Cell Biol 6:657–663
26. Kim I, Rodriguez-Enriquez S, Lemasters JJ (2007) Selective degradation of mitochondria by mitophagy. Arch Biochem Biophys 462:245–253
27. Maiuri MC, Criollo A, Tasdemir E, Vicencio JM, Tajeddine N, Hickman JA, Geneste O, Kroemer G (2007) BH3-only proteins and BH3 mimetics induce autophagy by competitively disrupting the interaction between Beclin 1 and Bcl-2/Bcl-X(L). Autophagy 3:374–376
28. Lemasters JJ (2005) Selective mitochondrial autophagy, or mitophagy, as a targeted defense against oxidative stress, mitochondrial dysfunction, and aging. Rejuvenation Res 8:3–5
29. Yen WL, Klionsky DJ (2008) How to live longer and prosper: autophagy, mitochondria, and aging. Physiology (Bethesda) 23:248–262
30. Chu CT, Zhu J, Dagda R (2007) Beclin 1-independent pathway of damage-induced mitophagy and autophagic stress. Autophagy 3:663–666
31. Clark SL Jr (1957) Cellular differentiation in the kidneys of newborn mice studied with the electron microscope. J Biophys Biochem Cytol 3:349–362
32. Takeshige K, Baba M, Tsuboi S, Noda T, Ohsumi Y (1992) Autophagy in yeast demonstrated with proteinase-deficient mutants and conditions for its induction. J Cell Biol 119: 301–311
33. Mizushima N (2004) Methods for monitoring autophagy. Int J Biochem Cell Biol 36:2491–2502
34. Twig G, Elorza A, Molina AJ, Mohamed H, Wikstrom JD, Walzer G, Stiles L, Haigh SE, Katz S, Las G, Alroy J, Wu M, Py BF, Yuan J, Deeney JT, Corkey BE, Shirihai OS (2008) Fission and selective fusion govern mitochondrial segregation and elimination by autophagy. EMBO J 27:433–446
35. Twig G, Hyde B, Shirihai OS (2008) Mitochondrial fusion, fission and autophagy as a quality control axis: the bioenergetic view. Biochim Biophys Acta 1777:1092–1097
36. Adam-Vizi V, Chinopoulos C (2006) Bioenergetics and the formation of mitochondrial reactive oxygen species. Trends Pharmacol Sci 27:639–645
37. Rodriguez-Enriquez S, He L, Lemasters JJ (2004) Role of mitochondrial permeability transition pores in mitochondrial autophagy. Int J Biochem Cell Biol 36: 2463–2472
38. Buetler TM, Krauskopf A, Ruegg UT (2004) Role of superoxide as a signaling molecule. News Physiol Sci 19:120–123
39. Scherz-Shouval R, Shvets E, Fass E, Shorer H, Gil L, Elazar Z (2007) Reactive oxygen species are essential for autophagy and specifically regulate the activity of Atg4. EMBO J 26:1749–1760
40. Galluzzi L, Vicencio JM, Kepp O, Tasdemir E, Maiuri MC, Kroemer G (2008) To die or not to die: that is autophagic question. Curr Mol Med 8:78–91
41. Kanki T, Klionsky DJ (2008) Mitophagy in yeast occurs through a selective mechanism. J Biol Chem 283:32386–32393

42. Kiššová I, Salin B, Schaeffer J, Bhatia S, Manon S, Camougrand N (2007) Selective and non-selective autophagic degradation of mitochondria in yeast. Autophagy 3:329–336
43. Tal R, Winter G, Ecker N, Klionsky DJ, Abeliovich H (2007) Aup1p, a yeast mitochondrial protein phosphatase homolog, is required for efficient stationary phase mitophagy and cell survival. J Biol Chem 282:5617–5624
44. Zhang Y, Qi H, Taylor R, Xu W, Liu LF, Jin S (2007) The role of autophagy in mitochondria maintenance: characterization of mitochondrial functions in autophagy-deficient *S. cerevisiae* strains. Autophagy 3:337–346
45. Narendra DA, Tanaka D, Suen F, Youle RJ (2008) Parkin is recruited selectively to impaired mitochondria and promotes their autophagy. J Cell Biol 183:795–803
46. McBride HM (2008) Parkin mitochondria in the autophagosome. J Cell Biol 183:757–759
47. Martinez-Vicente M, Sovak G, Cuervo AM (2005) Protein degradation and aging. Exp Gerontol 40:622–633
48. Cuervo AM (2004) Autophagy: in sickness and in health. Trends Cell Biol 14:70–77
49. Partin JC, Schubert WK, Partin JS (1971) Mitochondrial ultrastructure in Reye's syndrome (encephalopathy and fatty degeneration of the viscera). N Engl J Med 285: 1339–1343
50. Woodfin BM, Davis LE (1986) Liver autophagy in the influenza B virus model of Reye's syndrome in mice. J Cell Biochem 31:271–275
51. Yin XM, Ding WX, Gao W (2008) Autophagy in the liver. Hepatology 47:1773–1785
52. Komatsu M, Waguri S, Ueno T, Iwata J, Murata S, Tanida I, Ezaki J, Mizushima N, Ohsumi Y, Uchiyama Y, Kominami E, Tanaka K, Chiba T (2005) Impairment of starvation-induced and constitutive autophagy in Atg7-deficient mice. J Cell Biol 169:425–434
53. Koury MJ, Koury ST, Kopsombut P, Bondurant MC (2005) In vitro maturation of nascent reticulocytes to erythrocytes. Blood 105:2168–2174
54. Yoshida H, Kawane K, Koike M, Mori Y, Uchiyama Y, Nagata S (2005) Phosphatidylserine-dependent engulfment by macrophages of nuclei from erythroid precursor cells. Nature 437:754–758
55. Sandoval H, Thiagarajan P, Dasgupta SK, Schumacher A, Prchal JT, Chen M, Wang J (2008) Essential role for Nix in autophagic maturation of erythroid cells. Nature 454:232–235
56. Sivilotti ML (2004) Oxidant stress and haemolysis of the human erythrocyte. Toxicol Rev 23:169–188
57. Tinari A, Garofalo T, Sorice M, Esposito MD, Malorni W (2007) Mitoptosis: different pathways for mitochondrial execution. Autophagy 3:282–284
58. Zorov DB, Kinnally KW, Tedesci H (1992) Voltage activation of heart inner mitochondrial membrane channels. J Bioenerg Biomembr 24:119–124
59. Skulachev VP (1999) Mitochondrial physiology and pathology; concepts of programmed death of organelles, cells and organisms. Mol Aspects Med 20:139–184
60. Skulachev VP (2002) Programmed death phenomena: from organelle to organism. Ann NY Acad Sci 959:214–237
61. Karbowski M, Youle RJ (2003) Dynamics of mitochondrial morphology in healthy cells and during apoptosis. Cell Death Differ 10:870–880
62. Kundu M, Thompson CB (2005) Macroautophagy versus mitochondrial autophagy: a question of fate? Cell Death Differ 12:1484–1489
63. Lyamzaev KG, Nepryakhina OK, Saprunova VB, Bakeeva LE, Pletjushkina OY, Chernyak BV, Skulachev VP (2008) Novel mechanism of elimination of malfunctioning mitochondria (mitoptosis): formation of mitoptotic bodies and extrusion of mitochondrial material from the cell. Biochim Biophys Acta 1777:817–825
64. Deming PB, Rathmell JC (2006) Mitochondria, cell death, and B cell tolerance. Curr Dir Autoimmun 9:95–119

Chapter 7

Cellular Stress and Protein Misfolding During Aging

Rajiv Vaid Basaiawmoit and Suresh I.S. Rattan

Abstract

Cells are under constant onslaught from several intrinsic and extrinsic stressors, which lead to the occurrence and accumulation of molecular damage, functional impairment, aging, and eventual death. Protein misfolding is both a cause and a consequence of increased cellular stress. An age-related failure of the complex systems for handling protein misfolding results in the accumulation of misfolded and aggregated proteins, and consequent conformational diseases. However, some misfolded proteins have been found to be both toxic and, in some cases, protective, highlighting the various complex, dynamic, and interdependent mechanisms at play. Molecular mechanisms are being elucidated for the occurrence of protein misfolding and for its prevention by chaperones and various pathways of degradation. Insights from the knowledge about proteodynamics are likely to impact future interventional strategies to counter stress and to promote healthy aging by preventing and/or treatment of protein conformational diseases.

Key words: Abnormal proteins, Protein turnover, Proteasome, Lysosome, Proteostasis, Proteodynamics

1. Introduction

Aging at the molecular level is characterised by the progressive accumulation of molecular damage in DNA, RNA, lipids, and proteins (1). An age-related increase in the levels of structurally and functionally abnormal proteins is a universally observed phenomenon (2, 3). A variety of stressors have been implicated in the occurrence and increase of abnormal proteins during aging, of which oxidative stress is a major contributor. Chronic oxidative stress can lead to protein misfolding or unfolding, resulting in proteins unable to carry out their normal functions and thus initiating a breakdown of various cellular events. Mammalian cells have various maintenance, repair, and removal systems to counteract such events, and include the GroEL chaperone pathway to refold proteins (4), the ubiquitin-mediated proteosomal degradational

Peter Bross and Niels Gregersen (eds.), *Protein Misfolding and Cellular Stress in Disease and Aging: Concepts and Protocols*, Methods in Molecular Biology, vol. 648, DOI 10.1007/978-1-60761-756-3_7, © Springer Science+Business Media, LLC 2010

machinery to remove misfolded proteins (5), and lysosomal autophagy to sequester and degrade larger aggregates (6). Inefficient protein turnover by both proteasomal and lysosomal pathways are generally considered to be the main reason for the accumulation of abnormal proteins during aging (1). Recently, however, protein misfolding and its deleterious consequences have emerged as the crucial mediators of impaired cellular functions and increased cellular stress during aging. This is also underscored by observations that resistance to protein folding damage and proper protein folding over time correlate with species-specific longevity (7, 8). Thus, tight control of protein folding and misfolding seems to be the key in regulating cellular stress during aging.

It is well recognised that properly folded proteins are necessary for their efficient and accurate functions. Since the first demonstration by C. Anfinsen in 1973 that it is the primary amino acid sequence that determines the final protein structure (9), it is now well accepted that protein folding in vitro is many times simpler than what is known to happen in vivo. A remarkable fact in protein folding is the observation that proteins can fold in the millisecond timescale in a complex environment in which other proteins and macromolecules like DNA, RNA, and lipids populate the cellular space in concentration ranges of 300–400 mg/ml (10). This amazing biological feat clearly points out that protein folding is not an isolated event but is a "team event" with a number of molecules within this crowded interior assisting in the formation of the final correctly folded product. Thus, to attain proper function in the complex environment of a cell, appropriate mechanisms have evolved for folding proteins in the "right way," which overall comprise what is also known as the protein "quality control system."

The term "misfolded" proteins refers both to proteins that are unable to form a well-structured compact fold and to proteins that have attained a compact fold but a wrong one with the wrong amino acids resulting in aberrant or promiscuous functionality. What the term does not refer to is a new class of proteins that have been recently identified as functional proteins lacking a compact fold – the so-called "unfolded proteome" (11, 12). In this article, we review the basic mechanisms governing the process of protein folding, the occurrence of misfolding, and how it contributes towards increasing cellular stress during aging.

2. Protein Folding: A Team Event

Like all biological processes, protein folding requires energy, and is a mechanism that follows the basic thermodynamic principles. It has also been demonstrated that the folding process searches for the structural conformation with the lowest energy (13), again reaffirming Anfinsen's principle that the folding code is contained within the sequence itself. While Anfinsen's conclusions still hold for the most part, it has also become clear that the complex environment of a cell contains other "players," which are involved in the formation of the final, functional, and low-energy structure.

Many such macromolecules identified in this team-play are chaperones and folding catalysts. Molecular chaperones are the molecules whose primary function is to guide proteins to their "proper" fate but not to remain associated with the final "product" – a fully folded and functional protein (14). In comparison, folding catalysts are the molecules that accelerate the steps within the folding process, for example, disulphide isomerases which enhance the rate of formation and reorganisation of disulphide bonds within proteins (15). First set of evidence that chaperones prevent misfolding came from the observations that their levels were significantly increased under stress conditions such as heat, which cause protein denaturation and misfolding (16). Indeed, it is the chaperones that have emerged as the single largest class of molecules with an important role in the protein folding pathway, and an increasing diversity of their functions is still being uncovered. Not only do these molecules ensure that the proteins attain their native state, but also complex mechanisms have evolved to assist the folding of newly synthesised polypeptides and rescue existing proteins from partial stress-induced denaturation (17).

2.1. The Backstage Players: Multimodal Chaperones

A correctly folded protein is essential for carrying out functions within a cell. The mechanisms by which chaperones act differ depending on their nature and on their cellular location, but their active intervention in rescuing misfolded proteins is an energy consumptive process, and is thus largely driven by ATP (18). The GroEL-GroES chaperone system, for example, uses ATP to bind and release unfolded polypeptides (4). The binding stabilises the substrate in an unfolded state until its subsequent folding and release. The GroEL chaperone complex has been proposed to assist in protein folding via the Anfinsen Cage model (19, 20), whereby the cavity of the complex provides a stable microenvironment in which protein folding can occur unhindered and protected from aggregation. Another class of chaperones, the DnaK (Hsp70) family binds to unfolded proteins via exposed hydrophobic residues, thereby preventing misfolding and subsequent aggregation (21). The folding is then completed in the cytoplasm upon their release. In both cases, the chaperones partake indirectly without providing any folding information to the substrate.

An additional, albeit contrasting, function of chaperones has been found in the small periplasmic PapD-like chaperone family where protein information is supplied transiently to their substrate. This function was first observed in the subunit proteins of *E. coli* responsible for building up surface adhesive organelles whereby the chaperones supply steric information to their substrate proteins (22). An example of this is seen in the pilus subunits of gram-negative bacteria where PapD-like chaperones (of which more than 30 members exist in this superfamily) actually donate a beta strand of their own to the subunit to stabilise the complex (23). Interestingly, in the absence of the chaperone, a protease DegP, which recognises misfolded, denatured, and/or aggregated proteins, takes over, and degrades the now unstable subunits. DegP in itself has recently been shown not to be just a protease but more of a protease-chaperone. It has been observed that the binding of misfolded proteins transforms hexameric DegP into large active 12- or 24-mers where the highly flexible inner cavity serves opposite functions depending on the substrate (24), where outer membrane proteins are provided a safe transit while misfolded proteins are degraded within the same interior.

3. Stress and Protein Misfolding

Cells are under constant onslaught from several intrinsic and extrinsic stressors, which range from physical and metabolic stressors to environmental and genetic stressors. These also include changes in the microenvironment of the cell, up- or down-regulation of concentrations of metabolites and/or osmolytes (that can induce protein stability changes), temperature changes, pH changes, and so on. The consequent protein misfolding, if not rescued in time, can result in aberrant aggregate formation, whose accumulation can eventually lead to the onset of a disease phenotype. How cellular stress and protein misfolding overwhelm the protein quality control mechanisms to cause disease in the context of aging is discussed below.

3.1. Protein Homeodynamics in a Cell

"Proteostasis" has been suggested as a term to describe the control of the protein concentration, conformation, binding interactions, and cellular location by readapting the internal biology of the cell mostly through transcriptional and translational changes (25). Since the above processes are generally very dynamic, it may be more appropriate to call it as "proteodynamics," in line with increasing replacement of the term "homeostasis" with "homeodynamics" in biological systems (26). Indeed the protein quality control mechanisms within a cell, utilising various assistants in the protein folding pathway and other competing degradation pathways, indicate that proteodynamics plays a critical role in the cell. Evidence for the importance of the quality control is also highlighted by the fact that up to half of all the polypeptide chains fail to satisfy the quality control mechanism in the endoplasmic reticulum

(ER), and for some proteins the success rate is even lower (27). Like the HSP-response in the cytoplasm, the "unfolded protein response" (UPR) (28) in the ER is also up-regulated during stress and is strongly linked to the avoidance of protein misfolding diseases.

3.2. Natively Unfolded Proteins

Another recently observed parallel pathway in cells is that of the "natively unfolded proteome" (29). As the name suggests, these proteins are natively unstructured but are still functional, and a rather significant part of the eukaryotic genome actually codes for such proteins. The discovery of these proteins has challenged the paradigm of the need of a compact, ordered, and three-dimensional protein structure to carry out the function (30). Many of these proteins remain permanently unstructured under physiological conditions whereas many others interact with binding partners and then fold into a functional form making them ideal players in multiple targeting and cellular regulation. The presence of the natively unfolded proteome also suggests that chaperones and other macromolecular assistants probably bind to these proteins also and keep them from getting aggregated. One of the largest class of such natively unfolded proteins are the synucleins (31), of which the most extensively studied protein is α-synuclein for its role in Parkinson's disease (32). The cellular co-existence of parallel and seemingly antagonistic proteomes – the folded and unfolded proteome – is a clear example of proteodynamics in the cell. Imbalances in the proteodynamics, which could result from misfolding or aggregation, are sure to lead to functional impairment, disregulation, and other consequences including cell death, aging, and diseases.

3.3. Protein Misfolding and Aggregation as a Function of Stress

There are various factors in a cell – both intrinsic and extrinsic – that can raise cellular stress levels, and a cumulative action of these may overwhelm the quality control mechanisms, and result in the misfolding and/or aggregation of proteins. The main factors involved are spontaneous mutations due to intrinsic errors in DNA duplication, induced mutations due to reactive oxygen species (ROS), and other free radicals, errors in protein synthesis, and post-translational modifications due to damage by ROS, by other free radicals, and by nutritional metabolites such as glyoxal and methylglyoxal.

The role of mutations in protein misfolding is clearly identified and, as previously mentioned, this can be a severe form of stress in the cell. However, not all mutations have deleterious effects and thus it is important to distinguish the different types of mutations that can act as cellular stressors and those that cannot. Classification of mutations per se can be very complex as there may be many ways to define them based on pattern of inheritance, effect on the primary structure, phenotype, and function, and more importantly by their impact on protein sequence. However, generally mutations are classified based on their effect on the primary structure, and as such have three main types – point mutations, deletions, and insertions. Point mutations can be further subdivided into: (i) silent mutations – which code for similar amino acids; (ii) missense mutations – which code for different amino acids; and (iii) nonsense mutations – which code for a stop codon and can result in a truncated gene product.

For large deletions and, in most cases, stop mutations, the deleterious effect is obvious as the function of the affected polypeptide product is abolished. However, for missense mutations or short in-frame insertions and deletions, the disease-causing nature of the mutation is not directly evident (33). The Human Gene Mutation Database contains, at present, close to 88,000 entries on mutations in genetic diseases, of which approximately 49,000 entries are missense mutations (34). Missense mutations and short in-frame deletions and insertions often impair the propensity of the affected polypeptide to fold to the functional conformation, and/or decrease the stability of the functional conformation. Both effects lead to an increase in the proportion of mutant polypeptide present in nonfunctional conformations that are more susceptible to degradation or aggregation than the functional conformation.

Thus, diseases arising from misfolded or non-functional conformations were termed as conformational diseases (35), referring to conditions in which aggregation due to aberrant folding of a polypeptide appears to be the molecular pathological mechanism, as for example in Alzheimer's disease (AD) and Creutzfeldt-Jakob's disease. However, in a wider definition (36), conformational diseases were linked to disturbances of the folding process in general (37), thereby distinguishing two subgroups: one covered by the appearance of aggregating proteins and a second group of diseases in which impaired folding leads to rapid degradation of the affected polypeptides (e.g., cystic fibrosis and most forms of α-1-antitrypsin deficiency).

Over the years, a large number of human diseases have been linked to conformational upheavals or misfolding. The involvement of mutations in these diseases has also resulted in an alternate classification of these diseases by their effect on function. Inherited disorders like cystic fibrosis are due to amino acid substitutions that exhibit a loss-of-function pathogenesis because the aberrant protein is eliminated by the protein quality control system in the ER (38). However, not all aberrant proteins can be eliminated and the misfolded protein may accumulate and form toxic oligomeric and/or aggregated inclusions. In this case, the loss of function of a protein may be accompanied by a gain-of-function pathogenesis, which in many cases determines the pathological and clinical features, examples being Parkinson's and

Table 1
Representative protein misfolding diseases showing the proteins involved, the cellular location in which they fold, and the basic molecular pathogenesis

Protein involved	Location	Pathogenic mode	Disease
β-amyloid	C/E	Gain-of-function	Alzheimer's disease
Crystallin	C	Gain-of-function	Cataract
CFTR	ER	Loss-of-function	Cystic Fibrosis
Prion	ER	Loss-of-function	Creutzfeldt–Jakob disease or CJD
α-1-antitrypsin	ER	Loss-of-function	α-1-antitrypsin deficiency
Transthyretin (TTR)/ Lysozyme	ER	Gain-of-function	Familial amyloidosis
SCAD variants	M	Loss-of-function	Short-chain acyl-CoA dehydrogenase (SCAD) deficiency
LDL receptor	ER	Loss-of-function	Familial hypercholesterolemia

C cytoplasm, *E* extracellular, *ER* endoplasmic reticulum, *M* mitochondria

Huntington's disease. Table 1 lists some of the known protein misfolding or protein conformational diseases, the proteins involved in such disorders, their cellular location, and the suggested molecular pathogenesis.

3.4. Errors and Damage in Proteins

Since, the error frequency of amino acid misincorporation is generally considered to be quite high (3) as compared with nucleotide misincorporation (less than 10^{-6}), the role of protein error feedback in aging has been a widely discussed issue, and is the basis of the so-called error catastrophe theory of aging (3, 39). So far, no direct estimates of protein error levels in any aging system have been made, primarily due to the lack of appropriate methods to determine spontaneous levels of errors in a normal situation. However, several indirect estimates of the accuracy of translation in cell-free extracts, using synthetic templates or natural mRNAs have been made. Studies performed on various young and old animal tissues such as chick brain, mouse liver, and rat brain, liver, and kidney, and human cells undergoing aging in vitro did reveal some age-related increase in protein errors (3, 39). Furthermore, an induction and increase in protein errors has been shown to accelerate aging in human cells and bacteria (1, 40). It will be important to know if there is a direct link between increased protein errors and increased protein misfolding during aging.

A large number of post-translational modifications of proteins have been described that determine the activity, stability, specificity, transportability, and lifespan of a protein. Several of these modifications are highly specific and regulated involving various enzymatic pathways, for example, phosphorylation, methylation, ADP-ribosylation, glycosylation, and acetylation. However, there are several non-enzymatic modifications of proteins, which occur stochastically and are considered as damage, for example, oxidation, glycation, and racemization. Many of these alterations in proteins will lead to protein misfolding and to consequent cellular stress and aging. A large number of oxidatively damaged proteins have been reported to accumulate during aging in a wide variety of biological systems including cells, tissues, organs, and organisms (1, 40).

4. Aging, Diseases, and Protein Misfolding

The past decade has seen an explosion in the literature on misfolding diseases in general, and it has become evident that the largest group of misfolding diseases are those associated with the conversion of proteins into highly organised fibrillar aggregates. These structures are known as amyloid fibrils or plaques when referring to extracellular deposits, and as "intracellular inclusions" when observed inside the cell (41). Owing to the fact that these aggregates usually involve the formation of tissue deposits, either extracellular or intracellular, the diseases represented by them are also known as protein deposition diseases.

Whether all such deposits are harmful (42, 43) or may be even protective is still being debated (44), as both toxic and non-toxic effects have been reported (45, 46). Such a phenomenon of protective effects of potentially harmful conditions is also known as hormesis (47). A hitherto unexplored hypothesis has also been put forward that the cell itself may be targeting the protein to form non-toxic aggregates as a form of defence mechanism (48). This has indeed been seen in *E. coli* (48) where it was suggested that the function of protein aggregates is a type of "trash organelle" for cellular detoxification. While one may wonder at the cause and effect pattern of protein aggregates, it is clear that the presence and number of aggregates is heavily biased towards an aged cell (49). However, this is a result of damage that has started when the cell was young, and the toxic cascade was already at work during an apparently healthy looking cell. Indeed, it has been clearly shown that in many misfolding or deposition diseases, soluble oligomers (low molecular weight aggregates) are the major culprits responsible for toxicity (50) in contrast to higher molecular weight aggregates that do not correlate with toxicity (45). All these studies thus show that misfolding diseases and protein deposition diseases are two sides of the same coin. Furthermore, it has been shown that the same protein under pathological conditions can lead to the formation of fibrillar, pore-like, spherical, or amorphous aggregates with diverse biological consequences (51), but the conditions leading to misfolding and the formation of such abnormal complexes are unclear. The mechanisms underlying aggregation in vivo in biological systems are even less clearly understood due to the experimental difficulties in monitoring aggregates in their natural environment.

To summarise, a schematic representation of the occurrence and consequences of protein misfolding is given in Fig. 1. While accurate protein synthesis, followed by correct protein folding, is essential for the normal functioning of proteins, transcriptional and translational errors or stressful conditions can result in protein misfolding. Unless misfolding is counteracted by either chaperone-mediated refolding processes or proteasome-lysosome-mediated protein degradation, misfolded proteins can form aggregates with varying consequences. Whereas the so-called non-toxic aggregates may even have some beneficial hormetic effects by challenging the homeodynamic processes, large and toxic aggregates increase intracellular stress levels and consequent impairment in function, including aging, diseases, and death. With a deeper understanding of these phenomena, there is a hope for the development of efficient therapeutics capable of preventing or reversing the occurrence of protein misfolding and deposition diseases during the life-time of an individual.

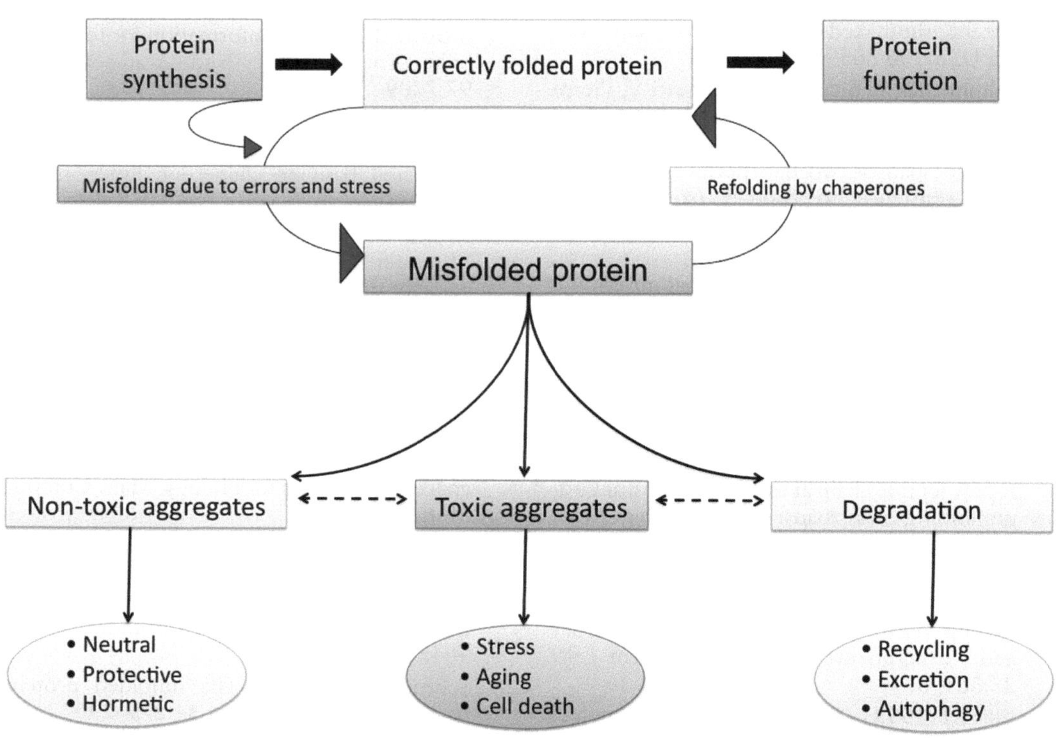

Fig. 1. Schematic representation of the occurrence of protein misfolding and the consequences of failed defences.

References

1. Rattan SI (2008) Increased molecular damage and heterogeneity as the basis of ageing. Biol Chem 389:267–272
2. Grune T, Jung T, Merker K, Davies KJA (2004) Decreased proteolysis caused by protein aggeregates, inclusion bodies, plaques, lipofuscin, ceroid, and "aggresomes" during oxidative stress, ageing, and disease. Int J Biochem Cell Biol 36:2519–2530
3. Hipkiss A (2006) Accumulation of altered proteins and ageing: causes and effects. Exp Gerontol 41:464–473
4. Fenton WA, Horwich AL (1997) GroEL-mediated protein folding. Protein Sci 6:743–760
5. Glickman MH, Ciechanover A (2002) The ubiquitin-proteasome proteolytic pathway: destruction for the sake of construction. Physiol Rev 82:373–428
6. Terman A, Gustafsson B, Brunk U (2007) Autophagy, organelles and ageing. J Pathol 211:134–143
7. Perez VI, Buffenstein R, Masamsetti V, Leonard S, Salmon AB, Mele J, Andziak B, Yang T, Edrey Y, Friguet B, Ward W, Richardson A, Chaudhuri A (2009) Protein stability and resistance to oxidative stress are determinants of longevity in the longest-living rodent, the naked mole-rat. Proc Natl Acad Sci USA 106:3059–3064
8. Salmon AB, Leonard S, Masamsetti V, Pierce A, Podlutsky AJ, Podlutskaya N, Richardson A, Austad SN, Chaudhuri AR (2009) The long lifespan of two bat species is correlated with resistance to protein oxidation and enhanced protein homeostasis. FASEB J 23:2317–2326
9. Anfinsen CB (1973) Principles that govern the folding of protein chains. Science 181:223–230
10. Ellis RJ, Minton AP (2003) Cell biology: join the crowd. Nature 425:27–28
11. Uversky VN (2002) What does it mean to be natively unfolded? Eur J Biochem 269:2–12
12. Uversky VN (2003) Protein folding revisited. A polypeptide chain at the folding-misfolding-nonfolding cross-roads: which way to go? Cell Mol Life Sci 60:1852–1871
13. Dinner AR, Sali A, Smith LJ, Dobson CM, Karplus M (2000) Understanding protein folding via free-energy surfaces from theory and experiment. Trends Biochem Sci 25:331–339
14. Ellis RJ (1993) The general concept of molecular chaperones. Philos Trans R Soc Lond B Biol Sci 339:257–261
15. Hartl FU, Hayer-Hartl M (2002) Molecular chaperones in the cytosol: from nascent chain to folded protein. Science 295:1852–1858
16. Pelham HR (1986) Speculations on the functions of the major heat shock and glucose-regulated proteins. Cell 46:959–961
17. Frydman J (2001) Folding of newly translated proteins in vivo: the role of molecular chaperones. Annu Rev Biochem 70:603–647
18. Parsell DA, Kowal AS, Singer MA, Lindquist S (1994) Protein disaggregation mediated by heat-shock protein Hsp104. Nature 372: 475–478
19. Ellis RJ, Hartl FU (1996) Protein folding in the cell: Competing models of chaperonin function. FASEB J 10:20–26
20. Shtilerman M, Lorimer GH, Englander SW (1999) Chaperonin function: folding by forced unfolding. Science 284:822–825
21. Teter SA, Houry WA, Ang D, Tradler T, Rockabrand D, Fischer G, Blum P, Georgopoulos C, Hartl FU (1999) Polypeptide flux through bacterial Hsp70: DnaK cooperates with trigger factor in chaperoning nascent chains. Cell 97:755–765
22. Barnhart MM, Pinkner JS, Soto GE, Sauer FG, Langermann S, Waksman G, Frieden C, Hultgren SJ (2000) PapD-like chaperones provide the missing information for folding of pilin proteins. Proc Natl Acad Sci USA 97:7709–7714
23. Sauer FG, Futterer K, Pinkner JS, Dodson KW, Hultgren SJ, Waksman G (1999) Structural basis of chaperone function and pilus biogenesis. Science 285:1058–1061
24. Krojer T, Sawa J, Schafer E, Saibil HR, Ehrmann M, Clausen T (2008) Structural basis for the regulated protease and chaperone function of DegP. Nature 453:885–890
25. Balch WE, Morimoto RI, Dillin A, Kelly JW (2008) Adapting proteostasis for disease intervention. Science 319:916–919
26. Yates FE (1994) Order and complexity in dynamical systems: homeodynamics as a generalized mechanics for biology. Math Comput Model 19:49–74
27. Schubert U, Anton LC, Gibbs J, Norbury CC, Yewdell JW, Bennink JR (2000) Rapid degradation of a large fraction of newly synthesized proteins by proteasomes. Nature 404:770–774
28. Schroder M (2006) The unfolded protein response. Mol Biotechnol 34:279–290
29. Fink AL (2005) Natively unfolded proteins. Curr Opin Struct Biol 15:35–41

30. Dyson HJ, Wright PE (2005) Intrinsically unstructured proteins and their functions. Nat Rev Mol Cell Biol 6:197–208
31. Surguchov A (2008) Molecular and cellular biology of synucleins. Int Rev Cell Mol Biol 270:225–317
32. Uversky VN (2008) Alpha-synuclein misfolding and neurodegenerative diseases. Curr Protein Pept Sci 9:507–540
33. Bross P, Corydon TJ, Andresen BS, Jorgensen MM, Bolund L, Gregersen N (1999) Protein misfolding and degradation in genetic diseases. Hum Mutat 14:186–198
34. Cooper DN, Ball EV, Krawczak M (1998) The human gene mutation database. Nucleic Acids Res 26:285–287
35. Carrell RW, Lomas DA (1997) Conformational disease. Lancet 350:134–138
36. Beissinger M, Buchner J (1998) How chaperones fold proteins. Biol Chem 379: 245–259
37. Thomas PJ, Qu BH, Pedersen PL (1995) Defective protein folding as a basis of human disease. Trends Biochem Sci 20:456–459
38. Amaral MD (2004) CFTR and chaperones: processing and degradation. J Mol Neurosci 23:41–48
39. Holliday R (1996) The current status of the protein error theory of ageing. Exp Gerontol 31:449–452
40. Rattan SIS (2006) Theories of biological ageing: genes, proteins and free radicals. Free Rad Res 40:1230–1238
41. Westermark P, Benson MD, Buxbaum JN, Cohen AS, Frangione B, Ikeda S, Masters CL, Merlini G, Saraiva MJ, Sipe JD (2005) Amyloid: toward terminology clarification. Report from the Nomenclature Committee of the International Society of Amyloidosis. Amyloid 12:1–4
42. Bucciantini M, Giannoni E, Chiti F, Baroni F, Formigli L, Zurdo J, Taddei N, Ramponi G, Dobson CM, Stefani M (2002) Inherent toxicity of aggregates implies a common mechanism for protein misfolding diseases. Nature 416:507–511
43. Isaacs AM, Senn DB, Yuan M, Shine JP, Yankner BA (2006) Acceleration of amyloid beta-peptide aggregation by physiological concentrations of calcium. J Biol Chem 281:27916–27923
44. Lansbury PT, Lashuel HA (2006) A century-old debate on protein aggregation and neurodegeneration enters the clinic. Nature 443:774–779
45. Cohen E, Bieschke J, Perciavalle RM, Kelly JW, Dillin A (2006) Opposing activities protect against age-onset proteotoxicity. Science 313:1604–1610
46. Malgaroli A, Vallar L, Zimarino V (2006) Protein homeostasis in neurons and its pathological alterations. Curr Opin Neurobiol 16:270–274
47. Rattan SIS (2008) Hormesis in ageing. Ageing Res Rev 7:63–78
48. Maisonneuve E, Fraysse L, Moinier D, Dukan S (2008) Existence of abnormal protein aggregates in healthy *Escherichia coli* cells. J Bacteriol 190:887–893
49. Maisonneuve E, Ezraty B, Dukan S (2008) Protein aggregates: an ageing factor involved in cell death. J Bacteriol 190:6070–6075
50. Caughey B, Lansbury PT (2003) Protofibrils, pores, fibrils, and neurodegeneration: separating the responsible protein aggregates from the innocent bystanders. Annu Rev Neurosci 26:267–298
51. Uversky VN (2003) A protein-chameleon: conformational plasticity of alpha-synuclein, a disordered protein involved in neurodegenerative disorders. J Biomol Struct Dyn 21:211–234

Chapter 8

Measuring Consequences of Protein Misfolding and Cellular Stress Using OMICS Techniques

Peter Bross, Johan Palmfeldt, Jakob Hansen, Søren Vang, and Niels Gregersen

Abstract

The ambition to measure all or at least a significant fraction of relevant molecules in a cell culture or tissue sample has reached possible realization with the development of the so-called OMICS technologies. We will here briefly review current technologies and give examples of their applications in investigations related to protein misfolding diseases. We will primarily cover the classical OMICS categories GENOMICS, TRANSCRIPTOMICS, METABOLOMICS, and with some more detail PROTEOMICS. These techniques are in most cases performed by dedicated core facilities or commercial services. We will give an assessment of uses as well as limitations of these technologies supported by examples of their application in research related to protein misfolding. We will further briefly discuss genome-wide RNA interference and finally touch on bioinformatics, because the huge amounts of data typically collected with OMICS techniques requires the application of specific software to handle and stratify the data sets. Today, most biologists using OMICS-techniques must, at least in part, be able to analyze their own data using user-friendly web-based tools.

Key words: Genomics, Transcriptomics, Metabolomics, Proteomics, Bioinformatics

1. Introduction

Starting with Walter Fiers' and Frederick Sanger's sequencing of entire virus genomes in the 1970s, the first "OMICS", genomics, was born. Genomics commenced its large-scale entry during the late 1980s, and was further developed during the 1990s and culminated in 2001 with the completion of the human genome sequence. Deriving from this we have seen a burst of OMICS disciplines with TRANSCRIPTOMICS, PROTEOMICS, and METABOLOMICS, and a wealth of further OMICS terms

Peter Bross and Niels Gregersen (eds.), *Protein Misfolding and Cellular Stress in Disease and Aging: Concepts and Protocols*, Methods in Molecular Biology, vol. 648, DOI 10.1007/978-1-60761-756-3_8, © Springer Science+Business Media, LLC 2010

(for a comprehensive listing and definition see http://omics.org). The rationale of all OMICS is to not only look at several selected molecules of a biological system, but to try to draw a global picture – as comprehensive as possible – of elements in a biological system. In this way, the research is data driven, that is, a careful review of the massive amount of data is expected to reveal the basis for the current condition of the biological system under study. With respect to diseases, this allows identifying the specific molecular differences between a disease condition and healthy controls. The hope is that this can produce hypotheses for further research into the pathogenesis and suggest diagnostic and therapeutic targets. This is the optimistic view; due to current limits of many of the technologies, hypotheses with preferred candidate molecules are in many cases underlying the design of the study. Hypothesis versus data-driven research is therefore not always sharply discernible.

1.1. Problems and Limits

The global approach of OMICS methods often has its limitations. For example, metabolites, transcripts, or proteins present in low amounts are frequently not detected or quantifiable due to limits of detection sensitivity. Although statistics are applied, the large amounts of data produce some accidental or technically caused errors. The challenge is then to obtain reproducible data and select the relevant parameters to eliminate false-positive and false-negative data as far as possible. To manage these difficulties bioinformatics has been developed strongly and as stated in (1), "*Bioinformatics has become too central to biology to be left to specialist bioinformaticians*". Molecular biologists have thus to become bioinformaticians in their own right to tackle these problems appropriately. There is no doubt that OMICS has come to stay and it will remain a challenge to filter the meaningful information (the needles in the haystack) out of the huge amounts of data and transform it into mechanistic understanding.

1.2. Scope of This Chapter

In the context of this book, we will in the following chapter discuss approaches, applications, and perspectives of OMICS techniques in connection with misfolding diseases and cellular stress. These techniques are rather equipment-demanding, and are subject to constant technological improvements. Therefore, these technologies in most cases require dedicated laboratories (core facilities) with advanced machineries and specifically trained specialists. If such core facilities are not available, such analyses can be performed by companies, who have specialized to perform such large-scale tasks. The commercial services, although expensive, spare the huge investments in equipment and trained staff. The follow-up of such analyses – data analysis, validation and generation and testing of hypotheses – is subsequently performed in the particular researcher's laboratory. Considering this, we have not included protocol

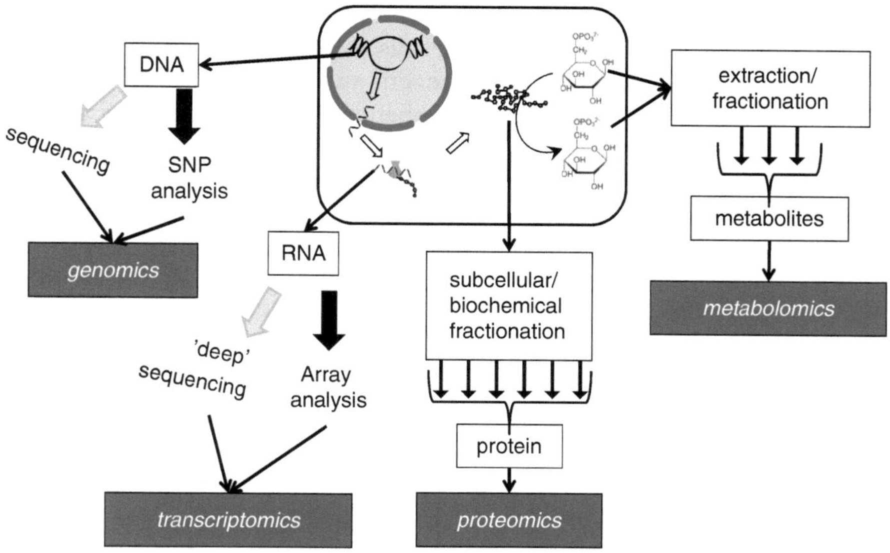

Fig. 1. Overview over major OMICS technologies. For details see text.

chapters for these methods in this volume. Instead, a summary of the approaches will be presented along with recent examples of applications pertinent to the contents of this volume. We will try to evaluate the uses and emerging perspectives of such methodology in this context. This will hopefully enable readers to make qualified decisions on when and how to perform such analyses and to develop strategies to explore the data and develop follow-up strategies. We will focus on four OMICS: GENOMICS, TRANSCRIPTOMICS, PROTEOMICS, and METABOLOMICS, which represent most molecules present in a cell (see Fig. 1). With the current state of the art of these methods, GENOMICS and TRANSCRIPTOMICS have the potential to cover a very high proportion of their targets. Typical PROTEOMIC analyses have a lower coverage, which can be partially overcome by fractionation and separation and joining together of many separate analyses. METABOLOMICS has so far the lowest coverage, also when combined with fractionation. Finally, we will briefly touch on genome-wide RNA interference (RNAi) studies, and, since the complexity of the data of all these analyses cannot be handled without the aid of computer-assistance, we will introduce the role and strategies of bioinformatics.

2. Genomics

2.1. Genome-Wide Association Studies

Genome-wide association studies involve rapid scanning for markers across the complete genomes of many individuals to find genetic variations associated with a particular disease. Such studies are particularly useful in finding genetic variations that contribute to common, complex diseases, such as asthma, cancer, diabetes, heart disease, and mental illnesses and can also be used to identify genetic variants which cause or influence the manifestation of the phenotype in protein misfolding diseases. Once new genetic associations are identified, researchers can use the information to develop better strategies to detect, treat, and prevent the disease. Technological developments in the recent past have made it possible and affordable to perform genome-wide association studies (GWAS) on large numbers of individuals from disease groups and controls. These studies typically use commercially available GeneChip services with typically approximately half-a-million validated single nucleotide polymorphism (SNP) markers spread across the genome to identify patterns of association with a specific condition. A wave of such studies has already been published (for a catalog of published studies see ref. 2).

One aim of GWAS in protein misfolding diseases is to detect novel disease loci/disease genes and susceptibility alleles by applying a much finer mesh of SNPs than in traditional association studies. However, GWAS typically utilize hundreds of case and control samples to achieve adequate statistical power. The low incidence of many diseases among the general population – Alzheimer's disease is a noteworthy exception – makes it difficult to collect enough patient material. One way to overcome this was realized in a recent GWAS on sporadic amyotrophic lateral sclerosis (ALS) (3). These authors exploited the availability of material from the Irish population, which, due to their geographic location, has been undisturbed by major demographic movements and is genetically rather homogenous thus giving a lower false-positive rate. Four hundred and thirty-two Irish individuals consisting of 221 patients with sporadic ALS and 211 control subjects were genotyped and 35 SNPs associated with ALS were identified with a *P*-value lower than 0.0001. A further discussion of the application of GWAS in neurological diseases can be found in (4).

Another objective of GWAS experiments is to identify susceptibility alleles which can modify the disease phenotype and contribute to the severity of disease expression. Such a study aiming to discover factors and susceptibility alleles in familial Parkinson's disease on 857 cases and a similar number of controls failed to detect significant candidate regions using stringent statistics (5). Regions comprising two previously implicated susceptibility factors scored high, but not significant; adding data from previous studies pinpointed two chromosome regions.

As in this example, the success rate of GWAS has in many cases not been very convincing and it is argued whether current GWAS, which are based on monitoring common variations, represent a good approach to reveal the genetic basis of common polygenic diseases (see, e.g., (6, 7)). If many very rare variations account for the phenotypes, these are overlooked when screening for common variations only.

2.2. Full Genome Sequencing

Full sequencing of individual genomes would get around many of the limitations pertinent to GWAS experiments. Although currently still a quite futuristic prospect due to still immense costs and technical challenges, sequencing of the first three individuals has shown that personal genome sequencing is feasible (8–10). The first of these personal genomes was performed by "classical" Sanger sequencing, whereas the other two have been determined by new technologies. These novel techniques referred to as "Next Generation Sequencing" (NGS), "Deep Sequencing," or with the more descriptive term "Massively parallel sequencing" allow to sequence in the range of 10^8–10^9 bases in a single experiment. A number of commercial platforms exist and we will not go into details with the different technologies, but just briefly sketch basic characteristics of these methods. For a more thorough discussion of the technical background, the interested reader is referred to the following reviews (11, 12). NGS is based on the sequencing of fragments from complex DNA mixtures like a whole genome or a full cDNA population. Each fragment is sequenced at a certain spot in an analysis cell and the sequence of bases is recorded for each of these spots. This setup allows to both sequence extremely large numbers of bases simultaneously (in parallel) and to get a quantitative record of the relative abundance of the sequences. The latter is especially relevant if cDNA produced from messenger RNA is sequenced as pointed out in Subheading 3.2 for a special application of these technologies termed "RNA-seq". Like all the other OMICS techniques, NGS also depends strongly on powerful bioinformatics tools to build large coherent contigs of the sequenced DNA fragments and extract all the information contained in it. These new technologies have a great potential and will likely be further developed (13) so that the price will fall in the near future.

Besides the technicalities, many challenges regarding the uses of such full genome sequence information have to be solved (14) before "personal genomics" holds its entrance into routine diagnostics. Likely the biggest challenge is to interpret the huge amount of variations present in different individuals, most of which are in no way characterized and, for those that are, there is in most cases only very limited information available. If only the variations in coding regions and promoter sequences discovered so far and to be discovered should be analyzed experimentally, this would engage scientists for a very long time. One possible solution to this may be the availability of thousands to millions of personal sequences tied to certain phenotypes and ethnic groups, which – with the use of smart bioinformatics – may enable scientists to find the needles in the haystack. Initiatives like the 1,000 Genomes Project (15), which will perform full sequencing of 1,000 genomes will provide an important basic catalog of human genome variation. Personal full genome sequencing will detect both common and rare variations, whereas genome-wide association studies are dependent on the informative quality of common variants. In the context of misfolding diseases and cellular stress, one may expect valuable information regarding modulating genetic factors from full genome sequencing of individuals with severe phenotypes versus those with mild ones who carry the same primary mutations in monogenic disease genes.

3. Transcriptomics

3.1. Microarray Analysis

Microarray analysis of the transcripts of cells or tissues is currently the most comprehensive, matured, and also affordable OMICS technique available. With the knowledge of the sequences of nearly all genes of humans and model organisms, it is possible to cover almost all transcripts. Recently, analyses have been further refined by the availability of so-called exon-arrays, which also give information on alternatively spliced transcripts (see, e.g., (16, 17)). This has further added to the information content of such analyses. Microarray analysis has been developed over the years and standardized, and, most importantly, microarray data have been made publicly available in the Gene expression omnibus (GEO; http://www.ncbi.nlm.nih.gov/geo/) for secondary analysis by researchers with similar interests. Great efforts have been made to ensure that raw and analyzed data are reported in a standardized manner (MIAME describes the Minimum Information About a Microarray Experiment) thus giving the best possibility for researchers to cross-compare and interpret data. Transcriptional changes are very sensitive markers for gene expression changes; however, transcript levels appear not to be very good indicators of protein levels (18).

A transcriptome study in connection with misfolding diseases was performed by Mandel et al. (19) to study Parkinson's disease. RNA samples from substantia nigra from a group of sporadic Parkinson's disease patients were compared to controls and a group of patients with familial Parkinsonism. This study revealed that similar patterns of transcript changes were observed for the sporadic and inherited patients, consistent with a convergent pathogenesis in both groups. The study also identified certain pathways, both formerly known ones like the proteasomal protein degradation pathway, whose expression was decreased, as well as previously unsuspected cellular processes. It could thus pinpoint possible targets for intervention strategies and biomarkers for diagnosis and follow-up of disease progression. For a more comprehensive overview of the methodological background of transcript profiling in especially neurodegenerative disease research, see (20, 21).

3.2. RNA-seq

An upcoming alternative to hybridization-based transcriptome analysis is RNA-seq (22), which makes use of "Next Generation Sequencing" (NGS) technology discussed in Subheading 2.2. The RNA-seq analyses the composition of a transcriptome by sequencing and counting cDNA molecules synthesized from mRNA preparations. Although different variants of the methods for sample preparation and analysis in RNA-seq approaches exist, the following steps are generally included: mRNA isolation, production of cDNA by reverse transcription, fragmentation (on cDNA or RNA level), cDNA sequencing and matching cDNA sequence reads back to the genome by alignment algorithms. From the distribution of sequence reads in genomic regions, transcript abundance can be estimated by calculating the number of sequence reads mapping to a specific region as for example, a particular gene or exon. Because of the inherent "single base resolution nature" of the sequence data, the genomic distribution of cDNA sequence reads can additionally be used to identify alternative start and termination sites as well as the presence and abundance of alternatively spliced transcripts. RNA-seq, principally in one step, gives a full record, both qualitatively and quantitatively, of the mRNA molecules in a cell and provides a level of information that would require data from a combination of different experimental techniques. The use of RNA-seq is currently restricted by the high cost and the limited number of facilities and laboratories with access to NGS techniques, but also by the fact that basic analysis of the huge amount of sequence data and extraction of biological relevant information is a massive challenge in terms of both man and computer power. Still, the number of published transcriptome studies using RNA-seq will likely increase, when next generation sequencing becomes more economical and broadly available.

4. Proteomics

As mentioned above, transcript levels are bad predictors of protein levels. Posttranscriptional processes have a large impact on protein levels as well as protein function and posttranslational modifications of proteins are very important for the regulation of activities. To get a trustworthy picture of a cell or tissue's proteome, it is therefore necessary to measure proteins directly.

Mass spectrometry (MS) is the main work horse in large-scale proteomics and this section therefore focuses on MS-based proteomics. Technological leaps ahead in MS instrumentation and bioinformatics mining of mass spectrometric peptide data have provided the basis for developing the current state of the art, which allows to address full qualitative and quantitative protein profiling of complex mixtures of proteins. For detection of a lower number of proteins and validation of MS proteome data, immunological techniques such as Western blot, enzyme linked immunosorbent assay (ELISA), multiplex Xmap-Luminex technology or antibody-based protein arrays can be used. The latter represents a valuable intermediate scale analysis technique in research areas, where the appropriate antibodies have been optimized for the array format. MS-based proteomics, in contrast to the immunological methods, is a truly discovery-based technique, which has the potential to also detect and analyze previously unidentified proteins. However, large-scale MS-based proteomics require access to genome (and thus proteome) sequence data and the establishment of comprehensive and freely accessible sequence databases during the last decade has been a prerequisite for the development of today's successful MS-based proteomics.

4.1. MS-Based Proteomics

Modern MS instruments have high capacity to analyze complex mixtures of proteins. In most current large-scale proteomics protocols, the MS detects and quantitates peptides produced by proteolytic digestion of the protein fraction to be analyzed. In one run, a fraction of peptides is separated by a reverse-phase liquid chromatography device, directly coupled to the mass spectrometer. The number of peptides that can be detected and quantitated in such a run is limited so that pretreatment of the samples at the protein or peptide level (or both) is applied to improve the coverage of the detected proteome. Basically, the pretreatments can be divided into (1) fractionation and (2) separation. Fractionation enables selection of a part of the proteome and thereby decreases sample complexity. It can be performed by, for example, dissection of tissue, centrifugation of organelles or specific biochemical isolation of proteins with certain properties. Through fractionation, one assures that the relevant biology is captured and removes non-relevant proteins, which could interfere with the analysis and decrease the analytical depth.

Separation technology is employed in order to divide the selected fraction of the total proteome, often termed subproteomes, into a series of samples that are analyzed separately one by one by MS. The classical approach running 2-dimensional gel electrophoresis of protein samples followed by MS identification represent a method where separation is performed down to single proteins, and where quantification is carried out by picture analysis of spot intensities on the gel. Each protein spot found to be altered in intensity, in the comparison of different samples, is then identified by MS, typically MALDI-TOF. Since modern tandem mass spectrometry (MS/MS) instruments are able to identify and quantitate more than hundred proteins in a mixture in one run, the labor-intensive 2D-gel electrophoresis has largely been replaced by liquid chromatography (LC) directly coupled (in-line) to the MS instrument. For both gel electrophoresis and LC, there exist several operation principles which can be applied at the protein or peptide level (Table 1) and often are two or more separation methods combined. When different LC separations are combined, it is referred to as multidimensional LC or Multidimensional Protein Identification Technology (MudPit) (23, 24). In multidimensional separations, the LC methods are chosen so that the separation methods are based on totally different principles (orthogonal principles) yielding efficient resolution of samples with high complexity. The most common combination is cation exchange followed by reverse phase chromatography, where the latter is coupled directly to the MS instrument.

The MS measures the mass to charge ratio (m/z) of peptides, and the most widespread method to prepare the peptide mixtures is by enzymatic cleavage of the proteins by trypsin which cleaves after lysine and arginine residues. Tandem mass spectrometry (MS/MS) consists of two steps in the instrument. The first step is a determination of the mass of the intact peptide, and the second is an amino acid sequencing of the peptide. Step two is done by fragmentation of the peptide and determination of the fragment's masses compared to a database of all proteins from the respective organism. On basis of the full genome sequence of the respective organism, the MS data confer identification of the peptides, and two peptides

Table 1
Examples of the most common separation methods applied prior to MS detection

	Method	Separation principle	References
In gel	SDS-PAGE	Size	
	Isoelectric focusing	Isoelectric point	
In liquid	Anion or cation LC	Charge	(50)
	Reverse phase LC	Hydrophobicity	(51)
	Free flow electrophoresis	Isoelectric point	(52)
	Capillary electrophoresis	Charge to mass ratio	(53)

Most of the methods can be adapted to separate either proteins or peptides, with the exception of SDS-polyacrylamide gel electrophoresis (PAGE), which cannot separate peptides. Within brackets is the biochemical property according to which the proteins/peptides are separated from the same protein are often sufficient for confident protein identification.

Relative quantification of proteins in a LC-MS/MS workflow – to obtain quantitative protein profiles – is possible through various strategies. Labeling at the protein or peptide level adds different mass tags to the peptides from the different samples to be compared. After labeling, the samples are mixed, and, after fractionation and separation steps, the relative signal intensities of the differentially labeled peptides give the relationship between the protein concentrations in the samples. There are several different labeling methods (25) out of which SILAC (Stable Isotope Labeling of Amino acids in Culture) (26) and iTRAQ (Isobaric Tag for Relative and Absolute Quantitation) (27) are the most commonly applied. SILAC requires cellular cultures to which two or three different isotopically labeled amino acids are added. Although not all cell types grow well in these specific cultivation media, the method has the advantage that labeling occurs very early in the analytical procedure resulting in low analytical variation. With iTRAQ labeling, up to eight analytical samples of peptides are amino-labeled chemically, yet after production of peptides by trypsin digestion. iTRAQ labeling has the advantage that it can be performed on all kinds of peptide mixtures irrespective of the origin of the peptides. Label-free methods are also being developed where software tools extract ion counts from the MS data or counts fragmentation events per peptide. These data can then be used for relative quantitation when proper internal standards and/or normalization procedures are applied (28).

The MS data obtained experimentally are then searched against sequence databases and the searching algorithms produce protein lists as output and quantitative information. Subsequent bioinformatic mining of the data is central for success (see below), as well as for careful statistical treatment of the protein identification and quantitation data to ensure a low error rate.

4.2. Proteomics in the Context of Protein Misfolding Diseases

PROTEOMICS has typically been assigned to be a large-scale discovery method, which is used to generate novel hypotheses instead of only answering specific questions. However, quantification of proteins to obtain protein profiles, might not only give the overview picture but can also be used in a more targeted mode. In the research area of protein misfolding, it is often desirable to map the global effects of misfolding as well as, for example, locating the fraction of proteins, which are specifically prone to misfold. During recent years, studies of both kinds have been performed and below are a few examples, which should serve to illustrate the principles and the possible outcomes of the proteomic methodologies.

4.2.1. Profiling Studies to Study the Effects of Misfolding

Large scale PROTEOMICS have been applied to map altered protein expression in neurological diseases such as Alzheimer's and Parkinson's disease. Analyses have been performed on samples from human cerebrospinal fluid, plasma, saliva, or urine on different platforms and consequently resulting in somewhat different protein lists from which candidate biomarkers can be selected ((29) and references therein). Animal models are used when proteomics analysis of specific tissues is required for elucidating the underlying disease mechanisms. PROTEOMICS of the striata in a mouse model of Parkinson's disease revealed mitochondrial dysfunction and oxidative stress (30). Another study used a mouse model of Alzheimer's disease, the 3xTgAD mice expressing a mutant form of the amyloid precursor protein presenilin-1 (31). They compared its hippocampal and cortical proteome with control mice using the iTRAQ technique and found changes in synaptic plasticity and several other pathways.

4.2.2. Novel Substrates of Chaperones

Previously unknown substrates of chaperones can be identified by gene knock-out or specific inhibition of the chosen chaperone followed by a comparison of the proteomes of treated versus unperturbed cells. This strategy has been applied in a study in which the Hsp90 chaperone was inhibited together with the proteasome, resulting in an overload of immature ubiquitinylated proteins and protein aggregates (32). Among 48 aggregated proteins, several were directly linked to the inhibition of Hsp90 and were thus identified as novel substrates of Hsp90. In another study, novel substrates of SecB, a chaperone for secretory proteins in *Escherichia coli*, have been identified through protein profiling of a SecB null mutant of *E. coli* (33). Secretory proteins were found to accumulate in the cytoplasm and could thus be linked to the SecB activity. Moreover, increased levels of protein quality control proteins such as DnaK, GroEL/ES, and ClpB, were described as the cellular protection response to the accumulation of the misfolded proteins.

4.2.3. Protein Carbonylation Related to Misfolding Diseases

A dominant form of oxidative protein modification is protein carbonylation (see also Chapter 1), which is an irreversible oxidative damage, often leading to loss of protein function (34). Carbonylation is highly relevant in biological systems with misfolded proteins since mistranslated and denatured proteins have been found to be more sensitive to carbonylation than native proteins (35, 36). Furthermore, carbonylated proteins have been reported to be more prone to degradation by cellular proteases than non-oxidized proteins (35, 37, 38). Intracellular proteases seem to be able to degrade mildly carbonylated proteins, whereas heavily carbonylated proteins appear to first aggregate and then to form covalent cross-links that make them highly resistant to proteolysis. The inability to degrade extensively carbonylated proteins may contribute to the accumulation of protein aggregates during diseases (39). The stable properties of carbonylated proteins make them interesting analytical targets and they have been used as markers of oxidative stress (see Chapter 1). Carbonylation is a consequence of a reaction between reactive oxygen species (ROS) and proteins and it keeps a record of bursts of oxidative stress in a specific cellular location, whereas ROS itself and downstream products like H_2O_2 are quickly metabolized. Measurement of carbonylation can be performed at different levels of sample complexity starting with a crude mixture of proteins. Overall measurement of the extent of protein carbonylation can be assayed on Western blots (see Chapter 17). Using MS, the carbonylated proteins can principally be identified directly in a liquid mixture. However, it might prove difficult since there are many combinations of carbonylation products, each at a relatively low level. Therefore, a common strategy is to increase the signal of carbonylated proteins by first enriching for the carbonylated proteins, which in some cases even enables identification of the amino acid site of carbonylation (40). Enrichment using affinity chromatography has been applied using different derivatization agents: DOTA (41), Girard P reagent (42) and biotin hydrazine (40, 43).

Large-scale protein carbonylation studies have been applied to, for example, human post mortem brain samples of Huntington disease (44) and on the mouse model PS1+AβPP of Alzheimer's disease (45), and have thus contributed to the understanding of neurogenerative diseases.

5. Metabolomics

METABOLOMICS, in some scientific circles also called METABONOMICS, is a growing OMICS branch, which has the aim to quantitate – as many as possible – metabolites from body fluids or cell extracts and thus describe the status of cells or organisms with respect to metabolic activity. Currently, the major employed methodologies are MS and nuclear magnetic resonance (NMR) spectroscopy. MS is able to detect a larger number of different metabolites, but laborious upstream fractionation strategies must be used focusing on fractions of metabolites with similar gross chemical properties (like hydrophobicity) and then combining the data to a more comprehensive picture. In contrast, NMR does not require upstream fractionation, but it covers clearly fewer metabolites, is less sensitive and identification is more challenging. METABOLOMICS is usually dealing with a limited fraction of the many metabolites present in biological systems, but the big step from analyzing one or a few metabolites to a more broad selection has definitely been made and methodologies are currently developing quickly (46). Estimates of the expected number of different small molecule metabolites range from 3,000 to 100,000 and current analysis is still coming short of the larger part of them. A further technical problem is the low chemical stability of many metabolites during sample preparation and the difficulty to preserve subcellular compartmentalization. This makes it difficult to distinguish between organellar, cytosolic, and extracellular pools of metabolites. Still, metabolomics has the potential to give unique information regarding cellular activity that cannot be obtained by the other OMICS techniques. Data on transcript and protein levels can sometimes be used to deduce alterations in metabolism, but the direct measurement of metabolites adds a further layer of useful information.

In a study of heat stress in the fruitfly, the metabolite profile was analyzed during recovery after exposure of the animals to different thermal stress treatments and compared with untreated controls using untargeted NMR metabolic profiling (47). Flies pretreated with moderate heat stress returned faster to normal metabolism than non-pretreated animals. Analysis of the metabolite profiles over time after heat shock showed good accordance between metabolite changes and gene expression changes.

6. Genome-Wide RNA Interference

By one-by-one knocking down of all genes using RNAi while following a specific phenotype, it is possible to identify genes that affect this phenotype. Although knocking down essential genes does not allow growth of the organism, the fact that knock down does not fully abolish expression of the gene, in fact a broad range of residual expression is observed, making it possible to grade the effect also of essential genes. The effort to identify factors that modulate phenotypes in misfolding diseases and in this way characterize molecular mechanisms as well as pinpoint targets for therapeutic intervention has been pursued by using genome-wide interference in the model organisms *Caenorhabditis elegans* and *Drosophila melanogaster*. *C. elegans* provides a technically easy handled target for genome-wide knock-down of a large proportion or all genes. Recently, genome-wide RNAi has also become available for cultured human cells (48); however, the costs are still extremely high, so that broader application has to await technical improvements and more wide application.

Genome-wide RNAi has become a quite common technique in studies related to protein misfolding. One example is a screen for factors influencing aggregation of α-synuclein in a *C. elegans* model of Parkinson's disease (49). This study identified both genes, which promote formation of α-synuclein inclusions and genes, which appear to counteract inclusion formation. Consistent with previous findings, quality control genes were overrepresented among aggregation suppressing genes and aging-associated genes were prominent among the genes that promote inclusion formation. The list of factors identified included both known and new factors which now may be further investigated and possibly targeted by therapeutic strategies.

7. Bioinformatics

A maybe not fully satisfactory, but simple and valid definition of bioinformatics is "*using computers to manage, organize, and analyze large amounts of biological information*" as stated in (1). Bioinformatics comes in three different flavors mainly performed by differently trained researchers: (1) highly specialized computer professionals, who develop databases and analysis tools allowing researchers to query with their datasets and often also to upload their datasets to contribute to the growth of the databases, (2) Janus-headed specialists, who are computer specialists as well as trained in the instruments used in the specific molecular biology field and work at the interface of front-line development of software for running the instruments and converting primary data to manageable and structured data sets, and (3) "ordinary" molecular biologists, who are clients to the former two and are able to take the output from database queries and software to produce meaningful and illustrative presentations of their complex experimental data as well as extract the information, which gives the rationale for experimental follow-up. The last group of researchers is steadily growing as seasoned molecular biologists have learned bioinformatics "by doing" and the younger generation has got bioinformatics as a natural part of the curriculum. This is facilitated by a quantum leap in usability and clever entanglement of the various databases. Also several meta-databases exist combining and presenting information in an intuitive manner. The field is very much in development and steady movement, as can be seen by websites often changing their setup, and favorite sites disappearing and new ones coming up. Due to this constant movement and the scope of this chapter we will in the following only mention some basic databases and a few examples of tools from our own practical experience as molecular biologist-bioinformaticians. These references may give some hints for beginners on where to start when dealing with large-scale data.

For functional classification of proteins and genes, the Gene Ontology (GO) terminology has been developed (http://www.geneontology.org/) giving descriptive terms to genes from many organisms. The terms are hierarchically structured and fall within the categories of molecular function, biological process, and cellular compartment. Two examples of web resources that utilize functional classification of data and that can be used for sorting long lists of proteins or genes into functionally related groups are DAVID (http://david.abcc.ncifcrf.gov/home.jsp) and PANTHER (http://www.pantherdb.org/). On NCBI pages, extensive access to sequence data and sequence analysis tools are found (http://www.ncbi.nlm.nih.gov/), and at ExPASy (http://www.expasy.org/), links to a wide range of proteomics tools can be found. Furthermore, organism-specific databases could be advantageous when searching for more specific gene descriptions.

Bioinformatics is also heavily used when combining datasets from several OMICS-techniques, a field termed systems biology; this, however, lies beyond the scope of this volume.

References

1. Stein LD (2008) Bioinformatics: alive and kicking. Genome Biol 9:114
2. Hindorff LA, Sethupathy P, Junkins HA, Ramos EM, Mehta JP, Collins FS, Manolio TA (2009) Potential etiologic and functional implications of genome-wide association loci for human diseases and traits. Proc Natl Acad Sci U S A 106:9362–9367
3. Cronin S, Berger S, Ding J, Schymick JC, Washecka N, Hernandez DG, Greenway MJ, Bradley DG, Traynor BJ, Hardiman O (2008) A genome-wide association study of sporadic ALS in a homogenous Irish population. Hum Mol Genet 17:768–774
4. Simon-Sanchez J, Singleton A (2008) Genome-wide association studies in neurological disorders. Lancet Neurol 7:1067–1072
5. Pankratz N, Wilk JB, Latourelle JC, DeStefano AL, Halter C, Pugh EW, Doheny KF, Gusella JF, Nichols WC, Foroud T, Myers RH (2009) Genomewide association study for susceptibility genes contributing to familial Parkinson disease. Hum Genet 124:593–605
6. Hardy J, Singleton A (2009) Genomewide association studies and human disease. N Engl J Med 360:1759–1768
7. Petsko GA (2009) Guilt by association. Genome Biol 10:104
8. Levy S, Sutton G, Ng PC, Feuk L, Halpern AL, Walenz BP, Axelrod N, Huang J, Kirkness EF, Denisov G, Lin Y, MacDonald JR, Pang AW, Shago M, Stockwell TB, Tsiamouri A, Bafna V, Bansal V, Kravitz SA, Busam DA, Beeson KY, McIntosh TC, Remington KA, Abril JF, Gill J, Borman J, Rogers YH, Frazier ME, Scherer SW, Strausberg RL, Venter JC (2007) The diploid genome sequence of an individual human. PLoS Biol 5:e254
9. Wheeler DA, Srinivasan M, Egholm M, Shen Y, Chen L, McGuire A, He W, Chen YJ, Makhijani V, Roth GT, Gomes X, Tartaro K, Niazi F, Turcotte CL, Irzyk GP, Lupski JR, Chinault C, Song XZ, Liu Y, Yuan Y, Nazareth L, Qin X, Muzny DM, Margulies M, Weinstock GM, Gibbs RA, Rothberg JM (2008) The complete genome of an individual by massively parallel DNA sequencing. Nature 452:872–876
10. Wang J, Wang W, Li R, Li Y, Tian G, Goodman L, Fan W, Zhang J, Li J, Zhang J, Guo Y, Feng B, Li H, Lu Y, Fang X, Liang H, Du Z, Li D, Zhao Y, Hu Y, Yang Z, Zheng H, Hellmann I, Inouye M, Pool J, Yi X, Zhao J, Duan J, Zhou Y, Qin J, Ma L, Li G, Yang Z, Zhang G, Yang B, Yu C, Liang F, Li W, Li S, Li D, Ni P, Ruan J, Li Q, Zhu H, Liu D, Lu Z, Li N, Guo G, Zhang J, Ye J, Fang L, Hao Q, Chen Q, Liang Y, Su Y, San A, Ping C, Yang S, Chen F, Li L, Zhou K, Zheng H, Ren Y, Yang L, Gao Y, Yang G, Li Z, Feng X, Kristiansen K, Wong GK, Nielsen R, Durbin R, Bolund L, Zhang X, Li S, Yang H, Wang J (2008) The diploid genome sequence of an Asian individual. Nature 456:60–65
11. Lister R, Gregory BD, Ecker JR (2009) Next is now: new technologies for sequencing of genomes, transcriptomes, and beyond. Curr Opin Plant Biol 12:107–118
12. Voelkerding KV, Dames SA, Durtschi JD (2009) Next-generation sequencing: from

basic research to diagnostics. Clin Chem 55:641–658
13. Bosch JR, Grody WW (2008) Keeping up with the next generation: massively parallel sequencing in clinical diagnostics. J Mol Diagn 10:484–492
14. Olson MV (2008) Human genetics: Dr Watson's base pairs. Nature 452:819–820
15. Kaiser J (2008) DNA sequencing. A plan to capture human diversity in 1000 genomes. Science 319:395
16. Nagalakshmi U, Wang Z, Waern K, Shou C, Raha D, Gerstein M, Snyder M (2008) The transcriptional landscape of the yeast genome defined by RNA sequencing. Science 320:1344–1349
17. Wold B, Myers RM (2008) Sequence census methods for functional genomics. Nat Methods 5:19–21
18. Gygi SP, Rochon Y, Franza BR, Aebersold R (1999) Correlation between protein and mRNA abundance in yeast. Mol Cell Biol 19:1720–1730
19. Mandel S, Grunblatt E, Riederer P, Amariglio N, Jacob-Hirsch J, Rechavi G, Youdim MB (2005) Gene expression profiling of sporadic Parkinson's disease substantia nigra pars compacta reveals impairment of ubiquitin-proteasome subunits, SKP1A, aldehyde dehydrogenase, and chaperone HSC-70. Ann N Y Acad Sci 1053:356–375
20. Papapetropoulos S, Shehadeh L, McCorquodale D (2007) Optimizing human post-mortem brain tissue gene expression profiling in Parkinson's disease and other neurodegenerative disorders: from target "fishing" to translational breakthroughs. J Neurosci Res 85:3013–3024
21. Kirby J, Heath PR, Shaw PJ, Hamdy FC (2007) Gene expression assays. Adv Clin Chem 44:247–292
22. Wang Z, Gerstein M, Snyder M (2009) RNA-Seq: a revolutionary tool for transcriptomics. Nat Rev Genet 10:57–63
23. Hu L, Ye M, Jiang X, Feng S, Zou H (2007) Advances in hyphenated analytical techniques for shotgun proteome and peptidome analysis – a review. Anal Chim Acta 598:193–204
24. Washburn MP, Wolters D, Yates JR III (2001) Large-scale analysis of the yeast proteome by multidimensional protein identification technology. Nat Biotechnol 19:242–247
25. Panchaud A, Affolter M, Moreillon P, Kussmann M (2008) Experimental and computational approaches to quantitative proteomics: status quo and outlook. J Proteomics 71:19–33
26. Ong SE, Blagoev B, Kratchmarova I, Kristensen DB, Steen H, Pandey A, Mann M (2002) Stable isotope labeling by amino acids in cell culture. SILAC, as a simple and accurate approach to expression proteomics. Mol Cell Proteomics 1:376–386
27. Ross PL, Huang YN, Marchese JN, Williamson B, Parker K, Hattan S, Khainovski N, Pillai S, Dey S, Daniels S, Purkayastha S, Juhasz P, Martin S, Bartlet-Jones M, He F, Jacobson A, Pappin DJ (2004) Multiplexed protein quantitation in Saccharomyces cerevisiae using amine-reactive isobaric tagging reagents. Mol Cell Proteomics 3:1154–1169
28. Kito K, Ito T (2008) Mass spectrometry-based approaches toward absolute quantitative proteomics. Curr Genomics 9:263–274
29. Shi M, Caudle WM, Zhang J (2009) Biomarker discovery in neurodegenerative diseases: a proteomic approach. Neurobiol Dis 35:157–164
30. Chin MH, Qian WJ, Wang H, Petyuk VA, Bloom JS, Sforza DM, Lacan G, Liu D, Khan AH, Cantor RM, Bigelow DJ, Melega WP, Camp DG, Smith RD, Smith DJ (2008) Mitochondrial dysfunction, oxidative stress, and apoptosis revealed by proteomic and transcriptomic analyses of the striata in two mouse models of Parkinson's disease. J Proteome Res 7:666–677
31. Martin B, Brenneman R, Becker KG, Gucek M, Cole RN, Maudsley S (2008) iTRAQ analysis of complex proteome alterations in 3xTgAD Alzheimer's mice: understanding the interface between physiology and disease. PLoS One 3:e2750
32. Falsone SF, Gesslbauer B, Rek A, Kungl AJ (2007) A proteomic approach towards the Hsp90-dependent ubiquitinylated proteome. Proteomics 7:2375–2383
33. Baars L, Ytterberg AJ, Drew D, Wagner S, Thilo C, van Wijk KJ, de Gier JW (2006) Defining the role of the *Escherichia coli* chaperone SecB using comparative proteomics. J Biol Chem 281:10024–10034
34. Dalle-Donne I, Rossi R, Giustarini D, Milzani A, Colombo R (2003) Protein carbonyl groups as biomarkers of oxidative stress. Clin Chim Acta 329:23–38
35. Dukan S, Farewell A, Ballesteros M, Taddei F, Radman M, Nystrom T (2000) Protein oxidation in response to increased transcriptional or translational errors. Proc Natl Acad Sci U S A 97:5746–5749
36. Fredriksson A, Ballesteros M, Dukan S, Nystrom T (2006) Induction of the heat shock regulon in response to increased

mistranslation requires oxidative modification of the malformed proteins. Mol Microbiol 59:350–359
37. Bota DA, Davies KJ (2002) Lon protease preferentially degrades oxidized mitochondrial aconitase by an ATP-stimulated mechanism. Nat Cell Biol 4:674–680
38. Rivett AJ (1985) Preferential degradation of the oxidatively modified form of glutamine synthetase by intracellular mammalian proteases. J Biol Chem 260:300–305
39. Grune T, Merker K, Sandig G, Davies KJ (2003) Selective degradation of oxidatively modified protein substrates by the proteasome. Biochem Biophys Res Commun 305:709–718
40. Mirzaei H, Regnier F (2005) Affinity chromatographic selection of carbonylated proteins followed by identification of oxidation sites using tandem mass spectrometry. Anal Chem 77:2386–2392
41. Lee S, Young NL, Whetstone PA, Cheal SM, Benner WH, Lebrilla CB, Meares CF (2006) Method to site-specifically identify and quantitate carbonyl end products of protein oxidation using oxidation-dependent element coded affinity tags (O-ECAT) and nanoliquid chromatography Fourier transform mass spectrometry. J Proteome Res 5:539–547
42. Mirzaei H, Regnier F (2006) Enrichment of carbonylated peptides using Girard P reagent and strong cation exchange chromatography. Anal Chem 78:770–778
43. Mirzaei H, Regnier F (2007) Identification of yeast oxidized proteins: chromatographic top-down approach for identification of carbonylated, fragmented and cross-linked proteins in yeast. J Chromatogr A 1141:22–31
44. Sorolla MA, Reverter-Branch G, Tamarit J, Ferrer I, Ros J, Cabiscol E (2008) Proteomic and oxidative stress analysis in human brain samples of Huntington disease. Free Radic Biol Med 45:667–678
45. Soreghan BA, Lu BW, Thomas SN, Duff K, Rakhmatulin EA, Nikolskaya T, Chen T, Yang AJ (2005) Using proteomics and network analysis to elucidate the consequences of synaptic protein oxidation in a PS1 + AbetaPP mouse model of Alzheimer's disease. J Alzheimers Dis 8:227–241
46. Blow N (2008) Metabolomics: Biochemistry's new look. Nature 455:697–700
47. Malmendal A, Overgaard J, Bundy JG, Sorensen JG, Nielsen NC, Loeschcke V, Holmstrup M (2006) Metabolomic profiling of heat stress: hardening and recovery of homeostasis in Drosophila. Am J Physiol Regul Integr Comp Physiol 291:R205–R212
48. Paddison PJ (2008) RNA interference in mammalian cell systems. Curr Top Microbiol Immunol 320:1–19
49. van Ham TJ, Thijssen KL, Breitling R, Hofstra RM, Plasterk RH, Nollen EA (2008) *C. elegans* model identifies genetic modifiers of alpha-synuclein inclusion formation during aging. PLoS Genet 4:e1000027
50. Motoyama A, Xu T, Ruse CI, Wohlschlegel JA, Yates JR III (2007) Anion and cation mixed-bed ion exchange for enhanced multidimensional separations of peptides and phosphopeptides. Anal Chem 79:3623–3634
51. Sandra K, Moshir M, D'Hondt F, Verleysen K, Kas K, Sandra P (2008) Highly efficient peptide separations in proteomics Part 1. Unidimensional high performance liquid chromatography. J Chromatogr B Analyt Technol Biomed Life Sci 866:48–63
52. Fonslow BR, Yates JR III (2009) Capillary electrophoresis applied to proteomic analysis. J Sep Sci 32:1175–1188
53. Malmstrom J, Lee H, Nesvizhskii AI, Shteynberg D, Mohanty S, Brunner E, Ye M, Weber G, Eckerskorn C, Aebersold R (2006) Optimized peptide separation and identification for mass spectrometry based proteomics via free-flow electrophoresis. J Proteome Res 5:2241–2249

Part II

Protocols

Chapter 9

Production of Cells with Targeted Integration of Gene Variants of Human ABC Transporter for Stable and Regulated Expression Using the Flp Recombinase System

Kanako Wakabayashi-Nakao, Ai Tamura, Shoko Koshiba, Yu Toyoda, Hiroshi Nakagawa, and Toshihisa Ishikawa

Abstract

The vector-mediated introduction of cDNA into mammalian cells by calcium phosphate co-precipitation or permeation with lipofectamine is widely used for the integration of cDNA into genomic DNA. Such integration, however, of cDNA occurs randomly at unpredictable sites in the host's chromosomal DNA, and the number of integrated recombinant DNAs is not controllable. To overcome this problem, we developed the Flp-In method to integrate one single copy of cDNA encoding the human ABC transporter ABCG2 into FRT-tagged genomic DNA. Examination of more than 20 metaphase spreads for both fluorescence *in situ* hybridization (FISH) mapping and multicolor-FISH analysis revealed that ABCG2 cDNA was incorporated into the telomeric region of the short arm on one of chromosomes 12 in Flp-In-293 cells. By using those cells, we investigated the effect of genetic polymorphisms and post-translational modifications of human ABC transporter ABCG2 on the protein expression and degradation. On the basis of our experience, it has been concluded that the Flp recombinase system provides a useful tool to quantitatively analyze the protein stability and endoplasmic reticulum (ER)-associated degradation of proteins like the ABC transporter. This system is also applicable for similar studies of the biogenesis of other proteins using the secretory pathway as well as proteins with other cellular localizations.

Key words: Flp recombinase, ATP-binding cassette (ABC) transporter, Genetic polymorphisms, Endoplasmic reticulum-associated degradation (ERAD), Proteasome, Aggresome, *N*-linked glycosylation

1. Introduction

The endoplasmic reticulum (ER) is the site of synthesis and maturation of proteins destined for the plasma membrane, for the secretory and endocytic organelles, and for secretion (1).

Peter Bross and Niels Gregersen (eds.), *Protein Misfolding and Cellular Stress in Disease and Aging: Concepts and Protocols*, Methods in Molecular Biology, vol. 648, DOI 10.1007/978-1-60761-756-3_9, © Springer Science+Business Media, LLC 2010

While the native conformation of a protein lies encoded in its primary amino acid sequence, the ER greatly enhances protein folding efficacy (2). During *de novo* synthesis in the ER, cysteine disulfide bonds are formed and oligosaccharides are added to asparagine (*N*-linked glycosylation), serine/threonine residues (*O*-linked glycosylation) of glycoproteins. The ER has a unique oxidizing potential that supports disulfide bond formation during protein folding (3). In addition to the well-known oxidoreductases, such as PDI and ERp57, many novel oxidoreductases have been identified over the past few years whose functions and substrates are unknown (4, 5). On the other hand, *N*-linked glycans are added *en blocc* to proteins as "core oligosaccharides" ($Glc_3Man_9GlcNAc_2$). Calnexin (CNX) is located near the translocon and can interact with nascent peptide chains of *N*-glycosylated proteins. *N*-linked glycans are subjected to extensive modification as glycoproteins mature and move through the ER via the *Golgi* apparatus to their final destination, for example, the plasma membrane.

Efficient quality control systems have evolved to prevent incompletely folded molecules from moving along the secretory pathway (1, 6–10). Accumulation of misfolded proteins in the ER would detrimentally affect cellular functions. Misfolded proteins are considered to be removed from the ER by retrotranslocation to the cytosol compartment where they are then degraded by the ubiquitin-proteasome system. This process is known as ER-associated degradation (ERAD). At present, however, it remains unclear how misfolded membrane proteins are selected and destroyed during ERAD. Chaperone proteins are considered to solubilize aggregation-prone motifs. In the case of the yeast ATP-binding cassette (ABC) transporter Ste6p, a 12-transmembrane protein, it has recently been shown that Hsp70/40s act before ubiquitination and facilitate the association of Ste6p with an E3 ubiquitin ligase (11). Furthermore, polyubiquitination was a prerequisite for retrotranslocation, which required the Cdc48 complex and ATP (11). Thus, it is becoming increasingly important to identify and to characterize multiple chaperone proteins that control the folding and degradation of ABC transporter proteins.

A total of 48 human ABC transporter genes have been identified and sequenced (12). On the basis of the arrangement of molecular structural components, that is, nucleotide binding domains and topologies of transmembrane domains, these reported human ABC transporters are classified into seven different sub-families (A to G). Human ABC transporter ABCG2 (BCRP/MXR/ABCP) belongs to the G subfamily of the ABC transporter proteins (13). This efflux pump is suggested to be responsible for protecting the body from toxic xenobiotics and for removing toxic metabolites (14). Human ABCG2 is a so-called "half ABC transporter" bearing a single ATP-binding fold at the NH_2-terminus and six transmembrane domains. Human ABCG2 exists in the plasma membrane as a homodimer bound through disulfide-bonded cysteine residues (15–19). The expression level of ABCG2 is regulated by both synthesis and degradation of the protein. By using Flp-In-293 cells, we have recently demonstrated that the formation of an intramolecular disulfide bond between Cys592 and Cys608 and *N*-linked glycosylation at Asn596 are critical check points for the stability and degradation of the *de novo* synthesized ABCG2 protein (20, 21). Furthermore, certain non-synonymous single nucleotide polymorphisms (SNPs), such as Q141K, F208S, and S441N, were also found to greatly affect the stability of ABCG2 in the ER and to enhance the protein degradation rate via ubiquitination and proteasomal proteolysis in Flp-In-293 cells (22–27) (Fig. 1). In this regard, we could successfully analyze the quality control of ABCG2 in the ER, namely, posttranslational modifications (intra- and inter-molecular disulfide bond formation and *N*-linked glycosylation) as well as ubiquitin-mediated proteasomal degradation of the ABC transporter protein (20, 21).

On the basis of our recent studies on the quality control of ABCG2 in the ER (19–27), we describe our methods for the production of Flp-In-293 cells with targeted integration of gene variants of the human ABC transporter for stable and regulated expression with the Flp recombinase system. The Flp-In-293 cell line appears to carry biologically important components required for post-translational modifications (*N*-linked glycosylation, disulfide bond formation, ubiquitination, etc.) of *de novo* synthesized proteins and also the machinery for protein degradation via endosome/lysosome or proteasome pathways. Taken together, it is concluded that the Flp recombinase system provides a useful tool to quantitatively analyze the protein stability and degradation of misfolded proteins. Expression of ABCG2 is used as an example and methods to validate successful integration, expression, and subcellular localization of this protein are described, which can be adapted to other proteins that can be expressed in the same way.

2. Materials

2.1. Flp-In™ System

1. Flp-In cell lines:
   (a) Flp recombinase-mediated site-specific integration and gene expression in mammalian cells was originally developed by S. O'Gorman, D. T. Fox, and G. M. Wahl at the Gene Expression Laboratory, The Salk Institute for Biological Studies, La Jolla, CA, in 1991 (28). The Flp recombinase system allows us to integrate one single copy of cDNA into the genomic DNA at a specific genome

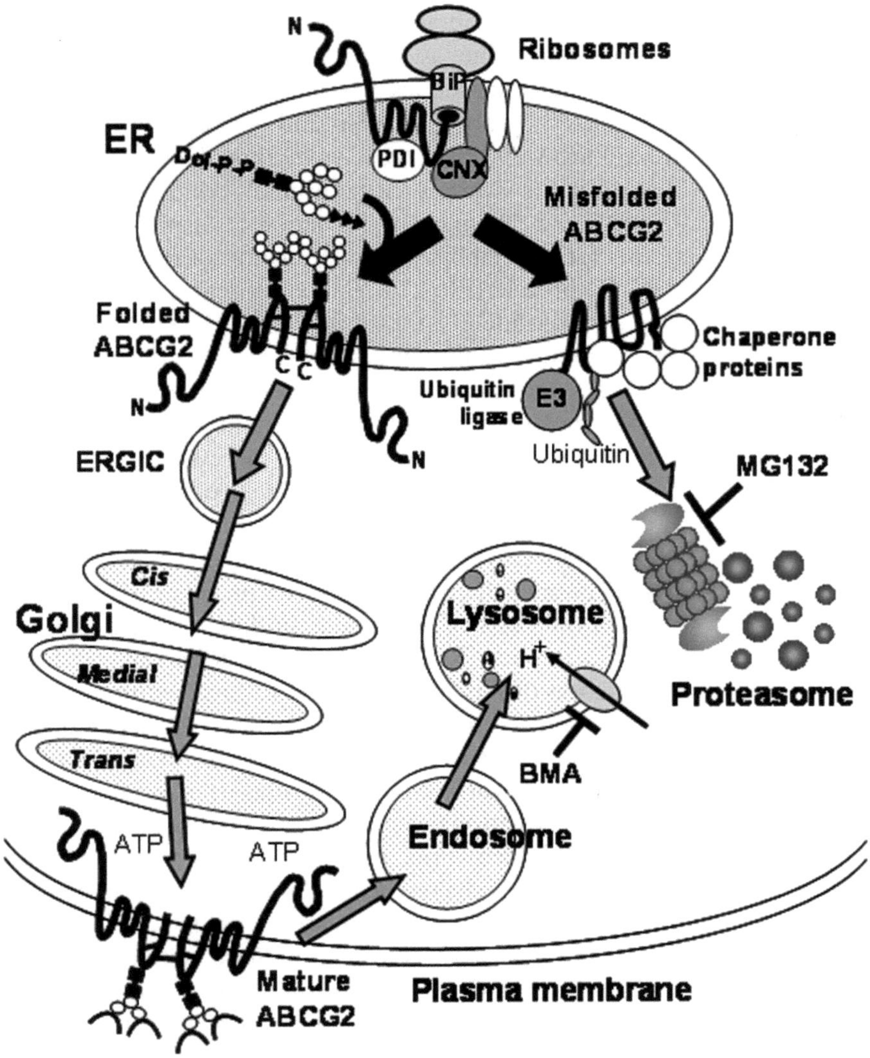

Fig. 1. Schematic illustration of protein folding and quality control in the ER and plausible pathways for degradation of ABCG2. BiP and CNX with their accessory proteins, for example, glucosidases I and II, and PDI, associate with the nascent peptide of ABCG2, as soon as the nascent peptide enters the ER lumen. When the ABCG2 protein has acquired its correctly folded structure with disulfide bonds and *N*-linked glycan, it is ready for exit from the ER and transfer to the ERGIC and *Golgi* for further processing. The ER-to-*Golgi* traffic may be mediated by COPII/Sar1p. The correctly processed ABCG2 WT is finally destined to reach the plasma membrane and is then degraded by the endosome-lysosome pathway after remaining in the plasma membrane domain for a certain period. In contrast, the misfolded ABCG2 protein undergoes ubiquitination-mediated proteasomal degradation. During the ERAD process, chaperones, such as Hsp90, Hsp70, Hsp40, CHIP, p23, Aha1, FKBP8, and/or Jab1/CSN5, may recruit E3 ubiquitin ligase and facilitate the ubiquitination of the misfolded ABCG2 protein. Bafilomycin A_1 (BMA) and MG132 inhibit lysosomal and proteasomal degradation, respectively.

location in mammalian host cells (28–34). At present, the Flp-In™ system is commercially available from Invitrogen (Carlsbad, CA, USA: www.invitrogen.com).

(b) Flp-In™ cell lines (Invitrogen, Carlsbad, CA, USA) were generated from the American Type Culture Collection

(ATCC) cell lines, for example, HEK293, CV-1, CHO-K1, BHK, NIH/3T3, or Jurkat, to stably express the *lac*Z-Zeocin fusion gene. Each cell line contains a single integrated **F**lp **R**ecombination **T**arget (FRT) site. The FRT site, originally isolated from *Saccharomyces cerevisiae*, serves as a binding site for Flp recombinase and has been well characterized (29, 35–37). The minimal FRT site consists of a 34-bp sequence containing two 13-bp imperfect inverted repeats separated by an 8-bp spacer that includes an *Xba*I restriction site (Fig. 2a). An

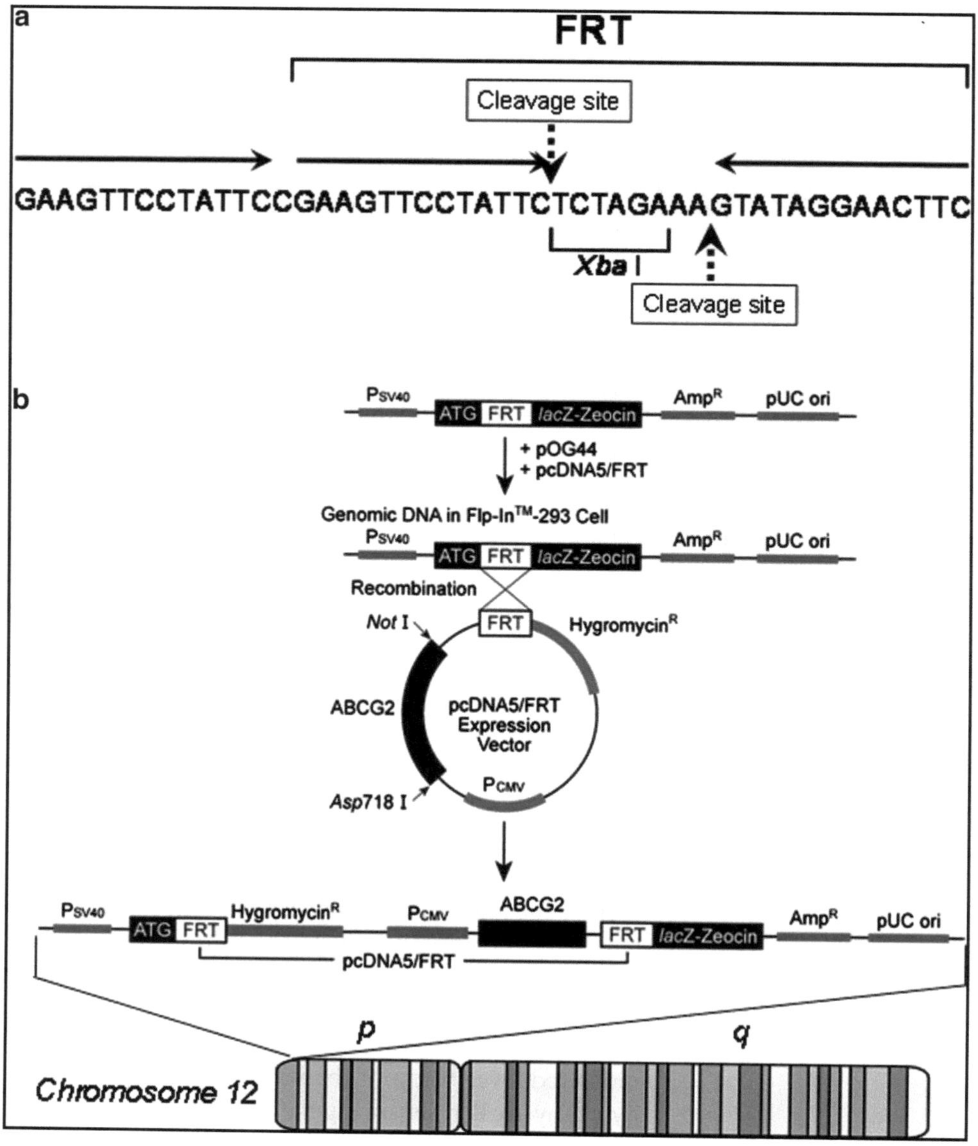

Fig. 2. DNA sequence of FRT site (**a**), schematic illustration of the stable expression of ABCG2 in Flp-In-293 cells (**b**), FISH mapping (**c**), and multicolor-FISH analysis (**d**). Flp-In-293 cells were co-transfected with the pcDNA5/FRT vector carrying

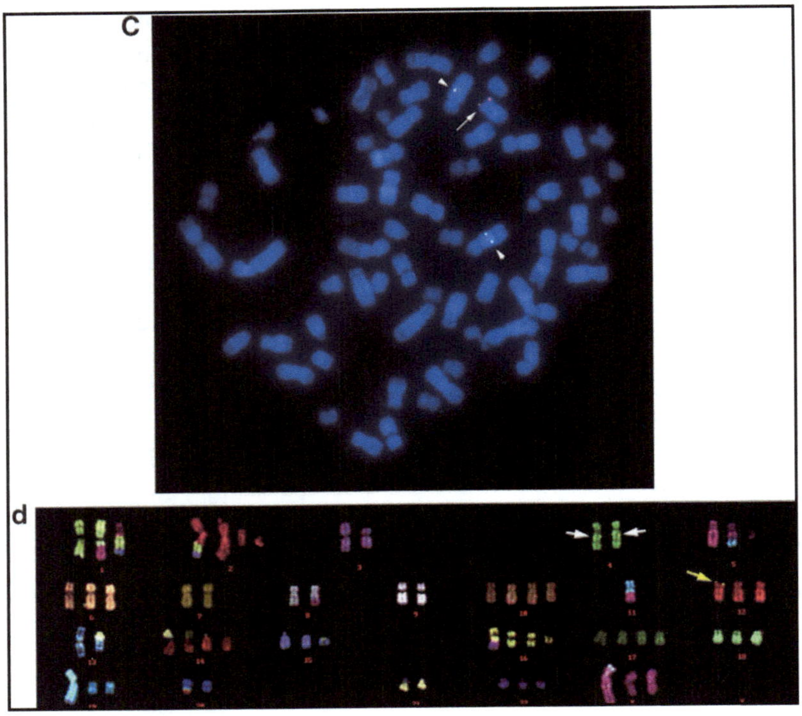

Fig. 2. (continued) the ABCG2 cDNA and the Flp recombinase expression plasmid pOG44. Flp recombinase mediates insertion of the expression construct with the ABCG2 cDNA into the genome at the integrated FRT site through site-specific DNA recombination. In the FISH mapping probed with the ABCG2-pcDNA5/FRT and Multicolor FISH-Human probe set, the hybridization signal of the transgene was detected in the telomeric region of the short arm of one of chromosomes 12 (*yellow arrow*) as overlapped signals of the biotin-labeled ABCG2-pcDNA5/FRT and digoxigenin-labeled pcDNA5/FRT, and two hybridization signals of the biotin-labeled ABCG2-pcDNA5/FRT (internal ABCG2 genes, *white arrows*) were detected in the q22 region on a pair of chromosomes 4.

additional 13-bp repeat is found in most FRT sites (38). While Flp recombinase binds to all three of the 13-bp repeats, strand cleavage actually occurs at the boundaries of the 8-bp spacer region (37, 38).

2. pcDNA5/FRT vector: pcDNA5/FRT (Invitrogen, Carlsbad, CA, USA) is a 5.1-kb expression vector designed for use with the Flp-In™ system. This vector contains the following elements: the human cytomegalovirus (CMV) immediate-early enhancer/promoter (39–41), multiple cloning sites with ten unique restriction sites, which can be used to introduce the cDNA sequence encoding the protein to be studied (in the present case ABCG2), the FRT site for Flp recombinase-mediated integration of the vector into Flp-In host cells; and hygromycin B-resistance gene for the selection of stable cell lines (42).
3. pOG44 vector: pOG44 is a 5.8-kb Flp recombinase expression vector (Invitrogen, Carlsbad, CA, USA). The *FLP* gene

was originally isolated from the *S. cerevisiae* 2-μm plasmid (43, 44) and encodes a site-specific recombinase that is a member of the integrase family of recombinases (45). The Flp recombinase mediates a site-specific recombination reaction between interacting DNA molecules via the pairing of interacting FRT sites (29, 46). The native *FLP* gene encodes a protein of 423 amino acids with a calculated molecular mass of 49 kDa. The *FLP* gene expressed from pOG44 encodes a temperature-sensitive Flp recombinase, which carries a point mutation (flp-F70L) that results in a change in amino acid 70 from Phe to Leu (47). The flp-F70L protein expressed from pOG44 exhibits increased thermo-stability at 37°C in mammalian cells when compared with the native Flp recombinase (47).

4. Dulbecco's modified Eagle's medium (D-MEM; Nacalai Tesque, Inc., Kyoto, Japan).
5. 10% (v/v) heat-inactivated fetal calf serum (FCS; Dainippon Pharmaceuticals, Osaka, Japan).
6. L-Glutamine (Wako Pure Chemical Industries, Ltd., Osaka, Japan), 2 mM.
7. Penicillin (100 U/ml) (Invitrogen, Carlsbad, CA, USA).
8. Streptomycin (100 μg/ml) (Invitrogen, Carlsbad, CA, USA).
9. Zeocin, 100 μg/ml (Invitrogen, Carlsbad, CA, USA).
10. Hygromycin B, 100 μg/ml (Invitrogen, Carlsbad, CA, USA).
11. Amphotericin B, 250 ng/ml (Invitrogen, Carlsbad, CA, USA).
12. 12- and 24-well culture plates (Becton Dickinson and Company, Franklin Lakes, NJ).
13. CO_2 incubator.
14. Trypan Blue dye (Nacalai Tesque, Inc., Kyoto, Japan).
15. Lipofectamine-2000 (Invitrogen, Carlsbad, CA, USA).

2.2. Validation of the System

2.2.1. FISH Analysis

1. Biotin 16 dUTP (Roche, Hague Road, IN, USA).
2. Digoxigenin-11-dUTP (Roche, Hague Road, IN, USA).
3. SSC (Sodium Citrate & Sodium Chloride, Wako Chemicals, Osaka, Japan).
4. Formamide (Wako Chemicals, Osaka, Japan).
5. Cy3-labeled streptavidin (Amersham Pharmacia Biotech, Uppsala, Sweden).
6. Cy5-labeled anti-digoxigenin (Roche, Hague Road, IN, USA).
7. Cooled CCD camera mounted on a microscope (e.g., Leica DMRA2).
8. Imaging software (e.g., CW4000 FISH application program of Leica Microsystems Imaging Solution Ltd.).

2.2.2. qPCR Analysis

1. NucleoSpin® RNA II kit (Macherey-Nagel GmbH & Co. KG, Duren, Germany).
2. SuperScript II RT (Invitrogen, Carlsbad, CA, USA) using random hexamers.
3. Thermal cycler (e.g., iCycler™ thermal cycler (BIO-RAD, Hercules, CA, USA)).
4. Primer sets: ABCG2 (5′-GATCTCTCACCCTGGGGCT TGTGGA-3′, 5′-TGTGCAACAGTGTGATGGCAAGGG A-3′), GAPDH (5′-ACTGCCAACGTGTCAGTGGTGGA CCTGA-3′, 5′-GGCTGGTGGTCCAGGGGTCTTACTCC TT-3′).
5. TaqMan® Fast Universal Master Mix (Applied Biosystems, Foster City, CA, USA).
6. TaqMan® probes (ABCG2; Hs00184979_ml, GAPDH; Hs99999905_ml) (Applied Biosystems, Foster City, CA, USA).
7. 7500 Fast Real Time-PCR System (Applied Biosystems, Foster City, CA, USA).

2.2.3. N-Linked Glycosylation

1. BXP-21 (ALEXIS Co., Lausen, Switzerland), a specific antibody to human ABCG2.
2. Anti-mouse IgG-horseradish peroxidase (HRP)-conjugate (Cell Signaling Technology, Beverly, MA, USA).
3. Phosphate-buffered saline (PBS).
4. Mouse monoclonal antibody against GAPDH (American Research Products, Inc., Belmont, MA, USA).
5. Lysis buffer, 50 mM Tris/HCl (pH 7.4), 1% (w/v) Triton X-100, 1 mM DTT, supplemented with protease inhibitor cocktail (Roche Ltd., Mannheim, Germany).
6. 27-gauge needle.
7. Endo H or PNGase F (New England Biosciences, Ipswich, MA).
8. β-Mercaptoethanol (ME).
9. SDS-PAGE equipment with 7.5% (w/v) polyacrylamide gels.
10. Hybond-ECL nitrocellulose membranes (GE Healthcare UK Ltd., Buckinghamshire, UK).
11. Western Lighting Chemiluminescent Reagent Plus (PerkinElmer Life And Analytical Sciences, Inc., Boston, MA, USA).
12. Lumino Imaging Analyzer FAS-1000 (TOYOBO Co., Ltd., Osaka, Japan).

2.2.4. Immunofluorescence for Subcellular Localization

1. MG132 (Sigma-Aldrich Co., St. Louis, MO, USA), 2 mM in DMSO.
2. Collagen type I-coated cover glasses (12-mm in diameter) (Asahi Techno Glass Corp., Chiba, Japan).
3. Paraformaldehyde 4% in PBS (Wako Pure Chemical Industries, Ltd., Osaka, Japan).
4. Triton X-100 (Nacalai Tesque, Inc., Kyoto, Japan).
5. Glycine (10 mg/ml, Nacalai Tesque, Inc., Kyoto, Japan) in PBS.
6. Bovine serum albumin (BSA, Nacalai Tesque, Inc., Kyoto, Japan) 0.5% (w/v) in PBS.
7. ABCG2-specific monoclonal antibody, BXP-21 antibody, and Alexa Fluor 488-conjugated anti-mouse IgG antibody (Invitrogen, Carlsbad, CA, USA).
8. Hoechst 33342 (1 μg/ml; Invitrogen, Carlsbad, CA, USA) in PBS containing 0.5% (w/v) BSA.
9. Confocal scanner unit (Yokogawa Electric Corp., Kanazawa, Japan) equipped with a CCD camera (Cool SNAP HQ, Roper Industries, Inc., Sarasota, Florida, USA) and an image intensifier unit (C9016-01, Hamamatsu Photonics K.K., Hamamatsu, Japan).

3. Methods

3.1. Principle of Flp Recombinase System

Generation of Flp-In expression cell lines requires co-transfection of Flp-In cells with the Flp-In expression vector (pcDNA5/FRT vector) containing the cDNA of interest and the recombinase expression plasmid pOG44. The pcDNA5/FRT vector contains a single FRT site immediately upstream of the hygromycin B-resistance gene for Flp recombinase-mediated integration and selection of the pcDNA5/FRT plasmid following co-transfection of the pOG44 vector into Flp-In mammalian host cells. Flp recombinase is expressed from the pOG44 plasmid. The FRT site serves as both the recognition and cleavage sites for the Flp recombinase and allows recombination to occur immediately adjacent to the hygromycin-resistance gene.

3.2. Expression of ABCG2 in Flp-In-293 Cells by Means of the Flp Recombinase System

Figure 2b illustrates the strategy by which we integrate one single copy of the human ABCG2 cDNA into the chromosomal DNA of Flp-In-293 cells by means of the Flp recombinase system. For this purpose, Flp-In-293 cells (see Note 1) are maintained in high-glucose Dulbecco's modified Eagle's medium (D-MEM) supplemented with 10% (v/v) heat-inactivated FCS, 2 mM L-glutamine, penicillin (100 U/ml), streptomycin (100 μg/ml), 250 ng/ml amphotericin B, and 100 μg/ml of Zeocin (Invitrogen, Carlsbad, CA, USA) at 37°C in a humidified atmosphere of 5% CO_2 in air. The number of viable cells should be determined from counts made in a hemocytometer with Trypan Blue dye exclusion.

Flp-In-293 cells are transfected with the ABCG2-pcDNA5/FRT vector, the Flp recombinase expression plasmid pOG44, and Lipofectamine-2000. Co-transfection of pOG44 and ABCG2-pcDNA5/FRT allows the expression of Flp recombinase and integration of the ABCG2-pcDNA5/FRT plasmid into the genome via the FRT site, as described above. The ABCG2-pcDNA5/FRT plasmid contains the hygromycin B resistance gene to allow selection of stable cell lines.

Flp-In-293 cells are pre-cultured in a 24-well culture plate (1×10^5 cells/well) for 24 h in 500 μl of D-MEM supplemented with 10% (v/v) heat-inactivated FCS and 2 mM l-glutamine, then the following solutions are added: total of 0.8 μg of plasmid DNA mixture (ABCG2-pcDNA5/FRT : pOG44 = 1 : 9 (w/w)) and 2.0 μl of Lipofectamine-2000 in 100 μl of serum- and antibiotics-free D-MEM. For preparation of the solution, Lipofectamine-2000 is diluted in 50 μl of D-MEM and incubated for 5 min, and then the diluted plasmid DNA solution in 50 μl of D-MEM is combined with 50 μl of the diluted Lipofectamine-2000 (final volume = 100 μl) and incubated for 20 min at room temperature. Following incubation at 37°C in a 5% CO_2 incubator for 24 h, the cells are cultured in a 12-well tissue culture plate containing fresh D-MEM supplemented with 10% (v/v) heat-inactivated FCS, 2 mM L-glutamine, penicillin (100 U/ml), and streptomycin (100 μg/ml). At 24 h after passage, hygromycin B is added to the cell culture at a final concentration of 100 μg/ml. For generating stable cell lines, cells are maintained in D-MEM containing 100 μg/ml of hygromycin B until foci can be identified.

Single colonies resistant to hygromycin B should be picked and subcultured. Selection of positive colonies can be performed by immunoblotting. The resulting cells are described as Flp-In-293/ABCG2 cells throughout this manuscript. Mock vector-transfected cells (Flp-In-293/Mock) are also prepared by transfecting Flp-In-293 cells with pcDNA5/FRT and pOG44 vectors in the same manner as described above.

3.3. Validation of the Flp Recombinase System

3.3.1. FISH Analysis to Detect the Chromosomal Site Where ABCG2 cDNA Has Been Integrated

The chromosomal site of ABCG2 cDNA that has been integrated into chromosomal DNA can be detected by FISH analysis (see Note 2). Here, we briefly describe the method and the corresponding results of our FISH analysis (see Note 3). The cell cultures are synchronized by thymidine blockage for 16 h, treated with 5-bromo-2′-deoxyuridine (BrdU; 30 μg/ml) for 5.5 h after the release of excessive thymidine, and harvested after treatment for 0.5 h. ABCG2-pcDNA5/FRT and pcDNA5/FRT plasmid

DNAs are labeled with biotin-16-dUTP and digoxigenin-11-dUTP, respectively, by nick translation kit (Roche, Hague Road, IN, USA) according to the manufacturer's protocol for use as probes. After hybridization, the slides are washed in 50% formamide/2 × SSC at 37°C and in 1 × SSC at room temperature for 20 min each. Detection of the probe signals is performed with Cy3-labeled streptavidin (3 μg/ml) and Cy5-labeled anti-digoxigenin (3 μg/ml) for biotin-labeled and digoxigenin-labeled probes, respectively. FISH images are captured by the CW4000 FISH application program of Leica Microsystems Imaging Solution Ltd. with a cooled CCD camera mounted on a Leica DMRA2 microscope. Multicolor FISH (M-FISH) with human chromosome-specific paints is performed after hybridization with biotin-labeled ABCG2-pcDNA5/FRT for assignment of the chromosomes where the transgenes are located. The M-FISH images are captured and merged with the images of the hybridization signals of the ABCG2-pcDNA5/FRT on the same metaphase spreads by the CW4000 FISH application program.

3.3.2. Quantitative Analysis of ABCG2 mRNA in Flp-In-293 Cells by RT-PCR

It is important to examine whether the genomic DNA-integrated ABCG2 cDNA is transcribed into mRNA. The transcript can be detected by conventional RT-PCR or real time-PCR methods (for ABCG2 results see Note 3). Total RNA is extracted from cultured cells with (for example with the NucleoSpin® RNA II kit). cDNA is prepared from the extracted RNA in a reverse transcriptase reaction with SuperScript II RT and random hexamers according to the manufacturer's instructions. The mRNA levels of ABCG2 and glyceraldehyde-3-phosphate dehydrogenase (GAPDH) are determined by PCR in an iCycler™ thermal cycler (BIO-RAD, Hercules, CA, USA) with the ABCG2 and GAPDH primer sets. The PCR reaction consists of hot-start incubation at 94°C for 2 min and 30 cycles of 94°C for 30 s, 59°C for 30 s, and 72°C for 30 s. After the PCR, products were separated by agarose gel electrophoresis and detected with ethidium bromide under UV light. The RNA levels of ABCG2 and GAPDH can be quantitatively determined by using the 7500 Fast Real Time-PCR System with TaqMan® Fast Universal Master Mix, and TaqMan® probes. The expression levels of ABCG2 should be normalized against those of GAPDH.

3.3.3. Detection of N-Linked Glycosylation and Disulfide Formation of ABCG2 by Immunoblotting

The ABCG2 protein expressed in Flp-In-293 cells can be detected by immunoblotting with BXP-21 (ALEXIS Co., Lausen, Switzerland), a specific antibody to human ABCG2 (see Note 4). The cells are rinsed with ice-cold PBS and subsequently treated with lysis buffer containing 50 mM Tris/HCl (pH 7.4), 1% (w/v) Triton X-100, 1 mM DTT, and a protease inhibitor cocktail (Roche Ltd., Mannheim, Germany). The samples are homogenized by passage through a 27-gauge needle and then centrifuged at 800×g for 10 min at 4°C. For glycosidase treatments, 20 μg of protein of the cell lysate sample is incubated with 20 U of Endo H or PNGase F at 37°C for 10 min.

Equal amounts of the resulting cell lysate (10 μg of protein) are subjected to SDS-PAGE in the presence or absence of β-mercaptoethanol (ME). Briefly, proteins are separated by electrophoresis on 7.5% (w/v) polyacrylamide gels and then electroblotted onto Hybond-ECL nitrocellulose membranes. Immunoblotting can be performed by using ABCG2-specific monoclonal antibody BXP-21 (1:200 dilution) as the first antibody and anti-mouse IgG-horseradish peroxidase (HRP)-conjugate (1:3,000 dilution) as the secondary antibody. HRP-dependent luminescence is developed by using Western Lighting Chemiluminescent Reagent Plus and detected in a Lumino Imaging Analyzer FAS-1000. To detect GAPDH, as an internal loading control, immunoblot detection is carried out in the same manner as described above, except for the use of a mouse monoclonal antibody against GAPDH (1:1,000 dilution) as the first antibody.

3.3.4. Detection of Cellular Localization of ABCG2 by Immunofluorescence Microscopy

ABCG2-expressing Flp-In-293 cells (4×10^4 cells) are seeded onto collagen type I-coated cover glasses (12-mm in diameter) (Asahi Techno Glass Corp., Chiba, Japan). Cells are pre-cultured for 24 h under the above-mentioned culture conditions and then incubated in the presence or absence of 2 μM MG132 for 24 h. Thereafter, cells are fixed with 4% paraformaldehyde in PBS at room temperature for 20 min, and then cell membranes are permeabilized by incubation with 0.02% Triton X-100 in PBS at room temperature for 5 min. To block free aldehyde groups in the formaldehyde, cells are treated with glycine (10 mg/ml) in PBS at room temperature for 10 min followed by a further incubation with 0.5% (w/v) BSA in PBS at room temperature for 1 h. To detect the ABCG2 protein, cells are treated with the ABCG2-specific monoclonal antibody BXP-21 antibody (1:1,000 dilution) as the first antibody and subsequently with Alexa Fluor 488-conjugated anti-mouse IgG antibody (1:1,000 dilution). In the same preparations, nuclear DNA is stained with Hoechst 33342 (1 μg/ml) in PBS containing 0.5% (w/v) BSA. Immunofluorescence of the Flp-In-293 cells is detected with a confocal scanner unit that is equipped with a CCD camera and an image intensifier unit to capture digital images. The fluorescence of Alexa Fluor 488 and Hoechst 33342 are observed with excitation laser light at 488 and 405 nm, respectively. It was of great interest to know how the inhibition of proteasomal protein degradation by MG132 affects the cellular localization of the F208S and S441N variant proteins (see Notes 5 and 6).

Figure 6a depicts the immunofluorescence images of Flp-In-293 cells expressing ABCG2 S441N that were incubated with or without 2 μM MG132 for 24 h. In the case of the S441N variant, MG132 treatments enhanced the localization of the ABCG2 variant protein at the plasma membrane and in intracellular compartments. By immunoelectron microscopy, we could detect ABCG2 aggresome formation adjacent to the nuclei when Flp-In-293 cells expressing the F208 variant were treated with 2 μM MG132 for 24 h (Fig. 6b, see Notes 7 and 8).

4. Notes

1. The Flp-In-293 cell line is considered to be a useful cell system for studying the molecular mechanism of protein misfolding and the subsequently occurring ERAD process. Flp-In-293 cells are not polarized cells. Therefore, for studying the apical or basolateral localization of membrane proteins, MDCKII (Madin-Darby canine kidney) and LLC-PK1 (porcine kidney) cells may be applicable (48).
2. The Flp-In method is based on the exchange of an expression cassette within a previously tagged FRT site. The targeting constructs harboring the gene of interest precisely replace the tagged reporter cassette making use of the Flp recombinase, and therefore the copy number of the recombinant DNA is well controlled. This criterion is of utmost importance to evaluate the effects of genetic variants on protein stability and expression level. If the copy numbers are not identical, it is impossible to quantitatively compare the genetic variants. In the FISH mapping experiment with the ABCG2-pcDNA5/FRT and pcDNA5/FRT (Fig. 3a), the signals of the ABCG2-pcDNA5/FRT and pcDNA5/FRT were detected on three chromosomes and one chromosome, respectively. M-FISH revealed that the signals of the ABCG2-pcDNA5/FRT were located in the telomeric region of the short arm on one of chromosomes 12 and in the q22 region on a pair of chromosomes 4 (Fig. 2d), while the hybridization signal of the pcDNA5/FRT was detected only in the telomeric region of the short arm on one of chromosomes 12 (Fig. 3b). Thus, it has been verified that ABCG2 cDNA was incorporated into the telomeric region of chromosome 12p. The number of chromosomes per cell ranged from 57 to 87 in the cell line, and the number of individual chromosomes was varied from one to four among cells. In addition, chromosome rearrangements, for example, translocations, have frequently occurred in the Flp-In-293 cell line (Fig. 2d).
3. Figure 3a depicts a schematic illustration of the human ABCG2 protein to indicate the sites of amino acid alteration in the SNP variants of F208S and S441N. ABCG2 WT,

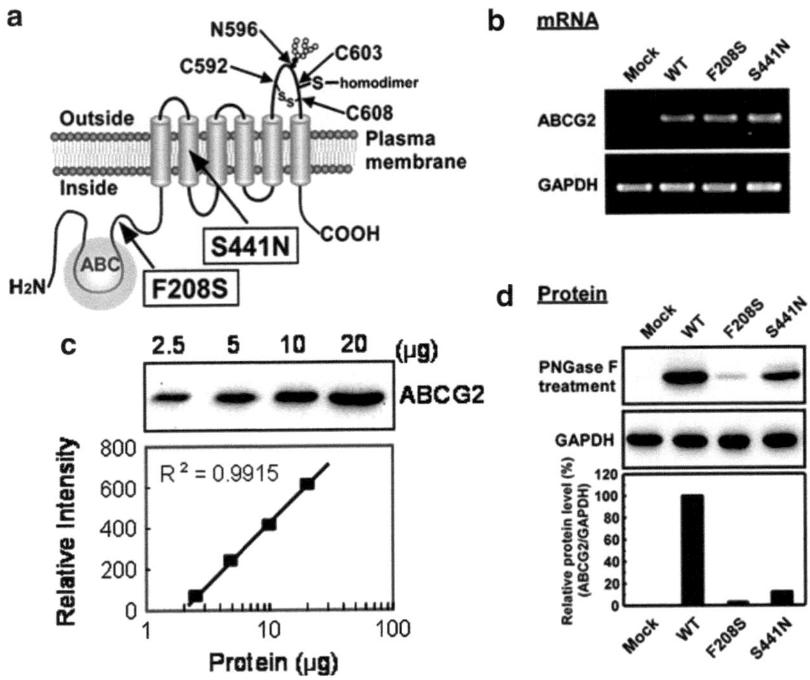

Fig. 3. Schematic illustration of human ABCG2 as well as the expression of ABCG2 WT, F208S, and S441N in Flp-In-293 cells at the mRNA and protein levels. (**a**) *Arrows* indicate the positions of amino acid substitutions (Phe208Ser and Ser441Asn) in the non-synonymous SNP variants of F208S and S441N. Cysteine residues, that is, Cys592, Cys603, and Cys608, are also indicated by *arrows*. Cys592 and Cys608 in the extracellular loop form an intra-molecular disulfide bond, whereas Cys603 is involved in homodimer formation via an inter-molecular disulfide bond. *N*-linked glycosylation reportedly occurs at Asn596 (N596). The *sphere labeled ABC* indicates the ATP-binding fold. (**b**) The mRNA level was analyzed by RT-PCR with total RNA extracted from Flp-In-293 cells expressing ABCG2 WT, F208S, or S441N. (**c**) The relationship between the relative intensity of ABCG2 immunoreactivity and the amount of cellular proteins. (**d**) For comparison of the protein levels, the cell lysate of each cell population was analyzed by immunoblotting with the ABCG2-specific monoclonal antibody (BXP-21) or the GAPDH-specific antibody after PNGase F treatment. The signal intensity ratio (ABCG2/GAPDH) was normalized to the WT level. Data are from our previous publication ref. 24 with permission.

F208S, and S441N were individually expressed in Flp-In-293 cells by using the Flp recombinase system. As shown in Fig. 3b, mRNA levels of ABCG2 WT as well as F208S and S441N were evenly represented in Flp-In-293 cells, where the mRNA levels of ABCG2 and GAPDH were measured by RT-PCR. On the other hand, ABCG2 WT, F208S, and S441N as well as GAPDH proteins were detected by immunoblotting, and their expression levels were quantified. For this purpose, we treated all of the samples with PNGase F and ME to remove glycomoieties and to break the cysteinyl disulfide bond forming a homodimer. Since there was a linear relationship between the signal intensity of immunoblotting and the logarithmic value of the amount of ABCG2 protein applied to the electrophoresis, the expression level of ABCG2 or GAPDH in cell lysate samples could be quantitatively estimated based on the linear relationship. The relative values of protein levels were then normalized to the ratio of ABCG2 WT/GAPDH. Although mRNA levels were almost the same in the WT and SNP variants (F208S and S441N), protein levels of those variants were markedly decreased (Fig. 3d). A standard curve plotting protein amounts versus relative scan intensity used to calculate the relative protein levels is given in Fig. 4c.

4. To gain insight into the molecular nature of ABCG2 WT and those two SNP variants, we performed immunoblot analysis experiments with cell lysate samples under reduced or non-reduced conditions. Figure 4a shows the effect of ME treatments on the SDS-PAGE migration of ABCG2 WT, F208S, and S441N proteins. When the cell lysate samples from Flp-In-293 cells expressing those proteins were applied to SDS-PAGE without ME treatments, one major band was

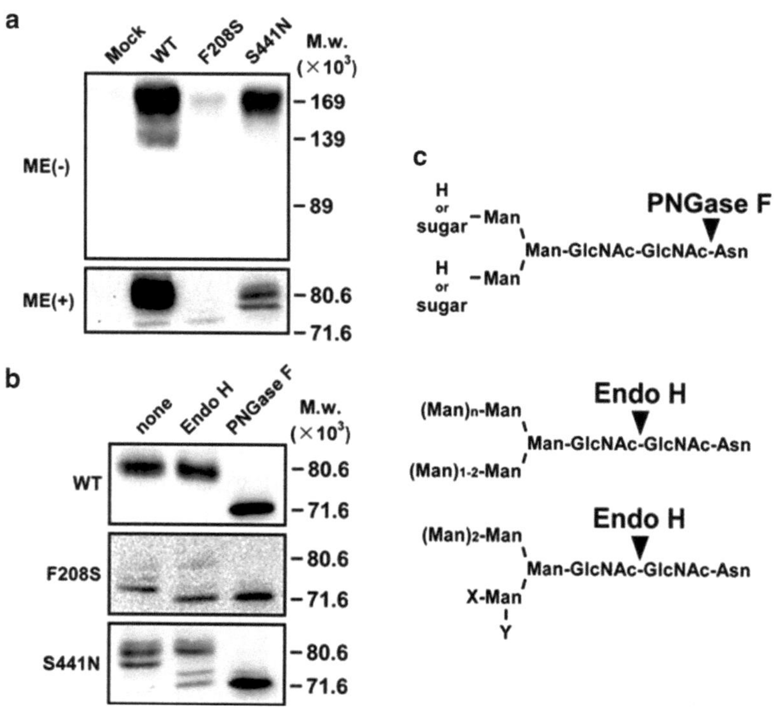

Fig. 4. Immunoblot detection of ABCG2 WT, F208S, and S441N proteins expressed in Flp-In-293 cells. (**a**) Effect of mercaptoethanol (ME) on the status (homodimer or monomer) of ABCG2 WT and the SNP variants. Cell lysate samples (20 μg of protein) were subjected to SDS-PAGE after treatments with or without ME; and, thereafter, ABCG2 WT, F208S, and S441N proteins were detected by immunoblotting with the BXP-21 antibody, as described in Subheading 3.3.3. By ME treatments, ABCG2 WT, F208S, and S441N proteins were changed from homodimers to monomers. (**b**) Effect of Endo H or PNGase F treatments on the glycosylated status of ABCG2 WT and the SNP variants. Cell lysate samples (20 μg of protein) were treated with Endo H or PNGase F, as described in Subheading 3.3.3. ABCG2 WT, F208S, and S441N proteins in the resulting samples were analyzed by immunoblotting with the BXP-21 antibody. The apparent molecular weights of mature and non-glycosylated ABCG2 WT were 81,000 and 72,000, respectively. (**c**) Schematic illustrations of *N*-linked oligosaccharides. Arrowheads indicate the sites of Endo H- or PNGase F-cleavage. X and/or Y = AcNeu-Gal-GluNAc. Data are from our previous publication ref. 24 with permission.

observed at the molecular weight of 162,000 for those ABCG2 proteins. In contrast, the ME treatment reduced the apparent molecular weight of those protein bands by half. These results provide evidence that F208S and S441N proteins form homodimers through a cysteinyl disulfide bond as does the ABCG2 WT protein.

5. It is important to note, however, that there were significant differences among the WT and those SNP variants with respect to the protein band patterns. As demonstrated in the ME (+) panel (Fig. 4a), two bands were detected at molecular weights of about 81,000 (major band) and 74,000 (minor band) in the cell lysate sample prepared from Flp-In-293/ABCG2 (WT) cells. On the other hand, for the S441N variant, two protein bands of ABCG2 were detected around the molecular weight of 81,000. In the case of the F208S variant, one faint band was detected at the apparent molecular weight of 74,000 by the same sample processing and immunoblotting experiment.
6. As demonstrated in Fig. 4b, PNGase F treatment of those cell lysate samples gave one major band with a molecular weight of 72,000, suggesting that glycosylation occurred at an asparagine moiety (Asn 596). Thus, the protein band (M.W. = 72,000) is regarded as the non-glycosylated form of ABCG2; however, Endo H treatments gave different results. Although the major band (M.W. = 81,000) of ABCG2 WT was not at all affected by the Endo H treatment, the apparent molecular weights of the smaller protein bands of F208S (M.W. = 74,000) and S441N (M.W. = 78,000) decreased after the Endo H treatment. Figure 5c shows the sites of PNGase F- and Endo H-cleavage in N-linked oligosaccharides. Among ABCG2 WT and those SNP variants, their *N*-linked oligosaccharide structures appear to be different. The matured *N*-linked oligosaccharide of ABCG2 WT may have a structure resistant to Endo H, whereas the immature *N*-linked oligosaccharides of the F208S and S441N variant proteins are susceptible to this endoglycosidase.
7. MG132 is a potent inhibitor of proteasomal proteolysis. Flp-In-293 cells expressing F208S or S441N were incubated in the presence of MG132 at different concentrations (0, 0.4, 2.0 μM) for 24 h, and then cell lysate samples were immediately prepared. Protein expression levels of the F208S and S441N variants were determined by immunoblotting after PNGase F treatments in the same way as described above. As shown in Fig. 5a, the protein levels of those ABCG2 variants were remarkably enhanced by the treatment with the proteasome inhibitor MG132 in a concentration-dependent manner. In contrast, the protein level of the ABCG2 WT was

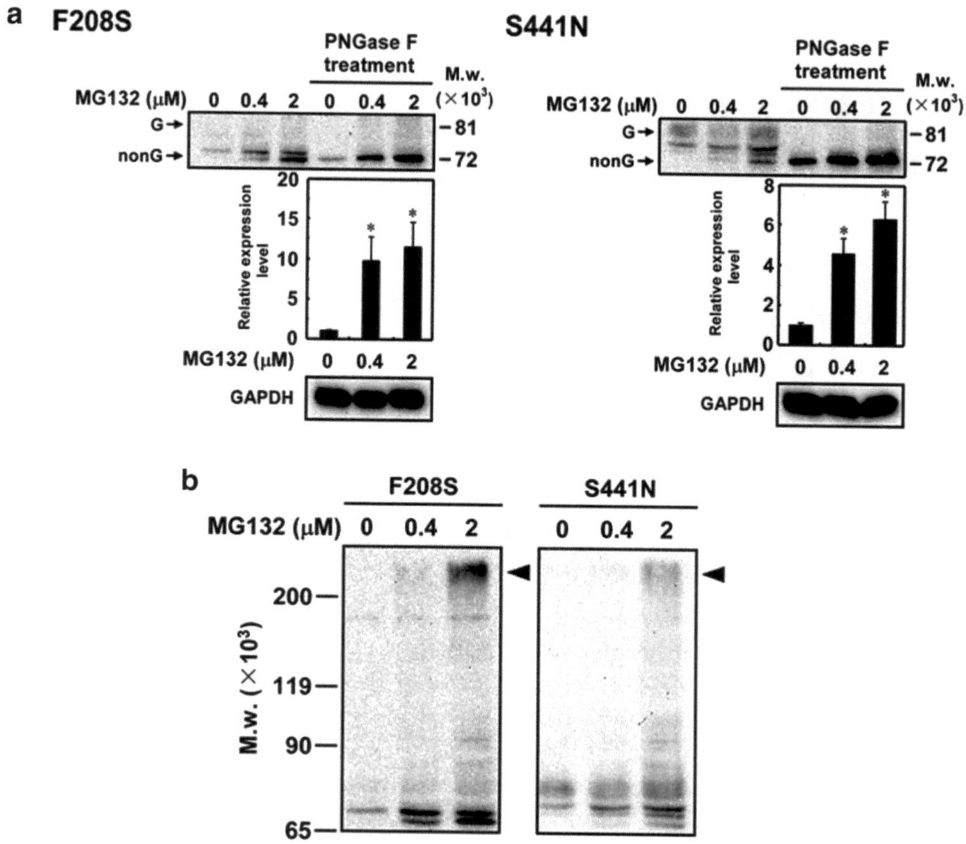

Fig. 5. Effects of MG132 on the glycosylation status and protein levels of ABCG2 F208S and S441N variants expressed in Flp-In-293 cells. (**a**) Flp-In-293/ABCG2 (F208S) and Flp-In-293/ABCG2 (S441N) cells were incubated with MG132 at concentrations of 0, 0.4, 2 μM for 24 h. Cell lysate samples (20 μg of protein) were prepared in the presence of ME. ABCG2 F208S and S441N variant proteins were analyzed by immunoblotting with the ABCG2-specific monoclonal antibody (BXP-21) either directly or after PNGase F treatment. The signal intensity of the non-glycosylated form of the ABCG2 F208S or S441N variant was measured after PNGase F treatment. The intensities are represented as ratios relative to the ABCG2 level in MG132-untreated cells. Data are expressed as mean values ± S.D. in triplicate experiments (*$p<0.05$). The expression levels of GAPDH in Flp-In-293 cells are shown as the loading controls. G, matured glycosylated form of ABCG2 (M.W. = 81,000); nonG, non-glycosylated form of ABCG2 (M.W. = 72,000) after PNGase F treatment. (**b**) Detection of aggregated forms (M.W. > 200,000) of ABCG2 F208S and S441N variants. Flp-In-293/ABCG2 (F208S) and Flp-In-293/ABCG2 (S441N) cells were incubated with MG132 at concentrations of 0, 0.4, 2 μM for 24 h. Cell lysate samples (20 μg of protein) were prepared in the presence of ME. ABCG2 F208S and S441N variant proteins were prepared from MG132-treated cells as described above and analyzed by immunoblotting with the BXP-21 antibody without PNGase F treatment. The aggregated forms of ABCG2 F208S and S441N are indicated by *arrowheads*. Data are expressed as mean values ± S.D. in triplicate experiments (*$p<0.05$). The expression levels of GAPDH in Flp-In-293 cells are shown as the loading controls. Data are from our previous publication ref. 24 with permission.

not significantly affected by MG132 treatment (data not shown). After the 24-h treatment with 2 μM MG132, F208S and S441N protein levels were increased 12- and 6-fold, respectively. MG132 treatments enhanced the levels of the

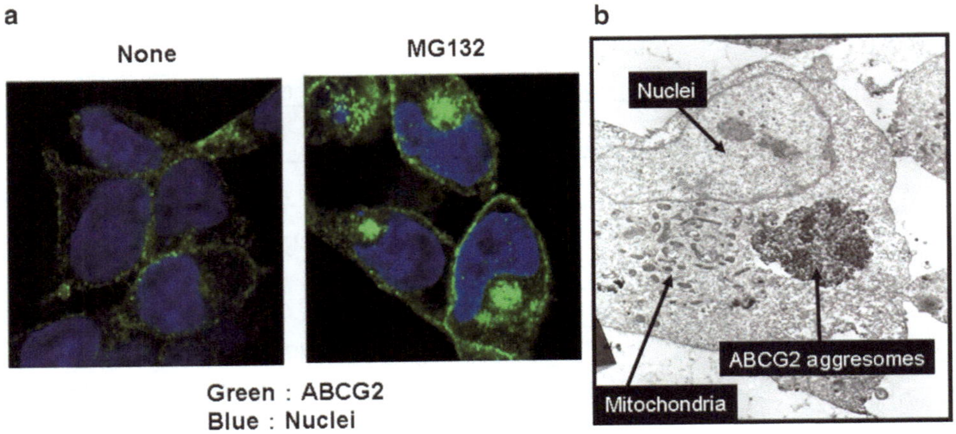

Fig. 6. Immunocytochemical staining of Flp-In-293 cells expressing ABCG2 S441N protein (**a**) and immunoelectron microscopic image of Flp-In-293 cells expressing ABCG2 F208S protein (**b**). Cells were incubated with or without 2 μM MG132 for 24 h. ABCG2 proteins were immunologically detected with an ABCG2-specific monoclonal antibody (BXP-21) and Alexa Fluor 488 (*green*). Cellular nuclei were stained by Hoechst 33342 (*blue*).

non-glycosylated form (M.W. = 72,000) for both F208S and S441N variants, where those cell lysate samples were not treated with PNGase F (Fig. 5a). More importantly, upon the immunoblot analysis without PNGase F treatments, aggregated forms (indicated by arrowheads in Fig. 5b) of both F208S and S441N variants were detected in the high molecular weight range of over 200,000.

8. It was of great interest to know how the inhibition of proteasomal protein degradation by MG132 affects the cellular localization of the F208S and S441N variant proteins. Figure 6a depicts the immunofluorescence images of Flp-In-293 cells expressing ABCG2 S441N that were incubated with or without 2 μM MG132 for 24 h. In the case of the S441N variant, MG132 treatments enhanced the localization of the ABCG2 variant protein at the plasma membrane and in intracellular compartments. By immunoelectron microscopy, we could detect ABCG2 aggresome formation adjacent to the nuclei when Flp-In-293 cells expressing the F208 variant were treated with 2 μM MG132 for 24 h (Fig. 6b).

Acknowledgments

The authors thank Dr. Masako Osumi (Integrated Imaging Research Support) and Dr. Hideaki Tamaki (Kitasato University) for their generous support for immunoelectron microscopy studies. Furthermore, they thank Dr. Yoichi Matsuda (Hokkaido

University) and Dr. Masahisa Tsuji (Chromosome Science Laboratory, Inc.) for the FISH analysis. The study performed in the authors' laboratory was supported by the NEDO International Joint Research Grant program "International standardization of functional analysis technology for genetic polymorphisms of drug transporters" as well as a Grant-in-Aid for Scientific Research (A) (No. 18201041) and a Grant-in-Aid for Exploratory Research (No. 19659136) from the Japanese Society for the Promotion of Science (JSPS). Kanako Wakabayashi-Nakao, Ai Tamura, Shoko Koshiba, and Yu Toyoda are JSPS Research Fellows.

References

1. Ellgaard L, Molinari M, Helenius A (1999) Setting the standards: quality control in the secretory pathway. Science 286:1882–1888
2. Anfinsen CB (1973) Principles that govern the folding of protein chains. Science 181:223–230
3. Tu BP, Weissman JS (2004) Oxidative protein folding in eukaryotes: mechanisms and consequences. J Cell Biol 164:341–346
4. Sargsyan E, Baryshev M, Szekely L, Sharipo A, Mkrtchian S (2002) Identification of ERp29, an endoplasmic reticulum lumenal protein, as a new member of the thyroglobulin folding complex. J Biol Chem 277: 17009–17015
5. Knoblach B, Keller BO, Groenendyk J, Aldred S, Zheng J, Lemire BD, Li L, Michalak M (2003) ERp19 and ERp46, new members of the thioredoxin family of endoplasmic reticulum proteins. Mol Cell Proteomics 2:1104–1119
6. Helenius A, Aebi M (2004) Roles of N-linked glycans in the endoplasmic reticulum. Annu Rev Biochem 73:1019–1049
7. Mori K (2000) Tripartite management of unfolded proteins in the endoplasmic reticulum. Cell 101:451–454
8. Ellgaard L, Helenius A (2001) ER quality control: towards an understanding at the molecular level. Curr Opin Cell Biol 13: 431–437
9. Hampton RY (2002) ER-associated degradation in protein quality control and cellular regulation. Curr Opin Cell Biol 14:476–482
10. Kleizen B, Braakman I (2004) Protein folding and quality control in the endoplasmic reticulum. Curr Opin Cell Biol 16:343–349
11. Nakatsukasa K, Huyer G, Michaelis S, Brodsky JL (2008) Dissecting the ER-associated degradation of a misfolded polytopic membrane protein. Cell 132:101–112
12. Klein I, Sarkadi B, Varadi A (1999) An inventory of the human ABC proteins. Biochim Biophys Acta 1461:237–262
13. Robey RW, To KK, Polgar O, Dohse M, Fetsch P, Dean M, Bates SE (2009) ABCG2: a perspective. Adv Drug Deliv Rev 61:3–13
14. Vlaming ML, Lagas JS, Schinkel AH (2009) Physiological and pharmacological roles of ABCG2 (BCRP): recent findings in Abcg2 knockout mice. Adv Drug Deliv Rev 61: 14–25
15. Kage K, Tsukahara S, Sugiyama T, Asada S, Ishikawa E, Tsuruo T, Sugimoto Y (2002) Dominant-negative inhibition of breast cancer resistance protein as drug efflux pump through the inhibition of S-S dependent homodimerization. Int J Cancer 97: 626–630
16. Mitomo H, Kato R, Ito A, Kasamatsu S, Ikegami Y, Kii I, Kudo A, Kobatake E, Sumino Y, Ishikawa T (2003) A functional study on polymorphism of the ATP-binding cassette transporter ABCG2: critical role of arginine-482 in methotrexate transport. Biochem J 373: 767–774
17. Kage K, Fujita T, Sugimoto Y (2005) Role of Cys-603 in dimer/oligomer formation of the breast cancer resistance protein BCRP/ ABCG2. Cancer Sci 96:866–872
18. Henriksen U, Fog JU, Litman T, Gether U (2005) Identification of intra- and intermolecular disulfide bridges in the multidrug resistance transporter ABCG2. J Biol Chem 280:36926–36934
19. Wakabayashi K, Nakagawa H, Adachi T, Kii I, Kobatake E, Kudo A, Ishikawa T (2006) Identification of cysteine residues critically involved in homodimer formation and protein expression of human ATP-binding cassette transporter ABCG2: a new approach using the flp recombinase system. J Exp Ther Oncol 5:205–222

20. Wakabayashi K, Nakagawa H, Tamura A, Koshiba S, Hoshijima K, Komada M, Ishikawa T (2007) Intramolecular disulfide bond is a critical check point determining degradative fates of ATP-binding cassette (ABC) transporter ABCG2 protein. J Biol Chem 282:27841–27846
21. Nakagawa H, Wakabayashi-Nakao K, Tamura A, Toyoda Y, Koshiba S, Ishikawa T (2009) Disruption of *N*-linked glycosylation enhances ubiquitin-mediated proteasomal degradation of human ATP-binding cassette transporter ABCG2. FEBS J. 276:7237–7252
22. Tamura A, Wakabayashi K, Onishi Y, Nakagawa H, Tsuji M, Matsuda Y, Ishikawa T (2006) Genetic polymorphisms of human ABC transporter ABCG2: development of the standard method for functional validation of SNPs by using the Flp recombinase system. J Exp Ther Oncol 6:1–11
23. Tamura A, Wakabayashi K, Onishi Y, Takeda M, Ikegami Y, Sawada S, Tsuji M, Matsuda Y, Ishikawa T (2007) Re-evaluation and functional classification of non-synonymous single nucleotide polymorphisms of the human ATP-binding cassette transporter ABCG2. Cancer Sci 98:231–239
24. Nakagawa H, Tamura A, Wakabayashi K, Hoshijima K, Komada M, Yoshida T, Kometani S, Matsubara T, Mikuriya K, Ishikawa T (2008) Ubiquitin-mediated proteasomal degradation of non-synonymous SNP variants of human ABC transporter ABCG2. Biochem J 411:623–631
25. Furukawa T, Wakabayashi K, Tamura A, Nakagawa H, Morishima Y, Osawa Y, Ishikawa T (2009) Major SNP (Q141K) variant of human ABC transporter ABCG2 undergoes lysosomal and proteasomal degradations. Pharm Res 26:469–479
26. Wakabayashi K, Tamura A, Saito H, Onishi Y, Ishikawa T (2006) Human ABC transporter ABCG2 in xenobiotic protection and redox biology. Drug Metab Rev 38:371–391
27. Wakabayashi-Nakao K, Tamura A, Furukawa T, Nakagawa H, Ishikawa T (2009) Quality control of human ABCG2 protein in the endoplasmic reticulum: ubiquitination and proteasomal degradation. Adv Drug Deliv Rev 61:66–72
28. O'Gorman S, Fox DT, Wahl GM (1991) Recombinase-mediated gene activation and site-specific integration in mammalian cells. Science 251:1351–1355
29. Sauer B (1994) Site-specific recombination: developments and applications. Curr Opin Biotechnol 5:521–527
30. Aladjem MI, Brody LL, O'Gorman S, Wahl GM (1997) Positive selection of FLP-mediated unequal sister chromatid exchange products in mammalian cells. Mol Cell Biol 17:857–861
31. Morschhauser J, Michel S, Staib P (1999) Sequential gene disruption in Candida albicans by FLP-mediated site-specific recombination. Mol Microbiol 32:547–556
32. Nakano M, Odaka K, Ishimura M, Kondo S, Tachikawa N, Chiba J, Kanegae Y, Saito I (2001) Efficient gene activation in cultured mammalian cells mediated by FLP recombinase-expressing recombinant adenovirus. Nucleic Acids Res 29:E40
33. Voziyanov Y, Konieczka JH, Stewart AF, Jayaram M (2003) Stepwise manipulation of DNA specificity in Flp recombinase: progressively adapting Flp to individual and combinatorial mutations in its target site. J Mol Biol 326:65–76
34. Wirth D, Hauser H (2004) Flp-mediated integration of expression cassettes into FRT-tagged chromosomal loci in mammalian cells. Methods Mol Biol 267:467–476
35. Gronostajski RM, Sadowski PD (1985) Determination of DNA sequences essential for FLP-mediated recombination by a novel method. J Biol Chem 260:12320–12327
36. Jayaram M (1985) Two-micrometer circle site-specific recombination: the minimal substrate and the possible role of flanking sequences. Proc Natl Acad Sci U S A 82:5875–5879
37. Senecoff JF, Bruckner RC, Cox MM (1985) The FLP recombinase of the yeast 2-micron plasmid: characterization of its recombination site. Proc Natl Acad Sci U S A 82: 7270–7274
38. Andrews BJ, Proteau GA, Beatty LG, Sadowski PD (1985) The FLP recombinase of the 2 micron circle DNA of yeast: interaction with its target sequences. Cell 40:795–803
39. Boshart M, Weber F, Jahn G, Dorsch-Hasler K, Fleckenstein B, Schaffner W (1985) A very strong enhancer is located upstream of an immediate early gene of human cytomegalovirus. Cell 41:521–530
40. Nelson JA, Reynolds-Kohler C, Smith BA (1987) Negative and positive regulation by a short segment in the 5'-flanking region of the human cytomegalovirus major immediate-early gene. Mol Cell Biol 7:4125–4129
41. Andersson S, Davis DL, Dahlback H, Jornvall H, Russell DW (1989) Cloning, structure, and expression of the mitochondrial cytochrome P-450 sterol 26-hydroxylase, a bile acid biosynthetic enzyme. J Biol Chem 264:8222–8229
42. Gritz L, Davies J (1983) Plasmid-encoded hygromycin B resistance: the sequence of

hygromycin B phosphotransferase gene and its expression in Escherichia coli and Saccharomyces cerevisiae. Gene 25: 179–188

43. Broach JR, Hicks JB (1980) Replication and recombination functions associated with the yeast plasmid, 2 mu circle. Cell 21:501–508

44. Broach JR, Guarascio VR, Jayaram M (1982) Recombination within the yeast plasmid 2mu circle is site-specific. Cell 29:227–234

45. Argos P, Landy A, Abremski K, Egan JB, Haggard-Ljungquist E, Hoess RH, Kahn ML, Kalionis B, Narayana SV, Pierson LS 3rd, Sternberg N, Leong JM (1986) The integrase family of site-specific recombinases: regional similarities and global diversity. EMBO J 5:433–440

46. Craig NL (1988) The mechanism of conservative site-specific recombination. Annu Rev Genet 22: 77–105

47. Buchholz F, Ringrose L, Angrand PO, Rossi F, Stewart AF (1996) Different thermostabilities of FLP and Cre recombinases: implications for applied site-specific recombination. Nucleic Acids Res 24:4256–4262

48. Mizuarai S, Aozasa N, Kotani H (2004) Single nucleotide polymorphisms results in impaired membrane localization and reduced ATPase activity in multidrug transporter ABCG2. Int J Cancer 109:238–246

Chapter 10

Stress Response Profiles in Human Fibroblasts Exposed to Heat Shock or Oxidative Stress

Christina Bak Pedersen and Niels Gregersen

Abstract

The upregulation of stress proteins is an important step in the cellular defense against various endogenous and exogenous stressors, and the stress response systems are therefore essential to cell maintenance and survival. In order to evaluate the cellular capability to cope with exogenous stressors, we investigated stress response profiles in cultured human fibroblasts exposed to short-term heat shock or oxidative stress. Quantitative PCR analysis as well as Western blotting demonstrated a specific upregulation of Hsp70 mRNA and protein after heat shock, while heme oxygenase 1 (HO1 or Hsp32) mRNA and protein was specifically upregulated following H_2O_2-induced oxidative stress. Hsp70 and HO1 should therefore be considered as useful biomarkers for acute heat stress and oxidative stress, respectively. Moreover, the results indicate that the expression of Hsp70 and HO1 in response to exogenous stressors may be regulated through different mechanisms.

Key words: Heat shock protein 70 (Hsp70), Heme oxygenase 1 (HO1), Heat shock, Oxidative stress, Quantitative PCR, Western blotting

1. Introduction

All living cells are recurrently subjected to a variety of chronic and acute stress conditions, including elevated temperature, exercise, caloric restriction, heavy metals, infection, genetic variations, and ageing (1–4). In order to adapt to such adverse circumstances, eukaryotic cells have developed a network of defense mechanisms which detect and respond to the different kinds of stressors (3, 5). The heat shock response, which is currently the most extensively studied cellular stress response system, results in the induction of genes encoding molecular chaperones, proteases and other proteins involved in the protection against stress-induced cellular

Peter Bross and Niels Gregersen (eds.), *Protein Misfolding and Cellular Stress in Disease and Aging: Concepts and Protocols*, Methods in Molecular Biology, vol. 648, DOI 10.1007/978-1-60761-756-3_10, © Springer Science+Business Media, LLC 2010

damage (1, 4). However, prolonged exposure to excessive stress or reduced cellular capability to cope with stress may initiate severe cellular dysfunctions. An efficient heat shock response is therefore essential for cell survival under various toxic conditions (3). The stress proteins induced during heat shock response are predominantly belonging to the highly conserved group of heat shock proteins (HSPs), including the HSP30, HSP40, HSP60, HSP70, and HSP90 families as well as the small HSPs (1, 3, 5, 6). The 70-kDa HSPs are probably the most important and abundant stress proteins and include two cytosolic forms, the constitutively expressed Hsc70 (or Hsp73) and the stress-inducible Hsp70 (or Hsp72). The HSP70s possess chaperone activity and assist in the stabilization of nascent polypeptides as well as in refolding of denatured proteins into native conformations, thus help preventing aggregation of misfolded proteins (3, 6, 7). For instance, overexpression of Hsp72 strongly decreased heat-induced aggregation of nuclear proteins (8, 9) as well as the formation of mutant Cu,Zn-superoxide dismutase (SOD1) containing proteinaceous aggregates in cultured primary motor neurons (10). Another important stress protein is the microsomal Hsp32, also known as heme oxygenase 1 (HO1), which is involved in regulation of the cellular redox state (11–14). Experiments in cells from HO1-deficient mice clearly demonstrated increased susceptibility to oxidative damage, thus illustrating the importance of HO1 expression in the protective mechanism against oxidant-induced stress (15). HO1 catalyzes the degradation of heme to biliverdin, carbon monoxide, and free iron, and biliverdin is subsequently converted into the potent antioxidant, bilirubin. The iron released by HO1 is rapidly incorporated into ferritin, which is also supposed to have an antioxidative capacity through sequestration of pro-oxidative free iron (11–14). In addition to the inducible HO1 stress protein, two constitutively expressed isoforms, HO2 and HO3, have been identified in mammals (5, 11, 16).

This chapter describes in detail molecular biological methods for evaluating the stress responses in cultured human fibroblasts exposed to short-term heat shock as well as oxidative stress. Expression profiles of Hsp70 and HO1 describe how fibroblasts respond to the specific stressors.

2. Materials

2.1. Cell Culture and Stress Treatments

1. Dulbecco's Modified Eagle's Medium (DMEM) (Gibco-BRL Life Technologies) supplemented with 10% fetal calf serum (FCS) (Gibco-BRL), 100 U/ml penicillin, 100 µg/ml streptomycin (both Leo Pharmaceutical Products Ltd.), and 0.29 mg/ml l-glutamine (Gibco-BRL).

2. Trypsin-EDTA solution 0.5% (v/v) (Gibco).
3. Phosphate-buffered saline (PBS, pH 7.4): 8 g NaCl, 0.2 g KCl, 1.15 g $Na_2HPO_4(2H_2O)$, 0.2 g KH_2PO_4 in 1 L H_2O.
4. Hydrogen peroxide (H_2O_2; 30% (w/w), Sigma-Aldrich) is diluted at 600 μmol/L in FCS-free DMEM.

2.2. RNA Isolation and cDNA Synthesis

1. TRIzol™ Reagent (Invitrogen).
2. Chloroform.
3. Isopropanol.
4. 70% (v/v) ethanol.
5. Diethylpyrocarbonate (DEPC)-treated water.
6. GeneQuant II RNA/DNA Calculator Spectrophotometer (Pharmacia Biotech).
7. 1st-strand™ cDNA synthesis Kit (Clontech) with oligo(dT) primers in a final volume of 20 ml according to the manufacturer's instruction.

2.2.1. Quantitative PCR Analysis

1. TaqMan® gene expression assays (20×) containing specific primers and a probe (Applied Biosystems) (details are summarized in Table 1).
2. TaqMan® Universal PCR Master Mix (2×) containing AmpliTaq Gold® DNA polymerase, dNTPs, and optimized buffer components (Applied Biosystems).
3. MicroAmp™ Optical 96-well Reaction Plate (Applied Biosystems).
4. MicroAmp™ Optical Adhesive Film (Applied Biosystems).
5. ABI PRISM® 7000 Sequence Detection System (Applied Biosystems).

2.3. Protein Extraction and SDS-Polyacrylamide Gel Electrophoresis

1. Cell lysis buffer: 50 mM Tris–HCl, pH 7.7, 5 mM EDTA pH 7.4, 1% (v/v) Triton X-100, 1 mM dithiothretiol (DTT), 0.5 μg/ml leupeptin, 10 μg/ml aprotinin, and 25 mM phenyl-methyl sulfonyl fluoride (PMSF) (see Note 1).
2. Sonicator (e.g. Branson 200).
3. Bradford protein assay (Bio-Rad).
4. Sodium dodecyl sulfate (SDS) sample buffer (6×): 650 mM Tris–HCl pH 6.8, 30% (v/v) glycerol, 10% (w/v) SDS, 9.3% (w/v) DTT, and 0.12 mg/ml bromphenol blue.
5. Tris–HCl Criterion polyacrylamide gels (12.5%) (Bio-Rad).
6. Running buffer (10×): 1% (w/v) SDS, 1.92 mol/L glycine and 250 mol/L Tris base.
7. Prestained molecular weight markers: Precision Plus Protein All Blue Standards (Bio-Rad).

Table 1
Details of TaqMan gene expression assays for QPCR

Assay	Forward primer	Reverse primer	Probe (exon–exon border)[a]	Fluorescence[b]
Hsp70 (Hs00271244_s1)[a]	No information	No information	No information (no introns)	6-FAM™ (NFQ)
HO1	5′-CCAGCGGGCCAGCAA-3′	5′-GGGAGCGGGTGTTGAGT-3′	5′-CAAAGTGCAAGATTCTG-3′(4-5)	6-FAM™ (NFQ)
β-actin (4310881E)[a]	No information	No information	No information	VIC™ (TAMRA™)

[a]ID number for a predeveloped assay designed by Applied Biosystems; primer/probe sequences are not available
[b]The probes are conjugated to a reporter dye (6-FAM™ or VIC™) in the 5′-end and to a quencher (non-fluorescent (NFQ) or TAMRA™) in the 3′-end

2.4. Western Blotting

1. Polyvinylidene difluoride (PVDF) membrane from Millipore and 3 MM Chr chromatography paper from Whatman.
2. Activation buffer: 96% (v/v) ethanol.
3. Transfer buffer: 50 mM borate, pH9.0 and 20% (v/v) ethanol.
4. Trans-blot SD Semi-dry Electrophoretic Transfer Cell blotting apparatus (BioRad).
5. Phosphate-buffered saline with Tween (PBS-T): PBS including 0.1% (v/v) Tween 20.
6. Blocking buffer: 5% (w/v) nonfat dry milk in PBS-T.
7. Primary antibodies: mouse monoclonal anti-Hsp70 (Hsp72) antibody (SPA-810, Stressgen) diluted 1:25,000 in PBS-T; mouse monoclonal anti-HO1 antibody (OSA-110, Stressgen) diluted 1:10,000 in PBS-T.
8. Secondary horseradish peroxidase (HRP) conjugated anti-mouse antibody (P0447, DAKO) diluted 1:25,000 in PBS-T.
9. Enhanced chemiluminescent (ECL Plus™) Western blotting detection reagent (GE Healthcare) and Phosphor-Imaging (STORM 840; Molecular Dynamics).

3. Methods

Fibroblasts from one healthy control person are cultured and used as model for studying stress responses following exposure to heat shock (1 h at 44°C) and oxidative stress (600 μmol/L H_2O_2). A recovery time of 24 h is used, and the expression of Hsp70 and HO1 is investigated in cell extracts using specific antibodies as well as by quantitative PCR. The stress response profiles are compared to fibroblasts which have not been stressed. The profiles clearly show that fibroblasts respond differently to heat and oxidative stress, suggesting that the expression of Hsp70 and HO1 may be regulated through different mechanisms. The stress treatments performed in this study can be regarded as an "acute" stress, which activates the appropriate adaptive stress response systems necessary to alleviate severe cellular damage. Accordingly, this approach may be useful for studying stress responses in patient cells, and in addition to Hsp70 and HO1 new potential stress response components can be investigated and selected as stress markers for other studies. Fibroblasts could also be exposed to the stress treatments for prolonged periods of time and combined with analysis of cell viability it may be possible to deduce if patient cells exhibited increased sensitivity to stress compared with control cells.

3.1. Cell Culture and Stress Treatments

1. Primary human skin fibroblasts are cultured in DMEM with 10% FCS (DMEM-10) at 37°C in humidified atmosphere of 5% CO_2 and passaged with trypsin-EDTA when approaching confluence. Six 95% confluent 75 cm^2 culture flasks are required for each stress treatment as well as for the "no stress treatment".
2. Heat shock treatment: Growth medium is changed on the day before the stress treatment (see Note 2). Seal all culture flasks and start the experiment by transferring cells to a 44°C heat incubator. After 1 h at 44°C, culture flasks are unsealed and moved back to the 37°C CO_2-incubator for 24 h recovery.
3. Oxidative stress treatment: Prior to the stress treatment, remove medium and rinse cells once in preheated (37°C) PBS. Add 10 ml H_2O_2 diluted in DMEM without serum (600 μmol/L final concentration) to each culture flask and incubate at 37°C (see Note 3). After 1 h, remove H_2O_2-medium and rinse with preheated PBS. Then add preheated DMEM-10 and move cells back to the 37°C CO_2-incubator for 24 h recovery.
4. "No stress treatment": Cells remain in DMEM-10 at 37°C for 24 h.
5. Rinse twice in PBS and scrape off cells from the six culture flasks at different times after the stress treatment: 0, 2, 4, 8, 12, and 24 h. Cells from each culture flask are split in two pellets (3 min at 1,500 rpm, 4°C) and stored at −80°C for RNA- and protein analysis.

3.2. RNA isolation and cDNA synthesis

1. Resuspend cell pellet corresponding to a half T75 culture flask in 1.8 ml TRIzol™ Reagent, add 180 μl chloroform, vortex 15 s, and incubate 5 min on ice (4°C).
2. Centrifuge at 12,000 × g for 5 min (4°C) and transfer the H_2O-phase (containing RNA) to a new 2 ml Eppendorf tube (see Note 4).
3. Add another 540 μl chloroform, vortex, and centrifuge 5 min at 12,000 × g (4°C). Transfer the H_2O-phase to a new 1.5 ml Eppendorf tube (see Note 4).
4. Precipitate RNA from the H_2O-phase by adding an equal volume of isopropanol (ca. 450 μl), mix by hand, and incubate at −20°C overnight.
5. Centrifuge at 15,000 × g for 15 min (4°C) and remove the supernatant.
6. Carefully wash the RNA pellet in 180 μl 70% (v/v) ethanol, centrifuge 1 min at 15,000 × g (4°C), and remove the supernatant.
7. Resuspend the RNA pellet in 25 μl DEPC-H_2O (mix by pipetting). RNA can be stored at −80°C.

8. RNA concentration and RNA purity (A_{260}:A_{280} ratio) can be measured by a GeneQuant RNA/DNA Calculator Spectrophotometer (see Note 5).
9. cDNA is synthesized using the 1st-strand™ cDNA synthesis Kit (Clontech): Transfer 1 μg RNA to an Eppendorf tube containing 1 μl Oligo$(dT)_{18}$ primer (20 μmol/L) and add DEPC-treated H_2O to a final volume of 13.5 μl. Incubate at 70°C for 2 min and place on ice (4°C).
10. Prepare 6.5 μl reaction mix containing 4.0 μl reaction buffer, 1.0 μl dNTP mix, 0.5 μl recombinant RNase inhibitor and 1 μl MMLV reverse transcriptase according to the manufacturer's instruction (see Note 6).
11. Add reaction mixture to and incubate 60 min at 42°C followed by subsequent enzyme inactivation (94°C for 5 min). cDNA can be stored at –20°C.

3.3. Quantitative PCR analysis

1. Quantitative PCR is performed using the real-time quantitative PCR-based TaqMan® fluorescence technology (Fig. 1). Relative quantification of Hsp70 (HSPA1B) and HO1 (HMOX1) mRNA is carried out using the "standard curve method" (17, 18) (User Bulletin #2, Applied Biosystems). β-actin (ACTB) has been selected as an endogenous reference gene (see Note 7).
2. Prepare a standard curve by pooling undiluted fibroblast cDNAs (1:1) and dilute 1:10, 1:100, and 1:1,000 with H_2O.
3. Dilute all fibroblast cDNAs 1:20 in H_2O.
4. The QPCR analyses are carried out in optical 96-well reaction plates. For each reaction use: 2 μl cDNA diluted 1:20 (sample or standard) as template, 12.5 μl TaqMan® Universal PCR Master Mix (2×), and 1.25 μl TaqMan® gene expression assay (20×) (see Table 1) containing probe (250 mM) and primers (900 nM) (see Note 8). Seal the 96-well plate with an optic adhesive film and centrifuge briefly. All samples and standards should be measured in triplicates!
5. Perform PCR in an ABI PRISM® 7000 Sequence Detection System using the following cycling conditions: 2 min at 50°C, 10 min at 95°C, and 40 cycles of 15 s at 95°C and 1 min at 60°C. See Hsp70 and HO1 mRNA expression profiles in Fig. 2.

3.4. Protein extraction and SDS-Polyacrylamide Gel Electrophoresis

1. Resuspend cell pellet corresponding to a half T75 culture flask in 75 μl cell lysis buffer (4°C) followed by 40 s in a sonication water bath and four cycles of freeze/thaw (2 min on dry ice and 2 min in a room temperature water bath).
2. Centrifuge 4 min at 14,800 ×*g* (4°C) and transfer the supernatant to a new Eppendorf tube.

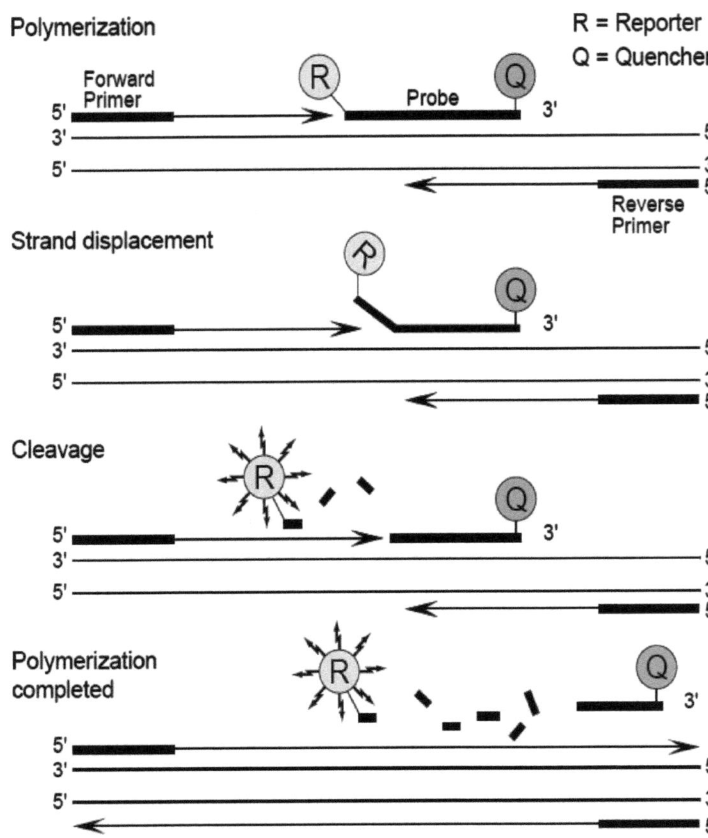

Fig. 1. Principle of the TaqMan chemistry®. The TaqMan® gene expression assay includes two unlabelled primers for amplification and one TaqMan® probe for detecting the sequence of interest. The probe contains a reporter dye (R) linked to the 5′-end and a quencher (Q) at the 3′-end, which anneals specifically to the target sequence between the two primer sites. When the probe is intact, the reporter fluorescence will be suppressed by the quencher. During the PCR, 5′–3′ exonuclease activity of the AmpliTaq Gold® DNA polymerase cleaves the probe and separates the reporter from the quencher, resulting in increased fluorescence of the reporter. The probe fragments are then displaced from the target, and elongation of the strand continues. Accumulation of PCR products can be detected directly by monitoring the increase in fluorescence of the reporter dye using an ABI PRISM® 7000 Sequence Detection System from Applied Biosystems.

3. Determine protein concentration of the supernatant fraction using the Bradford protein assay.
4. Dilute protein extracts to appropriate concentrations in 1:6 volumes of SDS sample buffer, mix, and boil 5 min in a heating block (95°C).
5. Centrifuge briefly and keep the samples at room temperature.
6. Separate the proteins by 12.5% SDS-polyacrylamide gel electrophoresis (PAGE) in cold 1× SDS running buffer. Use 10 µg protein for HO1 blots and 2 µg for Hsp70 blots. Include one well for prestained molecular weight marker (3 µl).
7. Run the gel at 200V for 60–70 min or until the dye front is reaching the bottom of the gel (4°C).

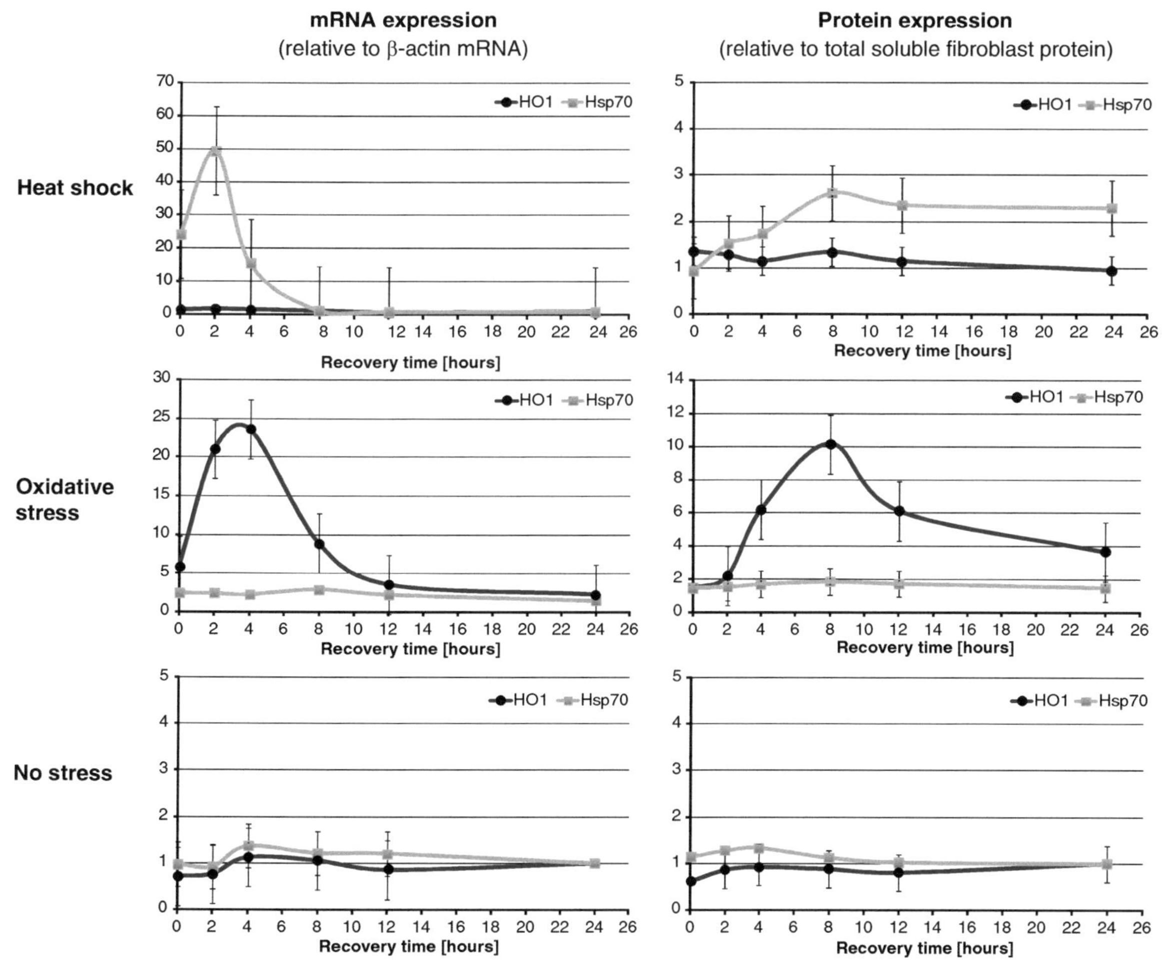

Fig. 2. Stress response profiles in human fibroblasts. Hsp70 and H01 mRNA and protein expression profiles in human fibroblasts exposed to heat shock (44°C), oxidative stress (600 mM H_2O_2), or no stress for 1 h followed by 24 h recovery at 37°C. The expression of Hsp70 and H01 mRNA is measured relative to β-actin mRNA, while Hsp70 and H01 protein expressions are measured relative to total soluble fibroblast protein. All expression profiles are normalized to unstressed cells at 24 h "recovery". The *curves* represent mean curves of two independent experiments and the standard deviation is estimated according to: $SD = \sqrt{(\Sigma d^2/2N)}$, d = difference between the two experiments; N = number of differences. Note different scales on the axes. —■— (Hsp70); —●— (H01).

3.5. Western Blotting

1. Transfer the separated proteins onto a PVDF membrane for 45 min at 220 mA using a Trans-blot SD Semi-dry Electrophoretic Transfer Cell blot apparatus from BioRad. Activate the PVDF membrane in 96% (v/v) ethanol (5–40 min) and wet gel, membrane and 3 MM chromatography paper in transfer buffer prior to blotting.
2. Once the transfer is completed, rinse the PVDF membrane in PBS-T and incubate 1 h in 25 ml blocking buffer.
3. Rinse twice in PBS-T (2 min) prior to addition of 25 ml of a 1:25,000 dilution of the anti-Hsp70 antibody in PBS-T or 25 ml anti-HO1 antibody diluted 1:10,000 in PBS-T. Incubate overnight at room temperature on a rocking platform. Use a sealed container!
4. Discard primary antibody and rinse the blot twice in PBS-T (2 min) followed by a 15-min wash and then three times 5-min wash in PBS-T.
5. Add 25 ml 1:25,000 secondary antibody solution in PBS-T and incubate 1 h at room temperature on a rocking platform.
6. Discard secondary antibody and rinse the blot twice in PBS-T (2 min) followed by a 15-min wash and then three times 5-min wash in PBS-T.
7. During the final wash, warm aliquots of ECL reagents A and B to room temperature. Once the final wash is removed from the blot, ECL reagents are mixed together, immediately added to the blot and rotated by hand to ensure even coverage. Use a lightproof container!
8. Incubate for 5 min and seal the blot in a plastic bag.
9. Use Phosphor Imaging (STORM840; Amersham Biosciences) for visualization and the Image Quant® software for quantification. See Hsp70 and HO1 protein expression profiles in Fig. 2.

4. Notes

1. The phenyl–methyl sulfonyl fluoride (PMSF) should be added to the lysis buffer immediately prior to use. As an alternative, leupeptin, aprotinin, and PMSF could be replaced by a protease inhibitor cocktail tablet, for example, Complete, Mini, EDTA-free from Roche Applied Science (Cat # 11836170001).
2. If you use a heat incubator without CO_2 for the heat shock treatment we recommend you to change the growth medium on the day before the stress treatment so that the medium is

buffered with CO_2, when sealing the culture flasks and starting the stress experiment.

3. Hydrogen peroxide (H_2O_2) is often used as an oxidative stress inducer. When applying H_2O_2 to cells in culture it should be dissolved in serum-free medium to avoid rapid detoxification of the H_2O_2 by enzymatic activities in the serum. It should be noticed that also the serum-free medium in itself contributes to oxidative stress, and you will get a probably additive effect of H_2O_2 and the serum free medium.
4. Be very careful when isolating the H_2O-phase containing the RNA to be purified. Avoid genomic DNA contamination from the intermediate phase.
5. RNA samples should be free of contaminating proteins, DNA, and other cellular material as well as phenol, ethanol, and salts from the RNA isolation procedure. Such impurities may lower the efficiency of reverse transcription and result in reduced amplification. In addition to estimate RNA purity from the A_{260}:A_{280} ratio (range: 1.7–2.1), it is important to assess the RNA integrity. This can be done by fractionating 0.5–1.0 μg of total RNA on a denaturing agarose gel (2%). Add 5 μl deionized formamide, 1 μl ethidium bromide (1:10), and H_2O to a final volume of 10 μl. Denature the RNA sample 3 min at 65°C and incubate on ice until loading. Ethidium bromide staining of the RNA reveals two bands – 18S and 28S ribosomal RNA (rRNA). If there is no smearing below each band and the 28S rRNA band is about twice as intense as the 18S rRNA band, then the mRNA quality is good.
6. It may be a good idea to prepare two identical cDNA synthesis reactions, but leave out the reverse transcriptase enzyme in one reaction. This reaction could then be used later in the quantitative PCR analysis as a control for contaminating genomic DNA. However, probe and primers may be selected to exclude amplification of genomic DNA (see Note 8).
7. One of the most important issues when performing relative gene expression is to select a "*good*" endogenous reference or control gene. A good control gene should be constantly expressed in all samples regardless of cell cycle, cell type, or cellular conditions. We investigated the expression of 11 different control genes in samples from human control fibroblasts exposed to stress (44°C, 1 h) followed by recovery at 37°C for 1, 8, and 48 h. We used a pre-developed TaqMan® human endogenous control plate from Applied Biosystems comprising the genes encoding: β-actin, β2-microglobulin, cyclophilin glyceraldehyde-3-phosphate dehydrogenase, β-glucronidase, hypoxanthine ribosyl transferase, phosphoglycerokinase, acidic ribosomal protein, transcription factor

IID TATA binding protein, transferring receptor, and 18S rRNA. For a good control gene, the stress treatment should not change the gene expression compared to that in unstressed cells. We found that the expression of eight of the genes changed in response to the stress treatment. Among the remaining three genes that appeared unaffected by the stress treatment, we selected β-actin as endogenous control gene.

8. You can buy pre-developed TaqMan® gene expression assays from Applied Biosystems or you can design it yourself. If you choose to design the assays yourself it is a very good idea to define the sequence of the probe to span an exon–exon junction. In this way, you exclude any possible amplification of genomic DNA which may be present in the RNA preparation.

Acknowledgements

The authors would like to thank lector Thomas J. Corydon at the Institute for Human Genetics, University of Aarhus, for advice, support and use of cell culturing facilities. This work has been supported by The Danish Medical Research Council; Aarhus University Hospital's Research Initiative; Institute of experimental Clinical Research, University of Aarhus; Faculty of Health Sciences, University of Aarhus.

References

1. Richter-Landsberg C, Goldbaum O (2003) Stress proteins in neural cells: functional roles in health and disease. Cell Mol Life Sci 60:337–349
2. Unno K, Asakura H, Shibuya Y, Kaihou M, Fukatsu H, Okada S, Oku N (2002) Stress response caused by chronic alcohol intake in aged rat brain. Alcohol Clin Exp Res 26:1017–1023
3. Verbeke P, Fonager J, Clark BF, Rattan SI (2001) Heat shock response and ageing: mechanisms and applications. Cell Biol Int 25:845–857
4. Westerheide SD, Morimoto RI (2005) Heat shock response modulators as therapeutic tools for diseases of protein conformation. J Biol Chem 280:33097–33100
5. Calabrese V, Lodi R, Tonon C, D'Agata V, Sapienza M, Scapagnini G, Mangiameli A, Pennisi G, Stella AM, Butterfield DA (2005) Oxidative stress, mitochondrial dysfunction and cellular stress response in Friedreich's ataxia. J Neurol Sci 233:145–162
6. Sonna LA, Fujita J, Gaffin SL, Lilly CM (2002) Invited review: Effects of heat and cold stress on mammalian gene expression. J Appl Physiol 92:1725–1742
7. Fourie AM, Peterson PA, Yang Y (2001) Characterization and regulation of the major histocompatibility complex-encoded proteins Hsp70-Hom and Hsp70-1/2. Cell Stress Chaperones 6:282–295
8. Gabai VL, Meriin AB, Yaglom JA, Volloch VZ, Sherman MY (1998) Role of Hsp70 in regulation of stress-kinase JNK: implications in apoptosis and aging. FEBS Lett 438:1–4
9. Stege GJ, Li GC, Li L, Kampinga HH, Konings AW (1994) On the role of hsp72 in heat-induced intranuclear protein aggregation. Int J Hyperthermia 10:659–674
10. Bruening W, Roy J, Giasson B, Figlewicz DA, Mushynski WE, Durham HD (1999)

Up-regulation of protein chaperones preserves viability of cells expressing toxic Cu/Zn-superoxide dismutase mutants associated with amyotrophic lateral sclerosis. J Neurochem 72:693–699
11. Elbirt KK, Bonkovsky HL (1999) Heme oxygenase: recent advances in understanding its regulation and role. Proc Assoc Am Physicians 111:438–447
12. Choi AM, Alam J (1996) Heme oxygenase-1: function, regulation, and implication of a novel stress-inducible protein in oxidant-induced lung injury. Am J Respir Cell Mol Biol 15:9–19
13. Stocker R (1990) Induction of haem oxygenase as a defence against oxidative stress. Free Radic Res Commun 9:101–112
14. Clark JE, Foresti R, Green CJ, Motterlini R (2000) Dynamics of haem oxygenase-1 expression and bilirubin production in cellular protection against oxidative stress. Biochem J 348(Pt 3):615–619
15. Poss KD, Tonegawa S (1997) Reduced stress defense in heme oxygenase 1-deficient cells. Proc Natl Acad Sci U S A 94:10925–10930
16. Scapagnini G, D'Agata V, Calabrese V, Pascale A, Colombrita C, Alkon D, Cavallaro S (2002) Gene expression profiles of heme oxygenase isoforms in the rat brain. Brain Res 954:51–59
17. Heid CA, Stevens J, Livak KJ, Williams PM (1996) Real time quantitative PCR. Genome Res 6:986–994
18. Winer J, Jung CK, Shackel I, Williams PM (1999) Development and validation of real-time quantitative reverse transcriptase-polymerase chain reaction for monitoring gene expression in cardiac myocytes in vitro. Anal Biochem 270:41–49

Chapter 11

Examining Ubiquitinated Protein Aggregates in Tissue Sections

Natalia A. Kaniuk and John H. Brumell

Abstract

In the cell, the binding of ubiquitin to abnormal or misfolded proteins marks them for degradation by the proteasome or lysosome via autophagy. Ubiquitinated-protein aggregates form when an increase in protein misfolding exceeds the degradation capacity of the cell. Many cellular stresses can cause an increase in the amount of ubiquitinated misfolded protein and failure to eliminate these proteins can disrupt cellular homeostasis and cause cellular toxicity. Ubiquitinated-protein aggregates accumulate in the cytosol and can be detected in tissues of patients with a variety of diseases, including Alzheimer's, Parkinson's, and Huntington's. Using a diabetic rat model, we have shown that ubiquitinated-protein aggregates form in pancreatic beta cells during diabetes-induced oxidative stress. Aggregates were also evident in the hippocampus, kidney, and liver of these animals. Our detailed protocol is provided here. Mounted tissue sections were first deparaffinized, then boiled in sodium citrate buffer to expose the antigen, followed by a specific staining procedure that allows for detection of ubiquitinated-protein aggregates. The ability to visualize ubiquitinated-protein aggregates in tissue sections can provide further understanding of the pathobiology of diseases associated with misfolded protein.

Key words: Ubiquitination, Immunofluorescence, Cellular stress, Aggregates, In vivo, Confocal microscopy, Tissue

1. Introduction

In disease situations such as diabetes and bacterial infection, cellular stress is particularly prevalent. As a result, the production of misfolded protein is greatly increased and cellular mechanisms that normally protect against this form of stress cannot keep up (1, 2). When the production of misfolded proteins exceeds degradation, ubiquitinated-protein aggregates form and accumulate in the cytosol of pancreatic cells during diabetes (3). Large cytosolic

Peter Bross and Niels Gregersen (eds.), *Protein Misfolding and Cellular Stress in Disease and Aging: Concepts and Protocols*, Methods in Molecular Biology, vol. 648, DOI 10.1007/978-1-60761-756-3_11, © Springer Science+Business Media, LLC 2010

ubiquitinated-protein aggregates termed DALIS (dendritic cell aggresome-like inducible structures) have also been observed in mature dendritic cells from murine lymphatic tissue (4). Several examples of pathological aggregation of ubiquitinated-proteins characteristic of amyloid diseases have been reported including Huntington's and Alzheimer's (5, 6). A model of ubiquitinated-protein aggregate formation and clearance can be seen in Fig. 1.

Paraffin-embedded tissues require deparaffinization and rehydration treatment to remove the wax, and this is done chemically with a series of xylene and ethanol washes. Following this, tissue sections may undergo an antigen retrieval treatment in order to visualize ubiquitinated-protein aggregates (7). Finally, a staining procedure has been suggested that allows fluorescent visualization of the aggregates using confocal microscopy. The microscopic visualization of individual proteins localizing within the cell is an important technique demonstrated in cell biology. Here, we describe a protocol to process diabetic pancreatic tissue in order to image ubiquitinated-protein aggregates within the cytosol of individual cells. Observing these aggregates within tissue is a significant step in characterizing this disease phenotype.

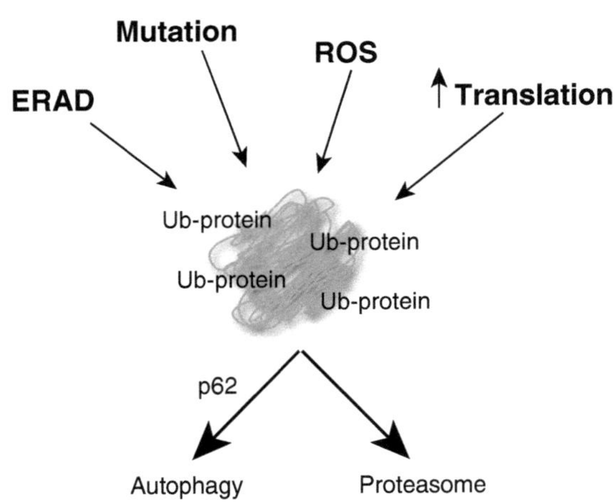

Fig. 1. Model of ubiquitinated-protein aggregate formation during cellular stress. Tissues are susceptible to many stresses during disease. Endoplasmic reticulum-association degradation (ERAD) targets damaged and misfolded proteins for proteasomal degradation by tagging them with ubiquitin (8). Mutated proteins can also accumulate into ubiquitinated aggregates (6). Reactive oxygen species (ROS) in the form of oxidative stress causes the formation of ubiquitinated-protein aggregates in the cytosol of β-cells during diabetes and plays a part in the accumulation of misfolded protein aggregates in neurodegenerative diseases (3, 9). It has also been documented that an increase in protein transcription and translation in dendritic cells can cause newly synthesized protein to be sequestered into aggregates where ubiquitination occurs (4, 10). Clearance may occur by proteasomal degradation or by the autophagy pathway, which degrades these ubiquitinated-protein aggregates by mediating their delivery to lysosomes (3, 11–14). Nbr1 and p62 adaptor proteins are thought to recruit autophagic machinery (15, 16).

2. Materials

2.1. Deparaffinization/ Rehydration of Sections

1. Xylenes certified A.C.S X5-1 (Fisher Chemicals).
2. Ethanol, anhydrous denatured, histological grade (100%, diluted with deionized water to 95% and 70%).
3. Deionized water (dH_2O).
4. 7 glass slide staining dishes, with covers and one glass slide rack (Fisher Scientific); 2 dishes contain xylenes bath, 2 dishes contains 100% ethanol bath, one dish contains 95% ethanol bath, one dish contains 70% ethanol bath, and one dish contains dH_2O bath (See Note 1).
5. Permeabilization buffer: Phosphate-buffered saline (PBS^{++}) (1× DPBS/Modified + Calcium + Magnesium, HyClone Thermo Scientific) with 0.1% Triton X-100.

2.2. Antigen Retrieval

1. Sodium citrate buffer: 10 mM Tri-sodium citrate dihydrate, 0.05% Tween 20, pH 6.0. To make 1 L, weigh 2.94 g of sodium citrate and mix with distilled water to dissolve. Adjust pH to 6.0 with 1 N HCl and then add 0.5 ml Tween 20 and mix well. Add distilled water to 1 L. Store at room temperature for 3 months.
2. Pyrex microwave-safe glass baking dish (7 × 5 × 1.5 in.) with a large glass plate used as a cover.
3. Forceps.
4. Plastic 5-place slide mailer containers, end opening (Fisher).
5. Washing buffer: PBS^{++}.

2.3. Staining

1. Blocking solution: 10% normal goat serum (Wisent) and 0.2% saponin (Calbiochem) in PBS^{++}. Make fresh as required. 10% saponin stock can be stored at −20°C and added to a final of 0.2%.
2. Primary antibody: Mouse monoclonal antibody FK2 (Biomol) (See Note 2).
3. Secondary antibody: Goat anti-mouse AlexaFluor 568 (Molecular Probes).
4. Nuclear stain: 300 nM DAPI (4,6-diamidino-2-phenylindole) in water.
5. Mounting media: Dako Fluorescent Mounting Media (Dako).
6. Microscope cover glass (circular, 25 mm, 0.13–0.17 mm thick) (Fisher).
7. Pap pen (Invitrogen).

3. Methods

Under diabetic stress, ubiquitinated-protein aggregates can be seen within a variety of different tissues (3). Depending on the tissue, antigen-retrieval steps may have to be performed in order to visualize the aggregates. The following method is specific for paraffin-embedded tissue. During the antigen-retrieval step, methylene bridges formed by formalin are broken and this in turn exposes antigenic sites.

3.1. Preparing Tissue Sections for Immunofluorescence

1. These instructions assume the samples are fixed paraffin-embedded tissue, which is mounted onto 25 × 75-mm glass microscope slides. A ventilated fume hood should be used when working with xylenes. Place slides in the glass rack and submerge the rack into the first glass dish containing xylenes. Incubate for 10 min. Once done, submerge the slides into the second glass dish containing xylenes for another 10 min. A third wash is done by submerging the slides into the first dish containing xylenes for a final 10 min.
2. Incubate slides in two washes of 100% ethanol for 5 min each.
3. Incubate slides in one wash of 95% ethanol for 5 min.
4. Incubate slides in one wash of 70% ethanol for 5 min.
5. Incubate slides in one wash of dH_2O for 5 min.
6. Incubate slides in permeabilization buffer within plastic slide mailer containers for 15 min. (See Note 3)
7. Pour off permeabilization buffer and wash sections twice with PBS^{++} for 2 min each, gently inverting with each wash.

3.2. Antigen Retrieval

1. Add the prepared sodium citrate buffer to the glass microwave-safe Pyrex dish covering approximately 3 cm from the bottom of the dish (See Note 4). Place the dish in the microwave and add a glass plate to cover the dish. Ideally, the microwave should be used inside a ventilated fume hood. Microwave the buffer until boiling. Once boiled, carefully remove the glass cover and very gently, using forceps, ease the slides (tissue side up) into the dish, allowing them to settle at the bottom.
2. Place the dish back in the microwave, add the cover and start a timer for 30 min. Begin to microwave the samples at a low power for a couple of minutes. Keep watching the samples, if you notice boiling in the dish, immediately stop the microwave and let the samples sit for 1 min (See Note 5). Then start the microwave again at a low setting for another 2 min, allowing the samples to sit for 1 min intervals. Two minutes is only suggested as a starting point. A control experiment is recommended

beforehand to evaluate how long it takes for boiling to occur and at what power setting this occurs. Every microwave is different and the time/power setting may have to be adjusted accordingly. The point of this step is to keep the samples at near boiling for 30 min. Do not allow the slides to dry out.

3. After 30 min at a high temperature, let the samples sit in the microwave for approximately 30 min to cool to room temperature.
4. Place slides in plastic mailer containers and wash sections in PBS^{++} 3 times for 5 min each, inverting gently at each wash.

3.3. Staining

1. Incubate slides in blocking solution within plastic slide mailer containers for 3 h at room temperature (See Note 6).
2. Remove slides one at a time with forceps from the plastic containers onto paper towel, tissue side up. Using Kimwipes or some filter paper, swab the slide and clean the area around the mounted tissue so that residual blocking solution is gone. Be careful not to touch or disturb the tissue when mopping up excess buffer.
3. Using the Pap pen, draw a circle around the tissue.
4. The freshly prepared primary antibody at a dilution of 1:200 (dilute antibody in blocking solution) is added to the circled tissue of each slide using a pipette (See Note 7).
5. Once the primary antibody is added, slides can be carefully transferred to a container lined with damp paper towels with a tight fitted lid and put at 4°C overnight.
6. Remove primary antibody by aspiration at one corner and transfer slides to plastic mailer containers. Wash sections in PBS^{++} 3 times for 10 min each, gently inverting 5 or 6 times with each wash.
7. Remove slides one at a time onto paper towel. Using a pipette, add fresh blocking solution to each circled tissue section. Let this incubate for 5 min (See Note 8).
8. Remove blocking solution by aspiration at one corner. The freshly prepared secondary antibody Goat anti-mouse AlexaFluor 568 at a dilution of 1:200 (dilute antibody in blocking solution) is added to the circled tissue in each slide using a pipette. Incubate the sections with secondary antibody for at least 2 h at room temperature in the dark (See Note 9).
9. Remove secondary antibody by aspiration at one corner and transfer slides to plastic mailer containers. Wash sections in PBS^{++} 3 times for 10 min each, gently inverting 5 or 6 times with each wash.
10. Wash sections in dH_2O 2 times for 2 min each, gently inverting 5 or 6 times with each wash.

11. DAPI is added for 10 min at room temperature to identify the nuclei.
12. Wash section in dH_2O 5 times at 10 min each and then aspirate dry from one corner.
13. Using Dako mounting media, drop a small amount onto a circular glass coverslip and apply directly onto the exposed tissue. Apply a small amount of pressure to the coverslip and then allow the slide to dry at room temperature in a dark drawer. The sample can be viewed under the microscope once the mounting media has dried. The sample can also be stored in the dark at −20°C for months.
14. The slides can be observed using confocal microscopy. Excitation at 578 nm induces the AlexaFluor 568 conjugate fluorescence exhibiting ubiquitinated aggregates as red structures in the tissue, whereas excitation at 364 nm induces DAPI fluorescence and the nucleus is seen in blue. OpenLab software can be used to analyze tissue and overlay fluorescent images. Ubiquitinated-protein aggregates occurring in diabetic pancreatic tissue can be seen by the arrows in Fig. 2 (3).

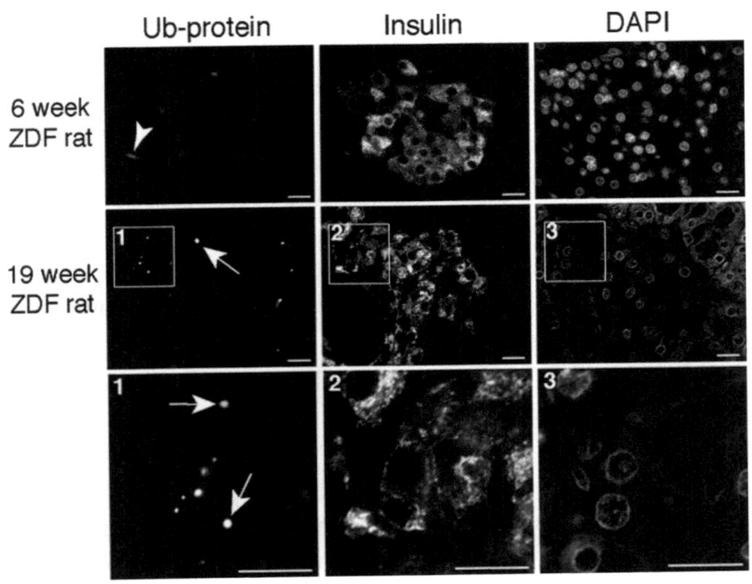

Fig. 2. Zucker diabetic fatty rat pancreatic tissue sections exhibit ubiquitinated protein aggregates in vivo. Representative images of pancreatic sections isolated from 19-week-old Zucker diabetic fatty rats (ZDF) and 6-week-old ZDF rats. Sections were stained using mAb FK2 (identifies mono-and polyubiquitin proteins) to examine ubiquitinated protein aggregates, and an antibody against insulin to identify β-cells within islets and a nuclear DAPI stain. Ubiquitinated-protein aggregates were observed in pancreatic β-cells of the 19-week-old ZDF rat as shown by the arrows. These aggregates were not evident in the control 6-week-old ZDF tissue. All pancreatic tissues contained a small number of infiltrating immune cells that stain nonspecifically with secondary antibodies as shown by the arrowhead. Scale bar, 10 μm.

4. Notes

1. Xylenes, ethanol, and ethanol dilution baths should be replaced after approximately 5 to 7 uses. Use a fresh dH_2O bath with every experiment.
2. The FK2 primary antibody recognizes mono- and poly-ubiquitinated proteins but not free ubiquitin. It is therefore ideal for detection of ubiquitinated-protein aggregates in the cell.
3. A deparaffinization step is not needed if frozen tissue sections are available. Once applied to a glass microscope slide, the frozen tissue sections can be simply incubated in the permeabilization buffer and the rest of the protocol can be carried out.
4. In some tissues, in order to unmask the ubiquitin antigen within aggregates, heat-induced epitope retrieval is done by boiling samples in sodium citrate buffer using a microwave. Use a sufficient volume of sodium citrate buffer to cover the slides by a least a few cm to allow for evaporation during boiling. The dish should be large enough to allow the slides to sit at the bottom.
5. Do not leave the microwave on continuously because the tissues may bubble off the slides.
6. In our experience, some tissues (i.e., pancreas, gut) do not require antigen unmasking in order to visualize ubiquitinated aggregates. In these cases, incubation in blocking solution will suffice. The blocking solution contains saponin, a mild detergent that further permeabilizes the tissue so that antigens can be revealed.
7. The amount of primary antibody added is just enough to cover the tissue sample, between 100 and 500 μl, depending on how big your tissue section is on the slide. The pap pen barrier around the sample will retain this volume. Since the dilution buffer contains saponin, a surfactant, the added primary antibody should spread over and cover the tissue section. Another antibody of different species lineage can also be added at this time, to compare localization of two proteins.
8. Adding blocking solution establishes a good meniscus on top of the tissue section.
9. Do not expose the secondary antibody to light for extended periods of time. Another secondary antibody specific to the primary can also be added at this time, to compare localization.

References

1. Carrell RW, Lomas DA (1997) Conformational disease. Lancet 350:134–138
2. Bukau B, Weissman J, Horwich A (2006) Molecular chaperones and protein quality control. Cell 125:443–451
3. Kaniuk NA, Kiraly M, Bates H, Vranic M, Volchuk A, Brumell JH (2007) Ubiquitinated-protein aggregates form in pancreatic beta-cells during diabetes-induced oxidative stress and are regulated by autophagy. Diabetes 56:930–939
4. Lelouard H, Gatti E, Cappello F, Gresser O, Camosseto V, Pierre P (2002) Transient aggregation of ubiquitinated proteins during dendritic cell maturation. Nature 417: 177–182
5. Kopito RR (2000) Aggresomes, inclusion bodies and protein aggregation. Trends Cell Biol 10:524–530
6. Gow A, Sharma R (2003) The unfolded protein response in protein aggregating diseases. Neuromolecular Med 4:73–94
7. Shi SR, Cote RJ, Taylor CR (1997) Antigen retrieval immunohistochemistry: past, present, and future. J Histochem Cytochem 45:327–343
8. Rutkowski DT, Kaufman RJ (2004) A trip to the ER: coping with stress. Trends Cell Biol 14:20–28
9. Nakamura T, Lipton SA (2009) Cell death: protein misfolding and neurodegenerative diseases. Apoptosis 14:455–468
10. Lelouard H, Ferrand V, Marguet D, Bania J, Camosseto V, David A, Gatti E, Pierre P (2004) Dendritic cell aggresome-like induced structures are dedicated areas for ubiquitination and storage of newly synthesized defective proteins. J Cell Biol 164:667–675
11. Yoshimori T (2004) Autophagy: a regulated bulk degradation process inside cells. Biochem Biophys Res Commun 313:453–458
12. Szeto J, Kaniuk NA, Canadien V, Nisman R, Mizushima N, Yoshimori T, Bazett-Jones DP, Brumell JH (2006) ALIS are stress-induced protein storage compartments for substrates of the proteasome and autophagy. Autophagy 2:189–199
13. Levine B, Klionsky DJ (2004) Development by self-digestion: molecular mechanisms and biological functions of autophagy. Dev Cell 6:463–477
14. Fortun J, Dunn WA Jr, Joy S, Li J, Notterpek L (2003) Emerging role for autophagy in the removal of aggresomes in Schwann cells. J Neurosci 23:10672–10680
15. Ding WX, Yin XM (2008) Sorting, recognition and activation of the misfolded protein degradation pathways through macroautophagy and the proteasome. Autophagy 4:141–150
16. Kirkin V, Lamark T, Sou YS, Bjorkoy G, Nunn JL, Bruun JA, Shvets E, McEwan DG, Clausen TH, Wild P, Bilusic I, Theurillat JP, Overvatn A, Ishii T, Elazar Z, Komatsu M, Dikic I, Johansen T (2009) A role for NBR1 in autophagosomal degradation of ubiquitinated substrates. Mol Cell 33:505–516

Chapter 12

Determination of Proteasomal Activities

Cristovao F. Lima and Suresh I.S. Rattan

Abstract

The proteasome is an important component of the intracellular system for the turnover of proteins. Themammalian proteasome is engaged to degrade a bulky fraction of soluble intracellular proteins both in an ubiquitin-dependent and independent manner. The proteasome is composed by a central catalytic core – the 20S proteasome – where three different proteases are located, whose activities can be measured. A detailed protocol for measuring accurately the three activities of the 20S proteasome in cell and tissue homogenates, using specific fluorogenic substrates and a microplate reader fluorometer, are described. Successful applications of this method include determining changes in the proteasomal activities during aging, anti-aging interventions, cell cycle analysis, and in various disease states including neurodegenerative diseases and cancers.

Key words: 20S Proteasome, Turnover of proteins, Chymotrypsin-like activity, Trypsin-like activity, Caspase-like activity, Microplate reader fluorometer

1. Introduction

In order to maintain cellular homeodynamics, macromolecules within a cell are submitted to constant repair and turnover. In the case of proteins, cells have several mechanisms of turnover, which include the ubiquitin and non-ubiquitin proteasome-mediated pathways, calpains or calcium-dependent proteases, and lysosome-mediated pathways, such as macroautophagy, microautophagy, and chaperone-mediated autophagy. The proteasome system represents about 1% of the total cellular protein, is present both in the cytosol and the nucleus of mammalian cells, and is responsible for the degradation of a large portion of soluble intracellular proteins (1, 2). It is a multicomponent enzymatic system incorporating different regulators and the catalytic core – the 20S proteasome.

Peter Bross and Niels Gregersen (eds.), *Protein Misfolding and Cellular Stress in Disease and Aging: Concepts and Protocols*, Methods in Molecular Biology, vol. 648, DOI 10.1007/978-1-60761-756-3_12, © Springer Science+Business Media, LLC 2010

Mammalian 20S proteasome is composed of two alpha and beta rings in the order αββα, arranged in a barrel-shaped structure of about 700 kDa. Each ring contains seven different subunits (α1-α7 or β1-β7) totalizing two each of different 14 types of subunits (3). The protease active site of β subunits in the inner rings is located inward to form the proteolytic chamber (4). The polypeptide substrate enters the opening formed by the outer rings, and it is degraded by the proteolytic chamber in small peptides (2–30 amino acids), which are further degraded by intracellular peptidases (3, 5). The 20S proteasome core alone is able to recognize and degrade oxidized and unfolded proteins in an ATP-independent and ubiquitin-independent manner, being important for the rapid removal of damaged proteins following oxidative stress (2, 3, 6, 7). It has been suggested that the exposure of hydrophobic amino acids at the surface of the proteins after being damaged may represent the recognition signal for the degradation by the 20S proteasome (5, 6).

The activity of the proteasome needs to be controlled in view of its abundance in the cytosol, and therefore, the 20S proteasome is strictly regulated. Different regulators influence the activity of the 20S proteasome, of which the 19S and the 11S regulators are the most understood (3). The 19S regulator is able to bind the 20S core proteasome on each side forming the 26S proteasome complex with a molecular weight of 2,000 kDa (4). The 26S proteasome is mainly responsible for the recognition and ATP-dependent degradation of poly-ubiquitinated proteins, although it also degrades non-ubiquitinated proteins. The 19S regulator is composed of a lid and a base-ring responsible for the recognition and ubiquitin recycling by poly-ubiquitin-cleaving hydrolases and the ATP-dependent unfolding of polyubiquitinated proteins, respectively, which are then degraded by the 20S core of the 26S proteasome (5, 7). Besides the removal of abnormal proteins, the ubiquitin system is believed to be responsible for the rapid degradation of normal short-lived proteins, such as cyclins, cyclin-dependent kinases, transcription factors, cell surface receptors, and structural proteins, important for the synchronized continuation of many cellular processes, including cell cycle progression, stress response, and cell differentiation (1, 7).

The 11S regulator is a hexa-or hepta-meric ring structure that also binds to either end (α-ring) of the 20S core proteasome forming an 11S-20S-11S structure. This complex is inducible by the interferon-γ and is specially expressed in tissues involved in immune response, and is therefore also known as the immunoproteasome (3, 8). Interferon-γ also induces three alternative subunits of the inner ring of the 20S proteasome core – βi subunits, which exchange constitutive catalytic active units in the β-rings, modifying and enhancing the catalytic activity of the proteasome. The immunoproteasome containing all three βi subunits produces peptides with about 8–10 amino acids that are suitable to be presented as antigen by the class I major histocompatibility complex (MHC-I) (3). The immunoproteasome is mainly located in the cytosol, but there is a special nuclear 11S regulator that is not induced by interferon-γ, which is particularly abundant in the brain (5, 8).

It was shown that the different proteasome forms might be distributed in the cells, and also a hybrid proteasome (11S-20S-19S) was identified (9). These studies have also shown that the free 20S proteasome levels exceed all other forms by two- to threefold. All forms with a regulator attached to the α-rings, where the 20S proteasome core and the regulators are superimposed opening a channel into the proteolytic chamber, result in increased activity (3).

In the eukaryotic 20S proteasome, only three β-subunits in each of the inner rings possess catalytic activities: β1 – a peptidyl-glutamyl-peptide-hydrolase or caspase-like activity that cleaves after acid residues such as glutamic acid; β2 – a trypsin-like activity that cleaves after basic residues like arginine or lysine; and β5 – a chymotrypsin-like activity that cleaves after large hydrophobic residues such as tyrosine or phenylalanine (4, 5, 10). The active center of each of these subunits is a threonine residue at the *N*-terminus of their protein chain, where the hydroxyl group acts as nucleophile defining a new type of protease as compared with serine and cystein proteases (10).

As stated above, the composition of β-subunits in the inner ring can be modified to form the immunoproteasome by incorporating inducible forms by de novo synthesis replacing their constitutive counterparts. These modify proteasome peptidase activities by increasing chymotrypsin-like and trypsin-like activities and decreasing caspase-like activity, which would be important to produce peptides with higher affinity for MHC class I complex (5).

Several specific proteasome inhibitors have been developed. Some widely used inhibitors are natural compounds such as epoxomycin and lactacystin, which act as irreversible inhibitors (5). Also, peptide aldehydes have been synthesized, such as Cbz-Leu-Leu-Leucinal (MG132), which are competitive inhibitors, and are able to enter cells. It was shown that MG132 is a more potent inhibitor against the activated 20S core than against the 26S proteasome. Nevertheless, it reduces the degradation of ubiquitin-conjugated proteins by the 26S complex without affecting its ATPase or isopeptidase activities (4). However, MG132 has also some affinity to inhibit lysosomal and calcium-activated proteases (4, 11).

1.1. Method-Background for Estimating Proteasome Activity

The three peptidase activities of the proteasome can be assayed using fluorogenic synthetic peptide substrates, which provide a convenient and sensitive way to monitor the proteasome activity in crude homogenates and purified fractions. These substrates consist of a bulky blocking group on the *N*-terminus followed by a three to four amino acid residue peptides with a fluorophore, usually amido-4-methylcoumarin (AMC), at the *C*-terminus. The proteasome cleaves an amido bond between an amino acid and the fluorophore, resulting in the release of a highly fluorescent product that can be followed by fluorometry (12). Depending on the amino acid sequence of the synthetic peptide, the three proteasome activities can be distinguished. Some of the most commonly used substrates are succinyl-Leu-Leu-Val-Tyr-AMC (Suc-LLVY-amc) for chymotrypsin-like activity, *N*-*t*-butyloxycarbonyl-Leu-Ser-Thr-Arg-AMC (Boc-LSTR-amc) for trypsin-like activity, and *N*-benzyloxycarbonyl-Leu-Leu-Glu-AMC (Z-LLE-amc) for caspase-like activity, respectively (10).

The different peptidase activities of the proteasome are optimal at neutral to weakly alkaline pH values depending on the substrates, but buffers at pH 7.5–8.0 work well for assaying proteasome activities (13). When the proteasome activity is determined in crude homogenates, and since the substrates are not 100% specific for the proteasome, other intracellular proteases may cleave the peptides giving overestimated results. Therefore, the non-specific proteolysis is commonly determined in the presence of a proteasome inhibitor, such as lactacystin or MG132, although care has to be taken in the interpretation of the results, since other proteases might be also inhibited by these reagents (11, 13).

To measure the ATP-dependent proteasome activity (26S proteasome), crude homogenates and proteasome activity have to be performed in the presence of ATP and 10–20% glycerol or 0.25 mM sucrose to isolate integral 26S proteasome and to stabilize the complex, respectively (12, 14). In addition, KCl is also needed to further decrease the contribution of the 20S proteasome to peptide cleavage by suppressing its spontaneous activation (14). It was demonstrated that different concentrations of ATP in the assay method affect differentially the proteasome activity, and, therefore, for each case, the concentration of ATP for optimal determination of 26S proteasome activity should be calculated (14). A detailed method for measuring 26S proteasome in purified proteasome fractions and in crude homogenates can be found in (12).

A detailed protocol for measuring different activities of the 20S proteasome in a microplate reader fluorometer using fluorogenic substrates is presented. This protocol is modified from combining protocols from other papers, and from our own experience (15–17).

2. Materials

2.1. Buffers and Stock Solutions

1. Proteasome Assay Buffer (PAB): 10 mM HEPES-NaOH, pH 8.0, 50 mM NaCl, 1 mM Na_2EDTA, 250 mM sucrose. Store solution at 4°C.
2. Proteasome Lysis Buffer (PLB): PAB with 0.2% Triton X-100 added fresh. Keep at 4°C.
3. Stock solutions of fluorogenic substrates: Commercially available fluorogenic substrates (e.g., Sigma-Aldrich, St. Louis, MO) dissolved in dimethyl sulfoxide (DMSO) at 10 mM: Suc-LLVY-amc, Boc-LSTR-amc and Z-LLE-amc. Store in aliquots at −20°C and protect from light.
4. Stock solution of MG132: dissolved in DMSO at 20 mM. Store in aliquots at −20°C.

2.2. Assay Solutions

Make the following assay solutions just before use in the 20S proteasome activity assays (see Subheading 3.2):

1. Assay Buffer (AB): Add DMSO to PAB to a final concentration of 2% (20 µl of DMSO to each 1 ml of PAB).
2. Inhibitor Buffer (IB): For each 1 ml of PAB, add 20 µl of stock solution of MG132 (20 mM in DMSO). Final concentration of the inhibitor in the assay will be ~35 µM.
3. Substrate Solutions (protect from light):
   1. *Chymotrypsin-like activity*: Add 20 µl of Suc-LLVY-amc stock for each 1 ml of PAB. Final concentration of the substrate in the assay will be ~21 µM.
   2. *Trypsin-like activity*: Add 32 µl of Boc-LSTR-amc stock for each 1 ml of PB. Final concentration of the substrate in the assay will be ~34 µM.
   3. *Caspase-like activity*: Add 100 µl of Z-LLE-amc stock for each 1 ml of PB. Final concentration of the substrate in the assay will be ~105 µM.

3. Methods

3.1. Sample Preparation

For measuring the three proteasomal activities in triplicate and for protein estimation, about 100–200 mg of tissue or $0.5–3 \times 10^6$ cells (depending on cell size) will be sufficient, following the procedures outlined below.

3.1.1. Tissues/Organs

1. Immediately after removing the organ/tissue rinse it with ice-cold PBS (phosphate buffer saline), mince with scissors (~2 mm^3 portions), and then immerse and rinse in ice-cold PBS.
2. Homogenize tissue with ice-cold PLB in a ratio 1:5 (5 ml PLB/g of tissue) using a Polytron homogeneizer (~10 passes).
3. Centrifuge the homogenate for 5 min at 500×*g* at 4°C and take the supernatant to a tube. Rehomogenize the pellet and centrifuge as above.
4. Combine both supernatants and centrifuge for 10 min at 10,000×*g* at 4°C and take the supernatant to a clean tube. Maintain the homogenate in ice for protein quantification.
5. After protein quantification measure proteasome activities (see Subheading 3.2 and Note 1).

3.1.2. Culture Cell Monolayers

1. Treat cells in culture as desired.
2. Wash cell monolayer twice with PBS or Hanks solution.
3. Add ice-cold PLB, spread it through the cell layer, and keep in ice for 5 min (see Note 2).
4. Scrap the cell layer with the PLB to a microtube and keep on ice.
5. Sonicate the cell homogenate twice for 6sec at lower amplitude with an interval of 2 min on ice between sonications.
6. Centrifuge the cell homogenate for 10 min at 10,000×*g* at 4°C and take the supernatant to a clean microtube. Maintain the homogenate in ice for protein quantification and then perform the proteasome activity assays (see Subheading 3.2 and Note 1).

3.1.3. Cell Pellets

1. Cell monolayers described above (see Subheading 3.1.2) can be scrapped in PBS or Hanks solution instead of PLB. In that case, centrifuge scrapped cells at 1,000×*g* at 4°C and keep the cell pellet. Cells that grow in suspension can also be washed with two cycles of centrifugation and resuspension using PBS or Hanks solution at 4°C to obtain a clean cell pellet (see Note 3).
2. Add ice-cold PLB to cell pellet and resuspend it. Maintain in ice for 5 min (see Note 4).
3. Continue the protocol as described above after Step 5 in Subheading 3.1.2.

3.2. 20S Proteasome Activity Assays

1. Set the microplate reader fluorometer to 37°C (see Note 5 and 6).
2. Pipette in triplicate for a flat-bottom black 96-well plate 10 μg of protein for each sample (see Note 7 and 8).

3. For 2 wells of each sample, pipette 200 μl of AB using a repeating pipette. To the other well for the same sample, pipette 200 μl of IB.
4. Incubate at 37°C for 10 min in the fluorometer (see Note 9).
5. Meanwhile, prepare the substrate solution to the proteasome activity that will be measured (see substrate solutions in Subheading 2.2).
6. After pre-incubation, take the plate from the fluorometer and using a repeating pipette, add 25 μl of the chosen substrate solution (e.g., Suc-LLVY-amc substrate solution for measuring the chymotrypsin-like activity).
7. Mix the plate for 5–10 s using the built-in automixing feature of the microplate reader fluorometer and read the fluorescence intensity along the time using ex_{380nm}/em_{460nm} (AMC) in the fluorometer for around 20 min in 1 min intervals at 37°C.
8. Take the slope (reaction kinetic) obtained from each sample with and without the proteasome inhibitor.

3.3. Calculation of the Proteasome Activity

To express any peptidase activity of the proteasome, the slopes of the linear kinetics of the reaction are taken by using the instrument software. The slope obtained in the replicate measured in the presence of MG132 is subtracted to the average of the other two replicates to eliminate the proteasome non-specific activity. This value can then be expressed as percentage of the sample considered as the control of the experiment.

Instead of the relative activity between the samples, the results can be expressed as specific activity of the enzyme by using the fluorescence intensity (FI) of one AMC standard obtained in the same final volume of the assay and the following equation:

Specific activity [nmol AMC/min/mg] = [slope (FI/min) × AMC standard concentration (μM) × volume of assay (ml)]/[fluorescence of AMC standard (FI) × quantity of protein (mg)].

An example is given for understanding how proteasome specific activity is calculated in an assay performed with 10 μg of protein and in a final volume of 230 μl:

1. Calculate proteasome activity:
   1.1. Slope of activity measured without MG132: 22 FI/min
   1.2. Slope of activity measured with MG132: 2 FI/min
   1.3. Proteasome activity will be: 22-2 = 20 FI/min
2. Measure fluorescence of an AMC standard:
   2.1. Fluorescence of 5 μM AMC standard: 330 FI
   2.2. Fluorescence of blank: 30 FI
   2.3. AMC standard fluorescence will be: 330-30 = 300 FI

3. Calculate proteasome specific activity using previous equation:

Proteasome specific activity = $(20 \times 5 \times 0.23)/(300 \times 0.01) = 7.67$ nmol AMC/min/mg

4. Conclusions

With the present protocol, the three different activities of the proteasome can be estimated in cell lysates using fluorogenic substrates. By measuring the 20S proteasomal activity in cell homogenates, many sources of interferences may occur, such as due to the use of the general proteasome inhibitor MG132 that may also inhibit other proteases, or due to the lack of 100% specificity of the proteasome substrates. However, if the protocol is performed as described above and by using the correct experiment controls run in parallel with the treatment conditions, it will give effective estimation of any changes in proteasomal activity. In addition, in opposition to the end-point measurements, estimation of proteasomal activities in a microplate reader fluorometer by following the kinetics of the reaction along the time will increase the accuracy of the assay, if care is taken to pipette precise volumes of the different solutions and protein.

Finally, it may be noted that the above protocols for the estimation of proteasomal activities have been successfully applied for determining changes during aging, anti-aging interventions, cell cycle analysis, and in various disease states, including neurodegenerative diseases and cancers (15–18).

5. Notes

1. Clean supernatant can be stored in aliquots at –80°C (or –20°C for shorter time) before proteasome activity assays, but it will decrease the activity by about 10% per each thawing. Avoid repeated freeze/thaw cycles of the sample.
2. As a guide, use 300 μl of PLB for each T75 flask or 60 μl per well of a 6-well plate. The final concentration of protein of the homogenate should be higher than 1 mg/ml, otherwise decrease the added PLB to the samples accordingly.
3. Cell pellets can be then either immediately used for preparation of cell homogenates, or stored at –80°C for later homogenization and proteasome activities assays.
4. As a guide, use 250 μl of PLB for a cell pellet taken from a T75 flask. The final concentration of protein of the homogenate

should be higher than 1 mg/ml, otherwise decrease the added PLB to the samples accordingly.

5. If a microplate reader fluorometer is not available, a kinetic assay can be replaced by end point measurements. For that, make the assay as mentioned in the Subheading 3.2 in microtubes at 37°C (protecting from light) and stop the reaction with 300 μl of an acidic solution (30 mM sodium acetate, 70 mM acetic acid, 100 mM sodium chloroacetate, pH 4.3). After addition of 2 ml of distilled water, read the fluorescence in a fluorometer with the proper excitation and emission filters. The time of incubation has to be previously tested and has to be within the linear response.
6. To calibrate the plate reader fluorometer, a calibration curve with a range of AMC concentrations (0.1–v10 μM) has to give a linear response.
7. The quantity of protein necessary for accurate results for each peptidase activity should be determined previously for each cell type or organ/tissue with preliminary assays. The correct protein load should give a linear response along the time (at least in the first 30 min) and not exceed during the assay 1.5× the signal generated by 5 μM of AMC standard. In addition, to increase accuracy, the chosen protein quantity should be pipetted in a similar volume between samples, that is, different samples should have similar protein concentration, preferably higher than 1 mg/ml. If the pippetted volume of sample differs much between samples, since the fluorescence intensity also depends on the height of the liquid in a well, the comparison of the results between different samples may be affected.
8. Pipet also a standard curve, or at least one standard within the linear range (e.g., 5 μM AMC), if the results are going to be given in nmol of AMC produced/min/mg protein (see Subheading 3.3). For the standards, one should take the fluorescent intensity in an endpoint measurement, and not the slope in a kinetic assay as in the samples.
9. This pre-incubation procedure will allow AB and IB to stabilize at 37°C and also to give time to the MG132 to completely inhibit proteasome.

References

1. Rivett AJ (1993) Proteasomes: multicatalytic proteinase complexes. Biochem J 291:1–10
2. Shringarpure R, Grune T, Davies KJA (2001) Protein oxidation and 20S proteasome-dependent proteolysis in mammalian cells. Cell Mol Life Sci 58:1442–1450
3. Jung T, Grune T (2008) The proteasome and its role in the degradation of oxidized proteins. IUBMB Life 60:743–752
4. Coux O, Tanaka K, Goldberg A (1996) Structure and functions of the 20S and 26S proteasomes. Annu Rev Biochem 65:801–807

5. Farout L, Friguet B (2006) Proteasome function in aging and oxidative stress: implications in protein maintenance failure. Antioxid Redox Signal 8:205–216
6. Grune T, Merker K, Sandig G, Davies KJA (2003) Selective degradation of oxidatively modified protein substrates by the proteasome. Biochem Biophys Res Commun 305:709–718
7. Poppek D, Grune T (2006) Proteasomal defense of oxidative modifications. Antioxid Redox Signal 8:173–184
8. Voss P, Grune T (2007) The nuclear proteasome and the degradation of oxidatively damaged proteins. Amino Acids 32:527–534
9. Tanahashi N, Murakami Y, Minami Y, Shimbara N, Hendil KB, Tanaka K (2000) Hybrid proteasomes. Induction by interferon gamma and contribution to ATP-dependent proteolysis. J Biol Chem 275:14336–14345
10. Friguet B, Bulteau AL, Conconi M, Petropoulos I (2002) Redox control of 20S proteasome. Methods Enzymol 353:253–262
11. Rodgers KJ, Dean RT (2003) Assessment of proteasome activity in cell lysates and tissue homogenates using peptide substrates. Int J Biochem Cell Biol 35:716–727
12. Kisselev AF, Goldberg AL (2005) Monitoring activity and inhibition of 26S proteasomes with fluorogenic peptide substrates. Methods Enzymol 398:364–378
13. Rivett AJ, Bose S, Pemberton AJ, Brooks P, Onion D, Shirley D, Stratford FLL, Forti K (2002) Assays of proteasome activity in relation to aging. Exp Gerontol 37:1217–1222
14. Powell SR, Davies KJ, Divald A (2007) Optimal determination of heart tissue 26S-proteasome activity requires maximal stimulating ATP concentrations. J Mol Cell Cardiol 42:265–269
15. Bulteau AL, Petropoulos I, Friguet B (2000) Age-related alterations of proteasome structure and function in aging epidermis. Exp Gerontol 35:767–777
16. Beedholm R, Clark BFC, Rattan SIS (2004) Mild heat stress stimulates 20S proteasome and its 11S activator in human fibroblasts undergoing aging in vitro. Cell Stress Chaperones 9:49–57
17. Bulteau AL, Moreau M, Saunois A, Nizard C, Friguet B (2006) Algae extract-mediated stimulation and protection of proteasome activity within human keratinocytes exposed to UVA and UVB irradiation. Antioxid Redox Signal 8:136–143
18. Tydlacka S, Wang CE, Wang X, Li S, Li XJ (2008) Differential activities of the ubiquitin-proteasome system in neurons versus glia may account for the preferential accumulation of misfolded proteins in neurons. J Neurosci 28:13285–13295

Chapter 13

Measurement of Autophagy in Cells and Tissues

Isei Tanida and Satoshi Waguri

Abstract

Two major proteolysis systems, the ubiquitin-proteasome system, and the autophagy-lysosome system, contribute to degradation of various types of protein and/or protein aggregates. In general, the autophagy-lysosome system is involved in bulk intracellular degradation of proteins and organelles, while the ubiquitin-proteasome system is selective. During autophagy, a cytosolic form of LC3 (LC3-I) is conjugated to phosphatidylethanolamine to form LC3-phosphatidylethanolamine conjugate (LC3-II), which is recruited to autophagosomal membranes, and LC3-II is degraded by lysosomal hydrolases after the fusion of autophagosomes with lysosomes. Therefore, lysosomal turnover of LC3-II reflects starvation-induced autophagic activity, and detection of LC3 by immunoblotting or immunofluorescence has become a reliable method for monitoring autophagy. When autophagy is impaired, the level of p62/SQSTM1, a ubiquitin- and LC3-binding protein, is increased in addition to the accumulation of ubiquitinated proteins. Here, we describe basic protocols to analyze endogenous LC3-II, p62, and autophagy-related proteins by immunoblotting, immunofluorescence, and electron microscopy.

Key words: LC3, Atg, p62, Ubiquitin, mTor

Abbreviations

Atg	Autophagy-related product
BSA	Bovine serum albumin
DMSO	Dimethylsulfoxide
IMPase	Myoinositol monophosphatase
PE	Phosphatidylethanolamine
PI3K	Phosphoinositide 3-kinase
LC3/MAP1LC3B	Microtubule-associated protein 1 light chain 3, a yeast Atg8 homolog
3-MA	3-Methyladenine
TBS	Tris-buffered saline

Peter Bross and Niels Gregersen (eds.), *Protein Misfolding and Cellular Stress in Disease and Aging: Concepts and Protocols*, Methods in Molecular Biology, vol. 648, DOI 10.1007/978-1-60761-756-3_13, © Springer Science+Business Media, LLC 2010

1. Introduction

Autophagy (macroautophagy) involves bulk lysosomal degradation of cytosolic proteins, lipids, and organelles, including mitochondria and the endoplasmic reticulum (Fig. 1) and is significantly related to neurodegenerative diseases, cancer, and programmed cell death (1–3). Autophagosomes, representative double-membrane structures involved in autophagy, accumulate in the brains of patients with a range of neurodegenerative diseases, including Alzheimer's disease, transmissible spongiform encephalopathy, Parkinson's disease, and Huntington's disease (1). Neuron-specific autophagy-deficient mice show progressive neurodegeneration associated with the accumulation of ubiquitinated protein aggregates and/or inclusion bodies (4, 5). Specific ablation

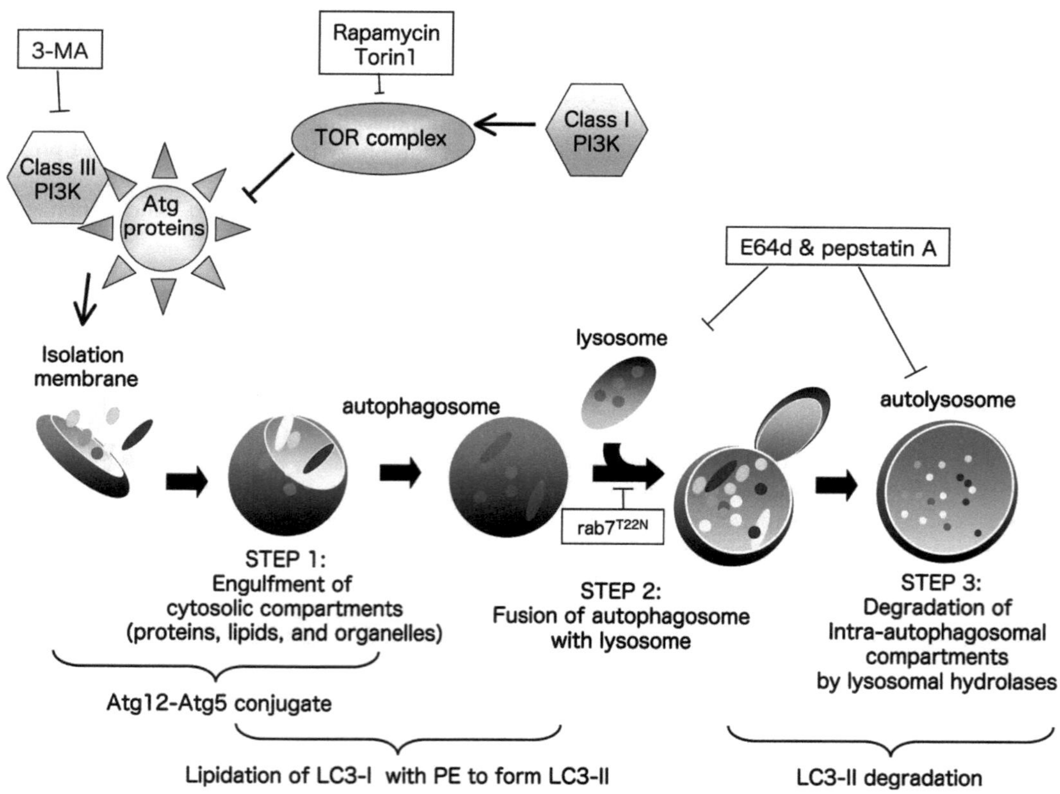

Fig. 1. Schematic representation of autophagy. Class I and Class III phosphoinositide 3-kinases (Class I and Class III PI3K, respectively) and the mTOR signaling pathway regulate autophagy. During autophagy, an isolation membrane appears in the cytosol, and cytosolic compartments including organelles are engulfed into a cup-shaped preautophagosome. Atg12–Atg5 conjugate contributes to these steps. Preautophagosomes finally close to form autophagosomes. LC3-II contributes to this step. Lysosomes fuse with autophagosomes, and intraautophagosomal contents are lysed. Rapamycin and Torin1 inhibit mTOR kinase, resulting in the induction of autophagy. 3-Methyladenine (3-MA) inhibits class III PI3K, resulting in inhibition of autophagy. As rab7 contributes to autophagosome – lysosome fusion, a dominant negative mutant of rab7 ($rab7^{T22N}$) inhibits autophagy. E64d and pepstatin A, inhibitors of cathepsins, inhibit the final step of autophagy, that is degradation of intraautophagosomal contents.

of autophagy in Purkinje cells in mice causes axonal swelling and degeneration of the axon terminals (6). The neurons in patients with neuronal ceroid lipofuscinosis (NCL) show accumulation of mitochondrial ATPase subunit c in lysosomes, indicating that mitochondria are degraded via the autophagy-lysosomal system in neurons, and that impairment of mitochondrial protein-degradation occurred in the patients. Mutation of cln3, one of the NCL-related genes, or a lack of cathepsin D, one of the major lysosomal proteases, leads to neuronal ceroid lipofuscinosis with autophagic vacuolization (7, 8).

During autophagy, unique double-membrane autophagosomes are formed for sequestration of intracellular components (AP in Fig. 2). According to their formation, cytosolic LC3 (LC3-I) is conjugated to phosphatidylethanolamine (PE) to localize the resultant LC3-II (a LC3-PE conjugate) on the autophagosomes (9). The autophagosomes then fuse with lysosomes to form autolysosomes (AL in Fig. 2), whereby sequestered intra-autophagosomal components are degraded by lysosomal hydrolases. Concomitantly, intra-autophagosomal LC3-II is also degraded by lysosomal proteases (Fig. 1). Therefore, LC3-II itself could be used as a marker of autophagosome formation, and its degradation as a monitoring indicator of the late period in autophagic flux (10). Monitoring of LC3-I and LC3-II by Western blotting and immunofluorescence analyses is essential to investigate the mechanism of mammalian autophagy (Figs. 3 and 4).

When mammalian cells and tissues are deficient in autophagy, the levels of ubiquitin and p62 are increased (11, 12). In addition to ubiquitin, p62/SQSTM1 is useful as a marker of autophagic

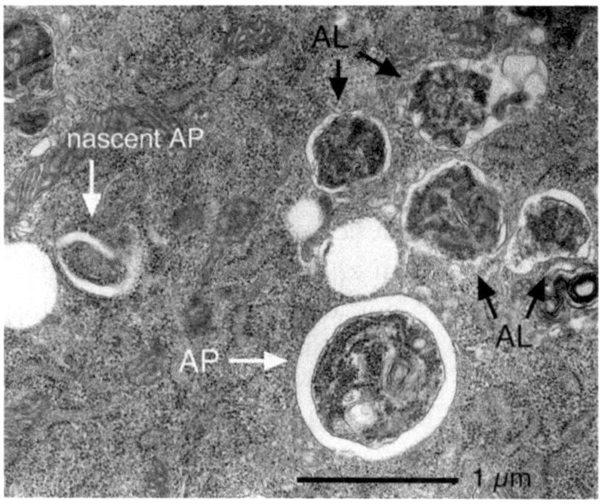

Fig. 2. Typical autophagosomes and autolysosomes on electron microscopy. Mouse embryonic fibroblasts were incubated in HBSS medium for 2 h and fixed as described in the text. Autophagosomes (AP), nascent AP, and autolysosomes (AL) can be detected as indicated. Note that the structure indicated as "AP" could also be nascent AP because a connection point between the segregated portion and surrounding cytosol may be present in the front or rear of this section. In autolysosomes, the segregated organelles lose their distinct shape and are usually replaced by electron-dense materials. Bar indicates 1 μm. Note that the diameter of the typical autophagosomes (AP) is 0.5-1 μm.

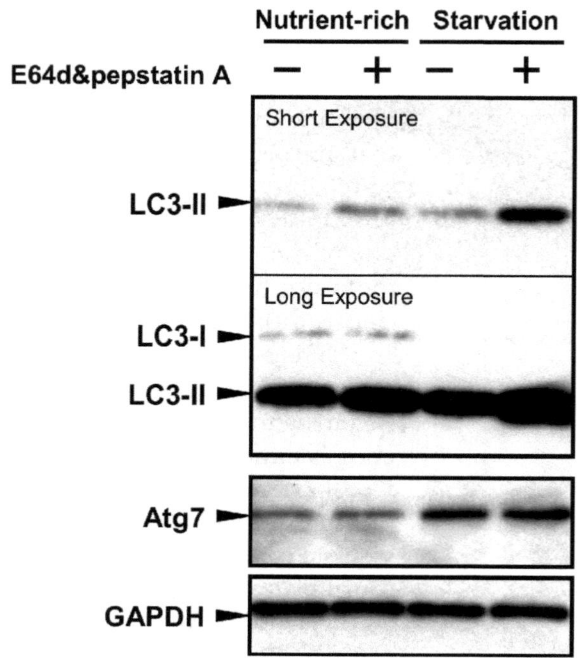

Fig. 3. LC3-II is degraded by lysosomal proteases during autophagy. For nutrient-rich conditions, cells were cultured in DMEM medium containing 10% FCS and non-essential amino acids. Where indicated, cells were treated with the protease inhibitors, E64d (10 μg/mL) and pepstatin A (10 μg/mL) (Inhibitors +) for 4 h, or with the solvent, dimethylsulfoxide, as a negative control (Inhibitors -). For starvation conditions, cells were incubated for 4 h in Krebs-Ringer medium in the presence (+) or absence (–) of protease inhibitors. The cells were lysed, total proteins (10 μg per lane) were separated by SDS-PAGE, and endogenous LC3 and Atg7 in the lysates were recognized by immunoblotting with anti-LC3 and anti-Atg7 antibodies, respectively. LC3-I, soluble form of LC3; LC3-II, membrane-bound form of LC3. GAPDH was employed as a loading control. Note that LC3-II accumulates under starvation conditions in the presence of the inhibitors.

defect in cells and tissues. p62 is a ubiquitin- and LC3-binding protein that contributes to inclusion body formation. Autophagy-deficient mouse brain and liver show accumulation of p62- and ubiquitin-positive inclusions. Therefore, the increase in p62-positive and ubiquitin-positive inclusions in tissues indicates the possibility of an insufficiency in autophagy.

Various reagents are used for studying autophagy (Table 1). mTor kinase negatively regulates autophagy via phosphorylation of ULK1-Atg13-FIP200 complex (13–16). As rapamycin (sirolimus) inhibits the activity of mTor kinase (mTORC1 complex) via binding to FKBP12, it induces autophagy. Torin1, another inhibitor of mTor kinase, induces autophagy more strongly, as it directly inhibits mTor kinase (both mTORC1 and mTORC2 complexes) (17). Lithium chloride (an inhibitor of IMPase), L-690 330 (another IMPase inhibitor), and trehalose induce autophagy via an mTor-independent pathway (18). Reagents that induce ER-stress (e.g. brefeldin A and tunicamycin) also induce autophagy.

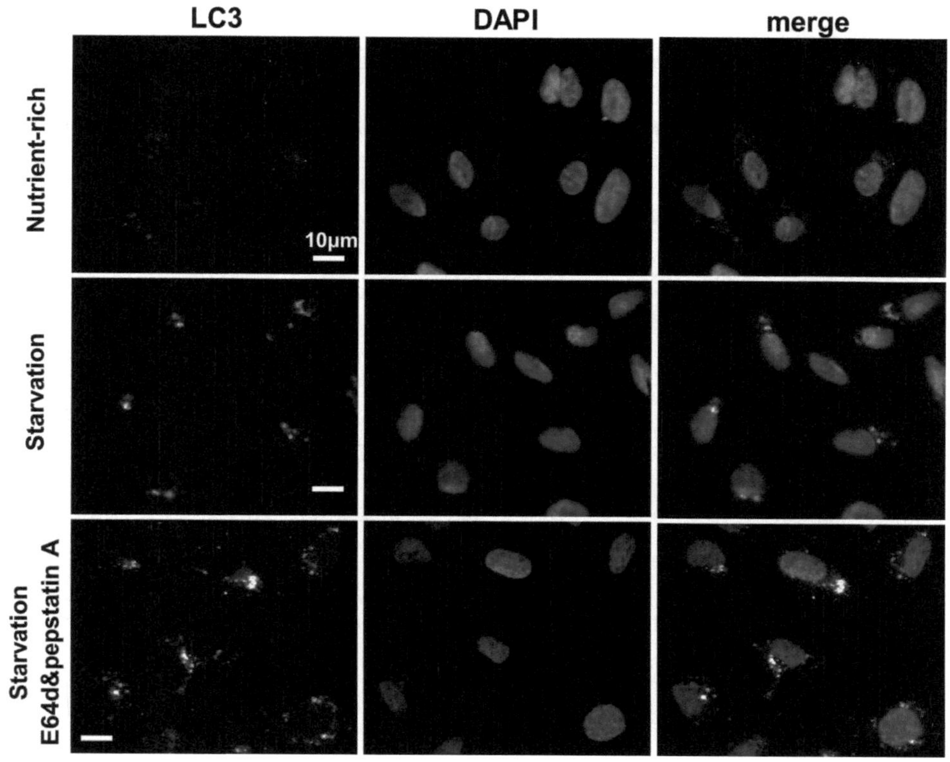

Fig. 4. Immunofluorescence analyses of endogenous LC3-puncta. Huh7 cells were incubated under nutrient-rich and starvation conditions in the absence and presence of E64d and pepstatin A as described in Fig. 3. Cells were fixed with 4% paraformaldehyde, and permeabilized with 50 μg/mL of digitonin. Endogenous LC3 was recognized using a rabbit polyclonal anti-human LC3 antibody and Alexa488-conjugated anti-rabbit IgG. Fluorescent images were obtained by fluorescence microscopy.

3-Methyladenine (3-MA) is used as an autophagy inhibitor. Class I and class III phosphoinositide 3 kinases (PI3Ks) positively regulate autophagy, and 3-MA inhibits class III PI3K. In this case, therefore, the formation of autophagosomes is blocked, that is little LC3-II, membrane-bound form of LC3, is formed. As 3-MA shows multiple metabolic effects and cytotoxicity, its off-target effects on the cells and tissues should be considered carefully.

Inhibition of lysosomal function by inhibition of cathepsins or by acidification of lysosomes inhibits the final step of autophagy. Cathepsins B, D, and L are major lysosomal proteases, and these cathepsins play indispensable roles in the final degradation step of autophagy. A mixture of E64d (but not E65 or E65c; See Note 1) (a membrane-permeable inhibitor of cathepsins B, H, and L) and pepstatin A (a membrane-permeable inhibitor of cathepsins D and E) is used for this purpose. Application of the inhibitor cocktail to the cells results in significant accumulation of autolysosomes (and LC3-II dots) because there is little intra-autolysosomal content degradation. This accumulation reflects

Table 1
Autophagy inducers, inhibitors, and media for starvation

Reagents that induce autophagy	Stock solution	Final conc.
Rapamycin	1 mM in DMSO or ethanol, –20°C	0.5–1 μM
Torin1	1 mM in DMSO, –20°C	1 μM
Lithium chloride	1 M in H_2O, 4°C	10 mM
L-690,330	1 M in H_2O, 4°C	100 μM
Trehalose	powder	100 mM
Reagents that inhibit autophagy	**Stock solution**	**Final conc.**
3-MA	powder	5–10 mM
E64d and pepstatin A	10 mg/ml each in DMSO, –20°C	10 μg/ml
Bafilomycin A1	0.1 mM in DMSO, –20°C	0.1 μM

E64d is membrane-permeable, while E64, E64c, and leupeptin are membrane-impermeable (see Note 1). When leupeptin is used instead of E64d and pepstatin A as cathepsin inhibitors, care should be taken to account for its membrane impermeability

the activity of the delivery process of LC3-II into lysosomes. Thus, these drugs can be used as tools for measurement of autophagy flux. Leupeptin, a membrane-impermeable inhibitor of cathepsin B, is used for investigation of autophagy in the liver (19), but only rarely in other tissues, as leupeptin accumulates effectively in the liver probably via endocytosis. Therefore, it is possible that leupeptin inhibits cathepsin B in some cells and tissues via the endocytic pathway. As leupeptin is a membrane-impermeable reagent, it should be confirmed whether this drug actually inhibits cathepsin B in cells and tissues before studying autophagy. Bafilomycin A1 is an inhibitor of V-ATPase that is essential for acidification of lysosomes. Therefore, this drug is also used for inhibition of autophagy and estimation of autophagic flux of LC3-II. As V-ATPase contributes to the acidification of the other organelles, including the Golgi and endosomes, bafilomycin A1 potentially shows multiple off-target effects (20). Rab7, one of the small GTPases, mediates the fusion of autophagosomes with lysosomes (21, 22). A dominant negative mutant of rab7, rab7^{T22N}, inhibits this fusion step. Overexpression of this rab7 mutant results in augmented accumulation of autophagosomes under starvation conditions. However, rab7 is not specific for autophagosome-lysosome fusion, as it also plays an indispensable role in endosome-lysosome fusion.

LC3 fused with green fluorescent protein (GFP-LC3) is a useful tool for *in vivo* imaging of autophagosome formation (23, 24). However, caution must be exercised due to the limitations of GFP-LC3 (25–28). GFP-LC3 tends to form puncta in cells independent of autophagy, especially when protein aggregates accumulate in the cells. Fluorescence of GFP in the lysosomes is observed even after degradation of the portion of LC3. Therefore, the number of autophagosomes detected in this method tends to be overestimated. To avoid these problems, it will be better to use a mutant GFP-LC3ΔG, which lacks the carboxyl terminal Gly essential for lipidation and localization to the autophagosomal membrane, as a negative control. GFP-LC3 transgenic mice are available to monitor autophagy in many tissues (http://www2.brc.riken.jp/lab/animal/detail.php?reg_no=RBRC00806). GFP-LC3 transgenic mice are good for studying autophagy in many tissues outside the brain where few puncta of GFP-LC3 are observed. The brain is probably one of the organs most protected against starvation, and therefore most endogenous LC3 in the brain is LC3-I, a cytosolic form. Some groups reported that GFP-LC3 works in cultured neurons (29, 30).

Tandem fluorescent protein-LC3 (tfLC3/mRFP-GFP-LC3) is used to distinguish between autophagosomes and autolysosomes (31). Intraluminal pH of autophagosomes is thought to be similar to cytosolic pH. On the other hand, autolysosomes are acidic, because they contain V-ATPase that is derived from lysosomes. Fluorescence of monomeric red fluorescent protein (mRFP) is stable, while that of GFP decreases at this acidic pH. When both green and red fluorescent puncta of tfLC3 are observed, they represent autophagosomes, and when their green signal decreases, this indicates delivery of tfLC3 into acidic compartments; that is, fusion occurs between autophagosomes and lysosomes. Although, tfLC3 is suitable for real-time (and short-term) monitoring of autophagy, care should be taken for use in long-term monitoring of autophagy, as it has been reported that fluorescence derived from GFP in the lysosomes is still observed even after degradation of LC3.

Although LC3 is an excellent marker for autophagosomes, LC3-positive dots observed at the light microscopic level could also represent autolysosomes or p62/ubiquitin inclusions under conditions where autophagy flux is partially or completely compromised (12). Moreover, recent findings indicated that a fraction of LC3-positive dots are also associated with lipid droplets (32, 33). Therefore, identification of typical autophagosomes by conventional electron microscopy is still useful and necessary in the field of autophagy research. However, the involvement of somewhat complex membrane dynamics in this phenomenon may lead to confusion.

2. Materials

2.1. Inducers, Inhibitors, Antibodies, and Other Reagents

1. Rapamycin stock solution: 1 mM Rapamycin (LC laboratories, Woburn, MA or Sigma-Aldrich, St. Louis, MO) in dimethylsulfoxide (DMSO) or ethanol. Stable at –20°C for at least 3 months.
2. Torin1: Provided by Dr. Sabatini's lab (http://web.wi.mit.edu/sabatini/pub/Torin1_requests.html).
3. Cathepsin inhibitor mixture: 10 mg/mL E64d (a membrane-permeable analog of E64c; Peptide Institute, Inc., Osaka, Japan) (see Note 1) and pepstatin A (Peptide Institute, Inc.) in dimethylsulfoxide (DMSO) (special grade for biochemical analysis) (Wako Pure Chemical Industries, Ltd., Tokyo, Japan) (see Note 2). These 1,000× concentrated stock solutions were stored at –20°C, at which temperature they are stable for at least 3 months.
4. Starvation media:
   (a) Earle's Balanced Salt Solution (EBSS) (GIBCO/Invitrogen, Carlsbad, CA or Hyclone/Thermo Fishier Scientific, Waltham, MA).
   (b) Krebs Ringer bicarbonate buffer (KRB) (Sigma-Aldrich) for nutrient starvation (see Note 3).
5. Phosphate-buffered saline (PBS): 20 mM sodium phosphate, pH 7.2, 150 mM NaCl (see Note 4).
6. Complete™ protease inhibitor cocktail (Roche Diagnostics, Basel, Switzerland).
7. Modified Laemmli buffer: 0.1 M Tris–HCl, pH 6.8, 2% (w/v) sodium dodecyl sulfate (SDS), 10% (w/v) glycerol, 6% (v/v) β-mercaptoethanol, and 0.03% (w/v) bromophenol blue.
8. SDS-lysis solution: 1% (W/V) SDS in PBS with complete protease inhibitor cocktail (Roche Diagnostics) (for option 2b).
9. TX100-lysis solution: 2% polyethylene glycol mono-*p*-isoo ctylphenyl ether (Triton X-100) in PBS with the complete protease inhibitor cocktail (Roche Diagnostics) (for option 2c).
10. Sucrose solution: 0.25 M Sucrose, 5 mM Tris–HCl, pH 7.5 (for preparing tissue lysates).
11. Potter-Elvehjem homogenizer for preparing tissue lysates.
12. BCA protein assay kit (Thermo Fisher/Pierce Biotechnology).

2.2. SDS-Polyacrylamide Gel Electrophoresis

1. Stock solutions for preparing 6–12.5% SDS-polyacrylamide gels: 1.5 M Tris–HCl, pH 8.8 (at 25°C); 0.5 M Tris–HCl, pH 6.8 (at 25°C); 30% acrylamide solution (29:1 (w/w) acrylamide: methylene-bisacrylamide); *N, N, N, N'*-tetramethyl-

ethylenediamide (TEMED) (see Note 5 for ready-made gel and recognition of LC3-I and LC3-II).

2. Ammonium persulfate (APS): 10% (w/v) in water, prepared immediately before use.
3. Water-saturated isobutanol: Equal volumes of water and isobutanol were shaken in a glass bottle and allowed to separate. The top layer was stored at room temperature.
4. Running buffer: 25 mM Tris, 192 mM glycine, and 0.1% (w/v) SDS. Do not adjust pH. Store at room temperature.
5. Prestained molecular weight markers, broad range (6–175 kDa) (New England Biolabs, Beverly, MA): The solution contained 0.01% (w/v) phenol red and 0.01% (w/v) bromophenol blue (see Note 6 for recognition of LC3-I and LC3-II).
6. Apparatus for SDS-gel electrophoresis: for example, NA-1030 (Nihon-Eido, Tokyo, Japan) for mini-wide gels (1.0 mm thick, 200 mm wide, 105 mm high, suitable for 16–20 samples per gel).

2.3. Western Blotting

1. Primary antibody solution: 0.1–1 μg/mL primary antibody (see Table 2), 1% (w/v) fraction V bovine serum albumin (BSA) (or 10% skim milk), 20 mM Tris–HCl, pH 7.5 at 25°C, 150 mM NaCl (TBS), and 0.1% NaN_3.
2. Bjerrum and Schafer-Nielsen's transfer buffer: 48 mM Tris, 39 mM glycine, 20% methanol (analytical grade; Wako), pH 9.0–9.4, depending on reagent purity. Do not adjust pH; if lower than pH 9.0, prepare again.
3. Polyvinylidene difluoride (PVDF) membrane (Durapore membrane GV, pore size 0.22 μm; Millipore, Bedford, MA) and chromatography paper (cat no 590; Advantec Japan, Tokyo, Japan).
4. Transblot SD Semi-Dry Transfer Cell (BioRad, Hercules, CA) for transferring proteins from polyacrylamide gels onto PVDF membranes. Transfer proteins from a single gel to a single PVDF membrane per cell. Avoid transferring proteins from two or more gels onto PVDF membranes per cell.
5. TTBS: 20 mM Tris–HCl, pH 7.5 at 25°C, 150 mM NaCl, and 0.05% (v/v) polyoxyethylene sorbitan monolaurate (Tween 20).
6. Blocking buffers: (a) 10% (w/v) nonfat dry milk in TTBS (or TBS: 20 mM Tris–HCl, pH 7.5 at 25°C, 150 mM NaCl), (b) 5% BSA (Fraction V; Sigma-Aldrich)-TBS, (c) 1% BSA-TBS, or (d) 1% casein-TBS. For preparing 10% casein solution, 100 mM maleic acid and 150 mM NaCl were dissolved and adjusted to pH 7.5 using solid NaOH. Thereafter, casein was added and dissolved with mild heating. This solution was

Table 2
Antibodies suitable for recognizing endogenous autophagy-related proteins by Western blotting

Name	Information/company/conditions for western blotting
Anti-Atg3	(1) Rabbit polyclonal antibody, Sigma-Aldrich, A3231; blocking, **5% BSA**-TBS, incubate at room temperature for 30 min; primary antibody reaction (dilution, 1:2,000), incubate at room temperature for 90–120 min
	(2) Mouse monoclonal IgG2b kappa, Medical & Biological Laboratories, Tokyo, Japan, #M133-3; blocking, 10% skim milk-TBS, incubate at room temperature for 30 min; primary antibody reaction (dilution, 1:2,000), incubate at **4°C for 15 h**
Anti-Atg4B	(1) Rabbit polyclonal antibody, Sigma-Aldrich, A2981; blocking, **5% BSA**-TBS, incubate at room temperature for 30 min; primary antibody reaction (dilution, 1:2,000), incubate at room temperature for 90–120 min
Anti-Atg5	(1) Rabbit polyclonal antibody, Sigma-Aldrich, A0731; blocking, **5% BSA**-TBS, incubate at room temperature for 30 min; primary antibody reaction (dilution, 1:2,000), incubate at room temperature for 90–120 min
	(2) Rabbit polyclonal antibody **specific for human Atg5**, Cell Signaling Technology, Danvers, MA, #2630; blocking, 1% BSA (or 10% skim milk)-TBS, incubate at room temperature for 30 min; primary antibody reaction (dilution, 1:2,000), incubate at room temperature for 90–120 min
	(3) Chicken polyclonal antibody, Abcam, Cambridge, MA, ab19130; blocking, 1% BSA (or 10% skim milk)-TBS, incubate at room temperature for 30 min; primary antibody reaction (dilution, 1:2,000), incubate at **4°C for 15 h**
Anti-Atg7	(1) Rabbit polyclonal antibody, Sigma-Aldrich, A2856; blocking, **5% BSA**-TBS, incubate at room temperature for 30 min; primary antibody reaction (dilution, 1:2,000), incubate at room temperature for 90–120 min
	(2) Rabbit polyclonal antibody, Cell Signaling Technology, #2631; blocking, 1% BSA (or 10% skim milk)-TBS, incubate at room temperature for 30 min; primary antibody reaction (dilution, 1:2,000), incubate at room temperature for 90–120 min

(continued)

Table 2
(continued)

Name	Information/company/conditions for western blotting
	(3) Goat polyclonal antibody, Santa Cruz Biotech, Santa Cruz, CA, APG7 (N-20) #sc-8668, **1% casein**-TBS, incubate at room temperature for 30 min; primary antibody reaction (dilution, 1:2,000), incubate at room temperature for 90–120 min
Anti-Atg10	Rabbit polyclonal affinity-purified antibody, Sigma-Aldrich, A9356; blocking, **5% BSA**-TBS, incubate at room temperature for 30 min; primary antibody reaction (dilution, 1:2,000), incubate at room temperature for 90–120 min
Anti-Atg16L	(1) Mouse monoclonal IgG1 kappa, Medical & Biological Laboratories #M150-3; blocking, 10% skim milk-TBS, incubate at room temperature for 30 min; primary antibody reaction (dilution, 1:2,000), incubate at **4°C for 15 h**
	(2) Rabbit polyclonal antibody, CosmoBio Tech, #TMD-PH-AT16; blocking, 1% BSA (or 10% skim milk)-TBS, incubate at room temperature for 30 min; primary antibody reaction (dilution, 1:2,000), incubate at room temperature for 90–120 min
Anti-beclin1	Rabbit polyclonal affinity-purified antibody, Cell Signaling Technology, #3738; blocking, 1% BSA (or 10% skim milk)-TBS, incubate at room temperature for 30 min; primary antibody reaction (dilution, 1:2,000), incubate at room temperature for 90–120 min
Anti-LC3	(1) Rabbit polyclonal affinity-purified antibody (anti-LC3B), Sigma-Aldrich, L7543; blocking, **5% BSA**-TBS, incubate at room temperature for 30 min; primary antibody reaction (dilution, 1:2,000), incubate at room temperature for 90–120 min
	(2) Rabbit polyclonal affinity-purified antibody (anti-LC3B), Cell Signaling Technology, Danvers, MA, #2775; blocking, 10% skim milk-TBS, incubate at room temperature for 30 min; primary antibody reaction (dilution, 1:2,000), incubate at room temperature for 90–120 min
	(3) Rabbit polyclonal antibody, Medical & Biological Laboratories, #PM036; blocking, 10% skim milk-TBS, incubate at room temperature for 30 min; primary antibody reaction (dilution, 1:2,000), incubate at room temperature for 90–120 min

(continued)

Table 2
(continued)

Name	Information/company/conditions for western blotting
Anti-p62/SQSTM1	Rabbit polyclonal antibody, Medical & Biological Laboratories, #PM045; blocking, 10% skim milk-TBS, incubate at room temperature for 30 min; primary antibody reaction (dilution, 1:2,000), incubate at room temperature for 90–120 min
Anti-ubiquitin	(1) Mouse monoclonal antibody (FK2 for multi-ubiquitin), Medical & Biological Laboratories, #D058-3; blocking, 10% skim milk-TBS, incubate at room temperature for 30 min; primary antibody reaction (dilution, 1:2,000), incubate at 4°C for 15 h
	(2) Rabbit polyclonal antibody, DAKO, Glostrup Denmark #Z0458; blocking, 10% skim milk-TBS, incubate at room temperature for 30 min; primary antibody reaction (dilution, 1:2,000), incubate at room temperature for 90–120 min

Bold characters are essential points for recognizing target proteins by these antibodies

stored at –20°C. The 10% casein solution was diluted tenfold with TBS (1% casein-TBS).

7. Secondary antibody solution: select appropriate secondary antibody (see Note 7) and conditions; (a) 0.005% (v/v) goat anti-rabbit IgG antibody (minimal cross-reactivity with other species) conjugated to horseradish peroxidase (HRP) and 0.15% (v/v) normal goat serum in TTBS, when using a rabbit polyclonal antibody as the primary antibody. (b) 0.005% (v/v) goat anti-mouse IgG antibody (minimal cross-reactivity with other species) conjugated to HRP and 0.15% (v/v) normal goat serum in TTBS, when using a mouse monoclonal antibody as the primary antibody. (c) 0.005% (v/v) rabbit anti-chicken IgG antibody (minimal cross-reactivity with other species) conjugated to HRP in TTBS, when using a chicken polyclonal antibody as the primary antibody. (d) 0.005% (v/v) donkey anti-goat IgG antibody (minimal cross-reactivity with other species) conjugated to HRP and 0.15% (v/v) normal donkey serum in TTBS, when using a goat polyclonal antibody as the primary antibody.
8. Enhanced chemiluminescent reagents: SuperSignal West Dura Extended Duration Substrate or SuperSignal West Pico Chemiluminescent Substrate (Pierce Biotechnology Inc./Thermo Fisher, Rockford, IL).

2.4. Immunofluorescence Analysis

1. Cell-fixative solution: 4% paraformaldehyde in PBS, pH 7.2, prepared just before use and warmed to 37°C.

2. Digitonin solution: 50 μg/mL digitonin (Wako) in PBS, pH 7.2.
3. Quenching solution: 50 mM NH_4Cl in PBS, pH 7.2.
4. Image-IT FX (Invitrogen/Molecular Probes) for blocking (see Note 8).
5. IF solution: 1% (w/v) BSA, 10 mM Tris–HCl, pH 7.5, 150 mM NaCl. Store at –20°C.
6. IF-primary antibody solution: 5 μg/mL of affinity-purified antibodies in IF solution.
7. IF-secondary antibody solution: 0.05% (v/v) anti-rabbit (or mouse) IgG conjugated to Alexa488, 594 or Q-dot 605 (Invitrogen) in IF solution.
8. Mounting Medium: SlowFade light or Prolong GOLD (Invitrogen).

2.5. Conventional Electron Microscopy for Cultured Cells

1. 10% paraformaldehyde (PFA) stock solution: paraformaldehyde powder (MERCK) was dissolved in water heated to approximately 70°C with stirring for a few minutes to make a homogeneous white suspension. Then, 1 N NaOH was added in a dropwise manner until the solution became clear. The solution was cooled, filtrated, and stored at 4°C.
2. 25% or 70% glutaraldehyde (GA) solution (TAAB).
3. PB: 0.2 M phosphate buffer, pH 7.4.
4. Fixative: 2% PFA-2% GA, 0.1 MPB, pH 7.4.
5. Wash buffer: 7.5% sucrose, 0.1 MPB, pH 7.4.
6. Osmium tetroxide (OsO_4): this reagent is highly poisonous even at low exposure levels, and must therefore be handled in a fume hood with appropriate precautions. OsO_4-fixative: 2% OsO_4, 7.5% sucrose in PB.
7. Propylene oxide: highly toxic flammable reagent.
8. Epon 812 resin kit (TAAB, Berkshire, UK,): to make ~100 g of epoxy resin (Epon 812 mixture), 55 g of Epon 812, 30 g of MNA, and 15 g of DDSA were mixed, stirred for 5 min, 1.5 g of DMP-30 was added, and stirred again for more than 10 min. After degassing, it could be stored at –30°C for 2 months.
9. Uranyl acetate (Wako): working solution was prepared by diluting saturated uranyl acetate solution (approximately 3% solution) with acetone at 1:2–5.
10. Lead citrate (EM grade: TAAB): 25–30 mg of lead citrate was dissolved in 10 mL of degassed water, to which 2 drops of 10 N NaOH are added. This solution could be stored for up to 1 month.
11. Electron microscope (for example JEM1200EX; JEOL, Tokyo, Japan).

3. Methods

To monitor autophagy, separation, and recognition of endogenous LC3-I (cytosolic form) and LC3-II (membrane-bound form) are critical in immunoblotting. Anti-LC3 antibodies, which are useful to recognize endogenous LC3-I and LC3-II, are now commercially available (Table 2). The listed anti-LC3 antibodies are also available for immunofluorescence analysis. There are some key points for sample preparation for SDS-polyacrylamide gel electrophoresis (SDS-PAGE) and immunofluorescence analysis. LC3-II (LC3-phosphatidylethanolamine conjugate) only dissolves partially in 1% Triton X-100, which is usually used for cell lysis due to its hydrophobicity. Therefore, when cells and tissues are lysed for SDS-PAGE, 2% Triton X-100 or 1% SDS should be used as the detergent. In contrast, structures of autophagosomes and autolysosomes are sensitive to MeOH- and Triton X-100 treatments. Therefore, for investigation of LC3 on autophagosomes and autolysosomes by immunofluorescence microscopy, digitonin or saponin is suitable for cell permeabilization.

There are two types of autophagy in mammals: inducible autophagy and constitutive autophagy. Inducible autophagy occurs under starvation conditions, and is observed mainly in the liver and muscles. This type of autophagy can be monitored by determining the levels of LC3-II and dephosphorylation of mTor-signaling pathway. On investigation of autophagic responses in tissues, the conditions for feeding and fasting should be considered. In general, after feeding, autophagy in the liver is suppressed to store nutrients in the liver. Autophagy in the liver is re-induced again several hours after feeding. When fasting continues for over 6 h, autophagy in the skeletal muscles occurs. Thereafter, autophagy in the heart occurs when fasting lasts longer.

Constitutive autophagy is the basal level of autophagy and functions mainly in maintenance of cells, especially neurons, and tissue-dependent physiological functions. Neuronal cell death by ablation of autophagy in the brain is caused by constitutive autophagy. In contrast to inducible autophagy, there is little increase of LC3-II during constitutive autophagy. However, as autophagic insufficiency even in these tissues leads to an increase of p62 and ubiquitin-positive aggregates, it is better to monitor p62 and ubiquitin by Western blotting, immunohistochemistry, and immuno-electron microscopy for estimation of autophagic insufficiency.

3.1. Cell Culture and Preparation of Cell Lysate

1. Pre-warmed KRB or EBSS buffer was prepared to investigate the autophagic responses of cells and cultured tissues under starvation conditions (see Note 3 when using HBSS instead). When lysosomal degradation of LC3-II is suspected, 1:1,000 dilutions of the stock solutions of E64d and pepstatin A

were added to the medium to yield final concentrations of 10 μg/mL. For the controls, the solvent DMSO was added to the medium. The cells were incubated at 37°C for 3–6 h to accumulate LC3-II; in general, LC3-II accumulates in cultured cells 3 h after addition of the inhibitors. As a control for nutrient-rich conditions, the cells were incubated in appropriate culture medium.

2. The cells were washed twice in PBS, harvested with Teflon cell scrapers, resuspended in 200 μL of ice-cold PBS with Complete™ protease inhibitor cocktail, and lysed with an ultrasonic generator for 10 s (0.5 s pulse with 50% duty) on ice.

– Option 2b for preparation of cell lysate without sonication: SDS-lysis solution was added to washed cells and incubated with mild shaking at room temperature for 1 min, followed by boiling for 10 min.

– Option 2c for preparation of cell lysate without sonication: TX100-lysis solution was added to washed cells and incubated with mild shaking at 4°C for 1 h.

To prepare tissue lysate, the tissue was homogenized in sucrose solution containing protease inhibitor cocktail with a Potter-Elvehjem homogenizer. The lysate was centrifuged at 700 × *g* for 5 min to remove debris.

3. The protein concentration of the lysate was measured using a BCA protein assay kit (see Note 9).
4. An appropriate volume of modified Laemmli buffer was added to the samples, followed by boiling for 5 min. (To recognize endogenous LC3, aliquots of 10–20 μg per lane were loaded onto SDS-polyacrylamide gels).
5. The samples were chilled on ice and centrifuged for 1 min at 10,000 × *g*.

3.2. SDS-Polyacrylamide Gel Electrophoresis

1. Glass plates were prepared for SDS-PAGE (for mini gels, spacer thickness was 1 mm). The inner plates were cleaned with 70% ethanol. (See Note 5 for use of a ready-made gel for Western blotting).
2. Gel running solution (12.5% gel: 3.2 mL of distilled water, 4.2 mL of 30% acrylamide solution, 2.5 mL of 1.5 M Tris–HCl, pH 8.8 (at 25°C), 0.1 mL of 10% SDS, 33 μL of 10% APS, and 5 μL of TEMED) was prepared and poured into the space between the glass plates.
3. The solution was gently overlaid with 150 μL of water-saturated isobutanol, and the gel was allowed to polymerize for 30–60 min at room temperature.
4. Stacking gel solution (2.89 mL of distilled water, 0.79 mL of 30% acrylamide solution, 1.25 mL of 1.5 M Tris–HCl,

pH 6.8 (at 25°C), 0.05 mL of 10% SDS, 17 μL of 10% APS, and 5 μL of TEMED) was prepared.

5. The isobutanol was removed from the polymerized running gel by washing the top with deionized water. Residual water was removed gently with a paper towel. The stacking gel solution was poured into the space between the glass plates, and a comb was inserted.
6. After polymerization of the stacking gel, the comb was removed, and the wells were washed.
7. The gel was mounted into the SDS-PAGE electrophoresis chamber, and running buffer was added.
8. Samples (10–20 μg per lane) were loaded onto the gel. An aliquot of 5 μL of prestained molecular weight marker was used as a molecular weight standard.
9. The gel was electrophoresed at a constant current of 1–1.5 mA per cm of gel width, until the pink color of the phenol red in the markers reached the bottom of the gel (see Notes 5 and 6).

3.3. Western Blotting

1. A PVDF membrane was soaked in methanol for 1 min, transferred to Bjerrum and Schafer-Nielsen's transfer buffer, and incubated for 5 min.
2. Four sheets of chromatography paper were soaked in the transfer buffer.
3. After removing the glass plates and stacking gel, the running gel was rinsed in the transfer buffer. Two sheets of chromatography paper were placed on the anode plate and overlaid with a PVDF membrane, the running gel, and two additional sheets of chromatography paper.
4. The stack was set onto a Transblot SD Semi-Dry Transfer Cell, and electrophoresed at 16–20 V for 1 h.
5. The gel and papers were discarded.
6. The PVDF membrane was incubated in blocking buffer at room temperature for 30 min (see Table 2 for blocking solution suitable for each antibody).
7. The membrane was washed three times in TTBS at room temperature for 2 min each time.
8. The membrane was incubated in primary antibody solution (see Table 2 for appropriate dilution, incubation time, and temperature).
9. The membrane was washed at least five times, for 5 min each time, in TTBS at room temperature.
10. Secondary antibody reaction: the PVDF membrane was incubated in secondary antibody solution for 60–90 min at room temperature (see Note 7).

11. The membrane was washed at least five times, for 10 min each time, in TTBS at room temperature.
12. The reaction of horseradish peroxidase with chemiluminescent reagent was started on the membrane, and the signal was detected on X-ray film or using a CCD camera (Fig. 3).

3.4. Immunofluorescence Analysis for Autophagosomes and Autolysosomes

1. After washing in TBS, cells were incubated in cell-fixative solution at 37°C for 5 min.
2. The cells were rinsed twice in TBS.
3. Cells were permeabilized in digitonin solution at 37°C for 5 min. The solution was then discarded by aspiration.
4. Excess digitonin was quenched by incubation in quenching solution at 37°C for 5 min. The solution was then discarded by aspiration.
5. The cells were rinsed twice in TBS.
6. Cells were incubated in a drop of Image-IT FX at 37°C for 30 min.
7. The cells were rinsed three times in TBS.
8. Cells were incubated in IF-primary antibody solution at 37°C for 60 min.
9. The cells were washed three times in TBS for 5 min each time at 37°C.
10. Cells were incubated in IF-secondary antibody solution at 37°C for 60 min.
11. The cells were washed five times in TBS for 5 min each time at 37°C.
12. The cells were mounted on glass slides in mounting medium.
13. Fluorescent images were obtained using a fluorescence microscope (Fig. 4).

3.5. Conventional Electron Microscopy for Cultured Cells

1. Cells were grown on round coverslips ~12 mm in diameter (see Note 10 for tissue preparations).
2. Immediately after removing the medium, 2% PFA-2% GA buffered with 0.1 M PB, pH 7.4 was added (see Note 11).
3. Cells were washed three times with 7.5% sucrose in PB.
4. Preparations were postfixed with OsO_4-fixative at 4°C for 30 min in dark chamber (e.g. refrigerator).
5. Preparations were then washed three times with pure water.
6. Dehydration was performed by successive immersion in 50%, 70%, 80%, and 95% ethanol for 5–10 min.
7. Coverslips were transferred into small glass dishes containing 100% ethanol, immersed for 10 min, and the ethanol was

exchanged with fresh 100% ethanol once and immersed for 10 min (see Note 12).

8. Immersion in propylene oxide twice for 5 min.
9. Immersion in a mixture of propylene oxide and Epon 812 (equal proportions) for 20 min.
10. Immersion in Epon 812 mixture overnight under weak negative pressure generated with a vacuum pump.
11. Embedded in fresh Epon 812 mixture.
12. Heated at 80°C for 2 days.
13. Coverslips were carefully removed while the Epon block was heated appropriately.
14. The side with cultured cells was confirmed, and ultrathin sections (approximately 70 nm thickness) were cut using a microtome (see Note 13).
15. Staining was performed with diluted uranyl acetate solution for 5 min.
16. Washing with water.
17. Staining with 0.25–0.3% lead citrate for 5 min.
18. Observation under an electron microscope.
19. If necessary, the number and size of autophagosomes (see Note 14) were determined using an imaging software.

4. Notes

1. E64d is a derivative of E64 and E64c. E64d is a membrane-permeable analog of E64c, while E64 and E64c are membrane-impermeable. Therefore, E64 and E64c have little inhibitory effect on cathepsins compared with E64d when added to culture medium. Leupeptin (Peptide Institute) is intraperitoneally injected into rats (or mice) when autolysosomes are prepared from the liver, as leupeptin effectively accumulates in the liver probably via endocytosis. As leupeptin is membrane-impermeable, care should be taken to determine whether leupeptin inhibits the activity of cathepsin B when used to treat cells and other tissues. Leupeptin is dissolved at 1 mg/mL in H_2O. The 1,000× concentrated stock solution is stored at −20°C, at which temperature it is stable for at least one month.
2. Storage of pepstatin A for over 3 months should be avoided. When pepstatin A is added to the medium, it should be mixed quickly, as it tends to precipitate.

3. If using Hanks' balanced salt solution (HBSS with calcium and magnesium) instead of EBSS or KRB as starvation medium, HBSS should be buffered by adding 10 mM HEPES-NaOH. Calcium and magnesium in HBSS are essential. When cells are incubated in HBSS, it may be better to incubate them at 37°C without control of CO_2.
4. PBS is an unsuitable solution for autophagy induction. PBS is too severe for cells and cultured tissues to incubate for long times, as it contains no magnesium, calcium, glucose, or amino acids.
5. One frequently asked question involves the separation and recognition of LC3-I and LC3-II. The ratio of acrylamide: methylene-bisacrylamide (29:1 (w/w)) is good for separation and detection of LC3-I and LC3-II on SDS-PAGE. While 30–40% acrylamide solution (30:1 (w/w) acrylamide:methylene-bisacrylamide) is available commercially, the separation of LC3-I and LC3-II is poor. A final concentration of acrylamide of 12.5% is good. Higher concentrations (>15%) are not suitable for separation of LC3-I and LC3-II. When using a ready-made gel for Western blotting, a NuPAGE system (4–12% or 12% Bis–Tris gel) using MES buffer (Invitrogen) works well.
6. For good separation between LC3-I and LC3-II, it is better to run phenol red (New England Biolabs's prestained marker) through a gel and to stop running just before bromophenol blue reaches the bottom of the gel.
7. All secondary antibodies conjugated to HRP were purchased from Jackson ImmunoResearch (West Grove, PA). If the chemiluminescence signal from HRP using other company's secondary antibody is insufficient to recognize a specific signal for the antibody, it is better to use the antibody from Jackson ImmunoResearch.
8. Image-IT FX markedly reduces background fluorescence from anti-rabbit, mouse, and chicken IgG conjugated to Alexa Fluor. IF-blocking solution (2% (w/v) BSA, 5% (v/v) normal goat serum, 20 mM Tris–HCl, pH 7.5, 150 mM NaCl) can be used as the blocking solution instead of Image-IT FX, but background signal is increased.
9. 0.25 M sucrose, 2% Triton X-100, and 1% SDS do not perturb the measurement of protein concentration using a BCA protein assay kit.
10. This method can be applied for cultured cells. For tissue preparations, please refer to Waguri and Komatsu (34).
11. Washing medium before fixation is usually unnecessary and may cause some morphological changes at the electron microscopic level. Some researchers adopted a method where cells

are pelleted before or after fixation, and then processed as tissue blocks. Although these procedures often work well, potential changes of cells before fixation or mechanical damage caused by scraping cells during collection should be taken into consideration.

12. Use glass dishes, because propylene oxide used in the next step is corrosive for plasticware. Transfer the coverslip at this step, because propylene oxide is so volatile that the solution may dry up when the transfer procedure is carried out after immersion in 100% ethanol.

13. In this method, sections are cut from the bottom of the cells. Therefore, cells are observed from the same direction as by light microscopy. Although the section number per block is limited depending on the height of cells, other fields on the coverslips can be selected.

14. During the formation of autophagosomes, the isolation membrane elongates or extends in three-dimensional space. Thus, attention should be paid to the two-dimensional features of autophagosomal structures (Fig. 2). To overcome this obstacle, serial sectioning and/or tomography technique may be useful.

    (a) First, check if the contents of autophagosome-like structures are not distinguishable from other parts of the cytosol; these could be autophagosomes or nascent autophagosomes.

    (b) It is difficult to distinguish real (closed) autophagosomes and nascent (open or closing) autophagosomes. A nascent autophagosome (nascent AP in Fig. 2) can be identified only when a connection is seen between the engulfed region and cytoplasm in the photograph.

    (c) Check the emptiness of the space (or cleft) between the outer and inner isolation membranes. The space has been proposed to be an artifact produced during fixation or later processes (35).

    (d) The presence of membranous materials in autolysosomal structures is not conclusive evidence that it is derived from autophagic mechanisms. Membranous structures may also occur in lysosomal structures through multivesicular body formation in the endocytic pathway, microautophagy, or as yet unknown mechanisms.

    (e) Tissue cells are often highly differentiated and show characteristic shapes, and "double membrane structures" may not belong to autophagosomes. For example, myelinated axons show invagination of the plasma membrane accompanied by the cytoplasm of adjacent oligodendrocytes, forming double-walled vesicles (36).

References

1. Komatsu M, Ueno T, Waguri S, Uchiyama Y, Kominami E, Tanaka K (2007) Constitutive autophagy: vital role in clearance of unfavorable proteins in neurons. Cell Death Differ 14:887–894
2. Uchiyama Y, Koike M, Shibata M (2008) Autophagic neuron death in neonatal brain ischemia/hypoxia. Autophagy 4:404–408
3. Sarkar S, Ravikumar B, Floto RA, Rubinsztein DC (2009) Rapamycin and mTOR-independent autophagy inducers ameliorate toxicity of polyglutamine-expanded huntingtin and related proteinopathies. Cell Death Differ 16:46–56
4. Komatsu M, Waguri S, Chiba T, Murata S, Iwata J, Tanida I, Ueno T, Koike M, Uchiyama Y, Kominami E, Tanaka K (2006) Loss of autophagy in the central nervous system causes neurodegeneration in mice. Nature 441:880–884
5. Hara T, Nakamura K, Matsui M, Yamamoto A, Nakahara Y, Suzuki-Migishima R, Yokoyama M, Mishima K, Saito I, Okano H, Mizushima N (2006) Suppression of basal autophagy in neural cells causes neurodegenerative disease in mice. Nature 441:885–889
6. Komatsu M, Wang QJ, Holstein GR, Friedrich VL Jr, Iwata J, Kominami E, Chait BT, Tanaka K, Yue Z (2007) Essential role for autophagy protein Atg7 in the maintenance of axonal homeostasis and the prevention of axonal degeneration. Proc Natl Acad Sci USA 104:14489–14494
7. Koike M, Shibata M, Waguri S, Yoshimura K, Tanida I, Kominami E, Gotow T, Peters C, von Figura K, Mizushima N, Saftig P, Uchiyama Y (2005) Participation of autophagy in storage of lysosomes in neurons from mouse models of neuronal ceroid-lipofuscinoses (batten disease). Am J Pathol 167: 1713–1728
8. Cao Y, Espinola JA, Fossale E, Massey AC, Cuervo AM, MacDonald ME, Cotman SL (2006) Autophagy is disrupted in a knock-in mouse model of juvenile neuronal ceroid lipofuscinosis. J Biol Chem 281:20483–20493
9. Kabeya Y, Mizushima N, Ueno T, Yamamoto A, Kirisako T, Noda T, Kominami E, Ohsumi Y, Yoshimori T (2000) LC3, a mammalian homologue of yeast Apg8p, is localized in autophagosome membranes after processing. EMBO J 19:5720–5728
10. Tanida I, Minematsu-Ikeguchi N, Ueno T, Kominami E (2005) Lysosomal turnover, but not a cellular level, of endogenous LC3 is a marker for autophagy. Autophagy 1:84–91
11. Bjorkoy G, Lamark T, Brech A, Outzen H, Perander M, Overvatn A, Stenmark H, Johansen T (2005) p62/SQSTM1 forms protein aggregates degraded by autophagy and has a protective effect on huntingtin-induced cell death. J Cell Biol 171:603–614
12. Komatsu M, Waguri S, Koike M, Sou YS, Ueno T, Hara T, Mizushima N, Iwata J, Ezaki J, Murata S, Hamazaki J, Nishito Y, Iemura S, Natsume T, Yanagawa T, Uwayama J, Warabi E, Yoshida H, Ishii T, Kobayashi A, Yamamoto M, Yue Z, Uchiyama Y, Kominami E, Tanaka K (2007) Homeostatic levels of p62 control cytoplasmic inclusion body formation in autophagy-deficient mice. Cell 131:1149–1163
13. Hara T, Takamura A, Kishi C, Iemura S, Natsume T, Guan JL, Mizushima N (2008) FIP200, a ULK-interacting protein, is required for autophagosome formation in mammalian cells. J Cell Biol 181:497–510
14. Ganley IG, Lam DH, Wang J, Ding X, Chen S, Jiang X (2009) ULK1-ATG13-FIP200 complex mediates mTOR signaling and is essential for autophagy. J Biol Chem 3:3
15. Jung CH, Jun CB, Ro SH, Kim YM, Otto NM, Cao J, Kundu M, Kim DH (2009) ULK-Atg13-FIP200 complexes mediate mTOR signaling to the autophagy machinery. Mol Biol Cell 20:1992–2003
16. Hosokawa N, Hara T, Kaizuka T, Kishi C, Takamura A, Miura Y, Iemura S, Natsume T, Takehana K, Yamada N, Guan JL, Oshiro N, Mizushima N (2009) Nutrient-dependent mTORC1 association with the ULK1-Atg13-FIP200 complex required for autophagy. Mol Biol Cell 20:1981–1991
17. Thoreen CC, Kang SA, Chang JW, Liu Q, Zhang J, Gao Y, Reichling LJ, Sim T, Sabatini DM, Gray NS (2009) An ATP-competitive mammalian target of rapamycin inhibitor reveals rapamycin-resistant functions of mTORC1. J Biol Chem 284:8023–8032
18. Sarkar S, Rubinsztein DC (2008) Small molecule enhancers of autophagy for neurodegenerative diseases. Mol Biosyst 4:895–901
19. Ueno T, Ishidoh K, Mineki R, Tanida I, Murayama K, Kadowaki M, Kominami E (1999) Autolysosomal membrane-associated betaine homocysteine methyltransferase. Limited degradation fragment of a sequestered cytosolic enzyme monitoring autophagy. J Biol Chem 274:15222–15229

20. Klionsky DJ, Elazar Z, Seglen PO, Rubinsztein DC (2008) Does bafilomycin A1 block the fusion of autophagosomes with lysosomes? Autophagy 4:849–950
21. Gutierrez MG, Munafo DB, Beron W, Colombo MI (2004) Rab7 is required for the normal progression of the autophagic pathway in mammalian cells. J Cell Sci 117:2687–2697
22. Jager S, Bucci C, Tanida I, Ueno T, Kominami E, Saftig P, Eskelinen EL (2004) Role for Rab7 in maturation of late autophagic vacuoles. J Cell Sci 117:4837–4848
23. Mizushima N, Yamamoto A, Matsui M, Yoshimori T, Ohsumi Y (2004) In vivo analysis of autophagy in response to nutrient starvation using transgenic mice expressing a fluorescent autophagosome marker. Mol Biol Cell 15:1101–1111
24. Mizushima N, Kuma A (2008) Autophagosomes in GFP-LC3 transgenic mice. Methods Mol Biol 445:119–124
25. Ciechomska IA, Tolkovsky AM (2007) Non-autophagic GFP-LC3 puncta induced by saponin and other detergents. Autophagy 3:586–590
26. Kuma A, Matsui M, Mizushima N (2007) LC3, an autophagosome marker, can be incorporated into protein aggregates independent of autophagy: caution in the interpretation of LC3 localization. Autophagy 3:323–328
27. Katayama H, Yamamoto A, Mizushima N, Yoshimori T, Miyawaki A (2008) GFP-like proteins stably accumulate in lysosomes. Cell Struct Funct 33:1–12
28. Tanida I, Yamaji T, Ueno T, Ishiura S, Kominami E, Hanada K (2008) Consideration about negative controls for LC3 and expression vectors for four colored fluorescent protein-LC3 negative controls. Autophagy 4:131–134
29. Wang QJ, Ding Y, Kohtz DS, Mizushima N, Cristea IM, Rout MP, Chait BT, Zhong Y, Heintz N, Yue Z (2006) Induction of autophagy in axonal dystrophy and degeneration. J Neurosci 26:8057–8068
30. Plowey ED, Cherra SJ, Liu YJ 3rd, Chu CT (2008) Role of autophagy in G2019S-LRRK2-associated neurite shortening in differentiated SH-SY5Y cells. J Neurochem 105:1048–1056
31. Kimura S, Noda T, Yoshimori T (2007) Dissection of the autophagosome maturation process by a novel reporter protein, tandem fluorescent-tagged LC3. Autophagy 3:452–460
32. Shibata M, Yoshimura K, Furuya N, Koike M, Ueno T, Komatsu M, Arai H, Tanaka K, Kominami E, Uchiyama Y (2009) The MAP1-LC3 conjugation system is involved in lipid droplet formation. Biochem Biophys Res Commun 382:419–423
33. Singh R, Kaushik S, Wang Y, Xiang Y, Novak I, Komatsu M, Tanaka K, Cuervo AM, Czaja MJ (2009) Autophagy regulates lipid metabolism. Nature 458:1131–1135
34. Waguri S, Komatsu M (2009) Biochemical and morphological detection of inclusion bodies in autophagy-deficient mice. Methods Enzymol 453:181–196
35. Kovacs AL, Palfia Z, Rez G, Vellai T, Kovacs J (2007) Sequestration revisited: integrating traditional electron microscopy, de novo assembly and new results. Autophagy 3:655–662
36. Li YC, Li YN, Cheng CX, Sakamoto H, Kawate T, Shimada O, Atsumi S (2005) Subsurface cisterna-lined axonal invaginations and double-walled vesicles at the axonal-myelin sheath interface. Neurosci Res 53:298–303

Chapter 14

Transcription Factor Sequestration by Polyglutamine Proteins

Tomoyuki Yamanaka and Nobuyuki Nukina

Abstract

In polyglutamine diseases including Huntington's disease, the causative gene products containing expanded polyglutamine form nuclear aggregates in neurons. Recent studies have identified several transcriptional factors, which interact with and are sequestered by expanded polyglutamine aggregates in neurons. Further, altered expression of many genes has been shown in several polyglutamine disease models. These observations suggest an involvement of transcriptional dysregulation in pathological process of these diseases. In this chapter, we introduce several methods to examine the interaction of transcriptional factors with and their sequestration by expanded polyglutamine proteins in vitro and in vivo.

Key words: Polyglutamine, Huntington's disease, Huntingtin, Transcriptional factor, Sequestration, Aggregate, Western blot analysis, Filter trap assay, Immunofluorescence microscopy, EMSA

1. Introduction

Polyglutamine diseases including Huntington's disease (HD), spinocerebellar ataxias (SCAs), dentatorubral and pallidoluysian atrophy (DRPLA) and spinobulbar muscular atrophy (SBMA) are autosomal-dominant, adult-onset neurodegenerative disorders. These diseases are caused by CAG repeat expansions in their causative genes. The gene products containing expanded polyglutamine form nuclear inclusions in neurons, leading to neuronal cell dysfunction and finally cell loss.

The causative gene of HD is huntingtin which contains more than 40 CAG repeats in its exon 1 (1). The product, mutant Htt containing expanded polyglutamine repeats forms nuclear aggregates in neurons in specific regions including the striatum and cortex, where severe neurodegenerations are observed (2). Recent studies have shown that mutant Htt interacts

Peter Bross and Niels Gregersen (eds.), *Protein Misfolding and Cellular Stress in Disease and Aging: Concepts and Protocols*, Methods in Molecular Biology, vol. 648, DOI 10.1007/978-1-60761-756-3_14, © Springer Science+Business Media, LLC 2010

with several transcriptional factors, such as CREB-binding protein (CBP), TBP, SP1, TAFII130, and p53 (3, 4). The interaction is suggested to suppress the functions of these factors by sequestration in neurons. Further, DNA microarray analyses have identified many genes whose expressions are altered in HD model mice brain or cells (5–7). Sequestrations of transcriptional factors as well as alterations of gene expressions are also reported in other polyglutamine diseases (8, 9). These observations suggest an involvement of transcriptional dysregulation induced by sequestration of transcriptional factors in the progression of these diseases (8–10).

Recently, we identified NF-Y transcriptional factor as a novel target of mutant Htt (11). NF-Y is composed of three subunits, NF-YA, NF-YB, and NF-YC, which binds to CCAAT in promoter regions to induce transcription. NF-YA interacts with and is sequestered by mutant Htt aggregates in cultured cells and in HD model mouse brain. Further, an electrophoretic mobility shift assay revealed the reduction of DNA binding of NF-Y in HD model mouse brain, suggesting the functional impairment of NF-Y during HD progression.

In this chapter, we introduce the methods used in our paper to examine transcriptional factor interaction with and its sequestration by mutant Htt in vitro and in vivo. For in vitro analysis, transfected cells co-expressing transcriptional factor with mutant Htt are used for the following three assays: Western blot analysis, filter trap assay and immunofluorescence microscopy. For in vivo analysis, brain samples prepared from HD model mice carrying mutant Htt transgene are used in the following two assays: immunofluorescence microscopy and electrophoretic mobility shift assay (EMSA). We also show example data of these assays.

2. Materials

2.1. Materials for Experiments Using Cultured Cells

2.1.1. Materials for Cell Transfection

1. 24 well plastic plate.
2. Dulbecco's modified Eagle's medium (DMEM) supplemented with 10% fetal bovine serum (FBS) and penicillin-streptomycin.
3. Lipofectamine 2000 reagent (Invitrogen).
4. Opti-MEM (Invitrogen).
5. pcDNA3.1 expression vectors (Invitrogen) for N-terminal Htt (Htt exon 1) containing a pathological (150Q) or normal length (18Q) of glutamine fused with EGFP and SV40 NLS (Nhtt150Q-EGFP-NLS or Nhtt18Q-EGFP-NLS, respectively) (*12*).

6. pcDNA3.1 expression vector with transcriptional factor cDNA tagged with V5.

2.1.2. Materials for Western Blot Analysis

1. Phosphate buffered saline (PBS).
2. 2× SDS sample buffer (125 mM Tris–HCl, pH 6.8, 4% sodium dodecyl sulfate (SDS), 20% glycerol, 0.01% bromphenol blue, 10% beta-mercaptoethanol).
3. Microsonicator.
4. Materials for SDS-PAGE and Western blotting (polyacrylamide gel, polyvinylidene difluoride (PVDF) membrane, running buffer, blotting buffer).
5. TBST (20 mM Tris–HCl, pH 8.0, 150 mM NaCl, 0.05% Tween20).
6. Dried skim milk.
7. Goat serum.
8. Mouse monoclonal antibodies against GFP (04363-66; nakalai tesque) and V5 tag (R960-25; Invitrogen).
9. Horseradish peroxidase (HRP) conjugated anti-mouse IgG.
10. Chemiluminescent reagent.
11. X-ray film or video camera system.

2.1.3. Materials for Filter Trap Assay

1. Phosphate buffered saline.
2. 2× SDS sample buffer (125 mM Tris–HCl, pH 6.8, 4% SDS, 20% glycerol, 0.01% bromphenol blue, 10% beta-mercaptoethanol).
3. Cellulose acetate membrane with 0.2 μm pore size.
4. Dot blotting apparatus.
5. Wash buffer (2% SDS, 50 mM Tris–HCl, pH 8.0, 10% Glycerol).
6. TBST (20 mM Tris–HCl, pH 8.0, 150 mM NaCl, 0.05% Tween20).
7. Dried skim milk.
8. Goat serum.
9. Mouse monoclonal antibodies against GFP (04363-66; nakalai tesque) and V5 tag (R960-25; Invitrogen).
10. HRP conjugated anti-mouse IgG.
11. Chemiluminescent reagent.
12. X-ray film or video camera system.

2.1.4. Materials for Immunofluorescence Microscopy

1. 4 well chamber slide.
2. Phosphate buffered saline.
3. 4% paraformaldehyde (PFA) in PBS.

4. TritonX-100.
5. Goat serum.
6. Mouse monoclonal antibody against V5 tag (R960-25; Invitrogen).
7. Anti-mouse IgG conjugated with Alexa 546 (Molecular Probes).
8. 4′-6-Diamidino-2-phenylindole (DAPI).
9. VECTASHIELD Mounting Medium (Vector Laboratories).
10. TBST (20 mM Tris–HCl, pH8.0, 150 mM NaCl, 0.05% Tween20).
11. Fluorescence microscope or confocal laser microscope.

2.2. Materials for Experiments Using HD Model Mouse Brain

2.2.1. Materials for Immunofluorescence Microscopy

1. Aminopropylsilane (APS) coated slide glass.
2. Phosphate buffered saline.
3. 4% PFA in PBS.
4. Bovine serum albumine (BSA).
5. Freezing microtome.
6. Mouse monoclonal anti-ubiquitin antibody (MAB1510; Chemicon International) or rabbit polyclonal anti-ubiquitin antibody (Z0458; DAKO).
7. Antibody against transcriptional factor.
8. Secondary antibodies conjugated with Alexa 488 or Alexa 546 (Molecular Probes).
9. 4′-6-Diamidino-2-phenylindole (DAPI).
10. VECTASHIELD Mounting Medium (Vector Laboratories).
11. TBST (20 mM Tris–HCl, pH8.0, 150 mM NaCl, 0.05% Tween20).

2.2.2. Materials for Electrophoretic Mobility Shift Assay

1. Sense and antisense oligonucleotides containing transcriptional factor binding site.
2. $\gamma[^{32}P]$-ATP (6,000 Ci/nmol).
3. T4 DNA kinase and kinase buffer.
4. Sepharose G-50 spin column (Amersham).
5. Lysis buffer (20 mM Hepes, pH7.9, 25% glycerol, 150 mM NaCl, 1.5 mM $MgCl_2$, and Complete protease inhibitor (Roche)).
6. Glass homogenizer.
7. Buffer C (20 mM Hepes, pH7.9, 25% glycerol, 420 mM NaCl, 1.5 mM $MgCl_2$, and Complete protease inhibitor).
8. 2× binding buffer (20 mM Tris–HCl, pH7.5, 100 mM NaCl, 2 mM EDTA, 10% glycerol).

9. Poly(dI-dC).
10. Bovine serum albumin.
11. Polyacrylamide gel apparatus.
12. 4% native polyacrylamide gel (2.7 ml of 30% acrylamide (29:1), 1 ml of 5× TBE, 0.2 ml of 10% ammonium sulfate (APS), 20 μl of N,N,N′,N′-Tetramethylethylenediamine (TEMED), 16.1 ml of H_2O for 20 ml gel).
13. 5× TBE (53.89 g of Tris, 27.5 g of Boric acid, 3.72 g of EDTA·2Na ($2H_2O$) and H_2O up to 1 L).
14. 0.25× TBE running buffer.
15. Loading dye (0.4% bromophenol blue, 0.2% xylene cyanol, 50% glycerol).
16. X-ray film or Imaging Plate (Fijifilm).

3. Methods

3.1. Analysis of Interaction of Transcriptional Factors with Polyglutamine Aggregates in Cultured Cells

The easiest way to analyze the interaction between polyglutamine aggregates and transcriptional factors is the use of transfected cultured cells. Here, we introduce three methods to examine the interaction of transcriptional factors with mutant Htt aggregates in cultured cells. For mutant Htt expression construct, pcDNA3.1 expression vector encoding Nhtt150Q-EGFP-NLS is used in the following methods (12). pcDNA3.1 expression vector encoding Nhtt18Q-EGFP-NLS is used as a control. By subcloning the cDNA for the transcriptional factor of your interest into pcDNA3.1-V5-His, you can detect the transcriptional factor by anti-V5 antibody. Alternatively, other tags such as Myc, HA are also available. In the following experiments, V5 tagged transcriptional factor is used as an example.

3.2. Neuro2a Cell Transfection

We often use Neuro2a cells, a mouse neuroblastoma cell line, for transfection experiments because they are easy to handle, grow very fast and are very efficiently transfected.

1. Seed 8×10^4 of Neuro2a cells on 24 well plate with 0.5 ml of 10% FBS/DMEM and incubate the cells overnight at 37°C in a CO_2 incubator (see Note 1).
2. Mix 0.4 μg of pcDNA3.1-Nhtt150Q-EGFP-NLS or Nhtt18Q-EGFP-NLS and 0.4 μg of plasmid encoding transcriptional factor with 50 μl of Opti-MEM in a tube (see Note 2). In another tube, mix 2 μl of lipofectamine 2000 reagent with 50 μl of Opti-MEM.
3. Incubate these mixtures at room temperature for 5 min.

4. Mix these two and incubate at room temperature for 20 min.
5. Add the mixed solution (100 μl) to cells.
6. Incubate the cells for 4–6 h at 37°C in a CO_2 incubator.
7. Change the medium to 0.5 ml of DMEM supplemented with 10% FBS and penicillin-streptomycin.
8. Incubate the cells for 20–24 h at 37°C in a CO_2 incubator (see Note 3).

3.3. Western Blot Analysis

Nhtt150Q-EGFP-NLS but not Nhtt18Q-EGFP-NLS forms SDS-insoluble aggregates in the transfected Neuro2a cells (*11, 12*). Because such insoluble materials do not migrate into the gel during SDS-PAGE, they can be detected in the gel top by Western blot analysis using antibody against the GFP-portion of the fusion protein. If the transcriptional factor is also incorporated into the aggregates and becomes SDS-insoluble, it can be detected in the gel top similar to mutant Htt aggregates.

1. Wash the transfected cells with PBS and harvest the cells with 150 μl of 2× SDS sample buffer.
2. Heat the cell lysates at 100°C for 5 min.
3. Sonicate the lysates with a microsonicator for about 10 s.
4. Resolve the proteins through SDS-PAGE and transfer them onto PVDF membrane using standard electrophoresis and blotting procedures.
5. Incubate the membrane in blocking buffer (5% skim milk/TBST) for 30 min at room temperature.
6. After brief washing with TBST, incubate the membrane with primary antibody against GFP or V5 diluted in 3% goat serum/TBST overnight at 4°C (see Note 4).
7. Wash the membrane with TBST three times for 10 min each.
8. Incubate the membrane with HRP conjugated secondary antibody diluted in 3% goat serum/TBST for 1 h at room temperature.
9. Wash the membrane with TBST three times for 10 min each.
10. Detect the protein by chemiluminescent reagent as recommended by the manufacturer (see Fig. 1).

3.4. Filter Trap Assay

Filter trap assay is another way to identify the SDS-insoluble aggregates (13, 14). In this assay, cellulose acetate membrane with 0.2 μm pore size is used. If the cell lysates expressing Nhtt150Q-EGFP-NLS are loaded, only aggregated proteins are trapped in the membrane and can be detected with anti-GFP. Incorporated transcriptional factor can be detected by anti-V5. This assay is often

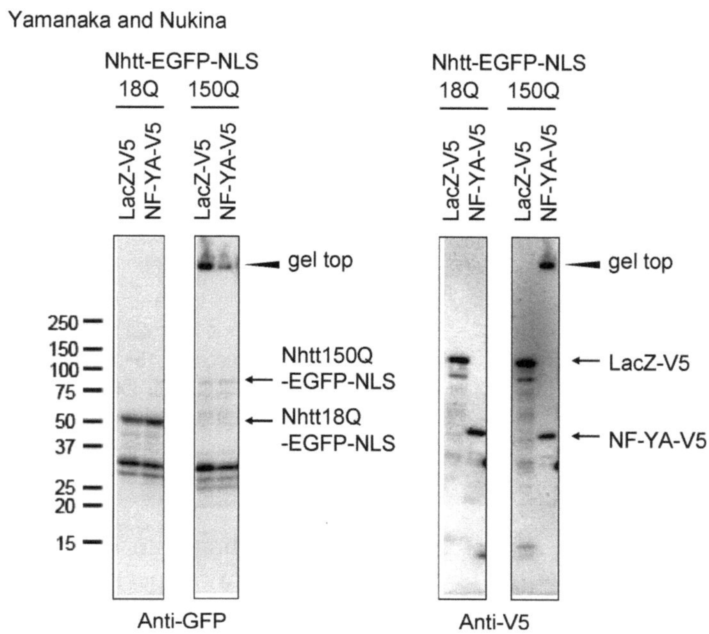

Fig. 1. Example data of Western blot analysis showing interaction of transcriptional factor with SDS-insoluble aggregates in Neuro2a cells. Neuro2a cells were transfected with expression vector for NF-YA or LacZ (control) tagged with V5 together with expression vector for Nhtt18Q-EGFP-NLS or Nhtt150Q-EGFP-NLS. After 24 h, cells were lyzed and subjected to SDS-PAGE and Western blot analysis using anti-GFP or anti-V5 antibody. Bands for expressed proteins are indicated by *arrows* and positions of the gel top are indicated by *arrowheads*. Note that gel top bands were detected by anti-GFP and anti-V5 staining in the cells expressing Nhtt150Q-EGFP-NLS but not ones expressing Nhtt18Q-EGFP-NLS.

more sensitive and quantitative than Western blot analysis described above to analyze the aggregate proteins in the cell lysates.

1. Wash the transfected cells with PBS and harvest the cells with 150 μl of 2× SDS sample buffer.
2. Heat the cell lysates at 100°C for 5 min.
3. Dilute 10 μl of cell lysates with 40 μl of 2× SDS sample buffer.
4. Pre-wet a cellulose acetate membrane and two filter papers with wash buffer.
5. Put the membrane on two filter papers and set them into the dot blotting apparatus.
6. Apply 50 μl of diluted lysates to each well and start suction (see Note 5).
7. Apply 300 μl of wash buffer twice to wash out soluble materials from the membrane.
8. Keep the suction for 20 min (see Note 6).

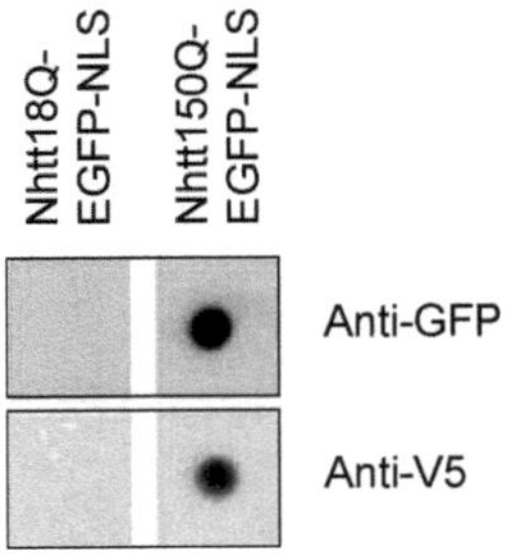

Fig. 2. Example data of filter trap assay showing interaction of transcriptional factor with SDS-insoluble aggregates in Neuro2a cells. Neuro2a cells were transfected with expression vector for NF-YA tagged with V5 together with expression vector for Nhtt18Q-EGFP-NLS or Nhtt150Q-EGFP-NLS. After 24 h, cells were lyzed and subjected to filter trap assay and stained with anti-GFP or anti-V5 antibody. Note that spots were detected by anti-GFP and anti-V5 staining in the cells expressing Nhtt150Q-EGFP-NLS but not ones expressing Nhtt18Q-EGFP-NLS.

9. Transfer membrane into blocking buffer and incubate it for 30 min at room temperature.
10. After brief washing with TBST, incubate the membrane with primary antibody against GFP or V5 diluted in 3% goat serum/TBST overnight at 4°C (see Note 4).
11. Wash the membrane with TBST three times for 10 min each.
12. Incubate the membrane with HRP conjugated secondary antibody diluted in 3% goat serum/TBST for 1 h at room temperature.
13. Wash the membrane with TBST three times for 10 min each.
14. Detect the protein using chemiluminescent reagent as recommended by the manufacturer (see Fig. 2).

3.5. Immunofluorescence Microscopy

The last method to analyze the interaction of transcriptional factor with mutant Htt aggregates in transfected cells is immunofluorescence microscopy. If the transcriptional factor is incorporated into mutant Htt aggregates, you can observe the co-localization of transcriptional factor with mutant Htt-EGFP inclusions. This method may be also useful for analysis of a weak and/or SDS-sensitive interaction between transcriptional factor and mutant Htt aggregates which may be hardly detected by Western blot analysis or filter trap assay described above. If the transcriptional factor is efficiently sequestered by mutant Htt aggregates, reduction of diffuse staining will be observed (see Fig. 3). In this experiment, you can use 4 well chamber slide instead of 24 well plastic plate for analysis with fluorescence microscope or confocal laser microscope.

1. Wash the transfected cells grown on a four well chamber slide with PBS, and fix the cells with 0.5 ml of 4% PFA/PBS for 10 min at room temperature (see Note 7).

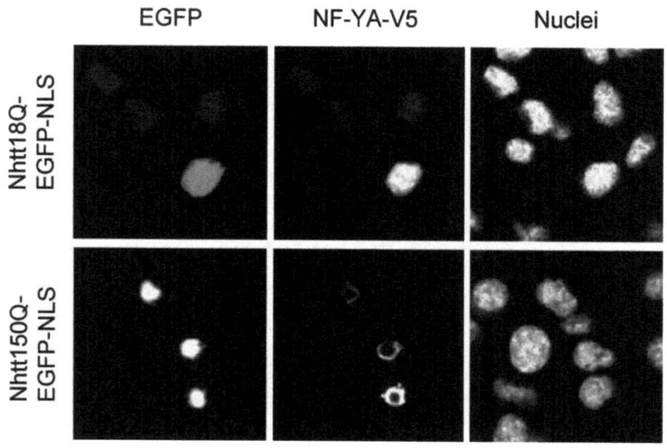

Fig. 3. Example data of immunofluorescence microscopy showing co-localization of transcriptional factor with mutant Htt inclusion in Neuro2a cells. Neuro2a cells were transfected with expression vector for NF-YA or LacZ (control) tagged with V5 together with expression vector for Nhtt18Q-EGFP-NLS or Nhtt150Q-EGFP-NLS. After 24 h, cells were fixed and stained with monoclonal anti-V5 antibody. Nhtt proteins were detected as GFP fluorescence. Nuclei were detected with DAPI. Note that Nhtt150Q-EGFP-NLS but not Nhtt18Q-EGFP-NLS, formed nuclear inclusions and NF-YA-V5 preferentially localized to Nhtt150Q-EGFP-NLS inclusions in Neuro2a cells.

2. Wash the cells with PBS, and permeabilize the cells with 0.5 ml of 0.1% TritonX-100/PBS for 10 min at room temperature.
3. Wash the cells with PBS, and incubate the cells with 0.5 ml of 10% goat serum/TBST (see Note 8).
4. After washing with TBST, incubate the cells with anti-V5 antibody diluted with 0.1% BSA/TBST for 1 h at 37°C.
5. Wash the cells with TBST three times for 10 min each.
6. Incubate the cells with anti-mouse IgG conjugated with Alexa 546 diluted with 0.1% BSA/TBST for 1 h at 37°C.
7. Wash the cells with TBST three times for 10 min each.
8. Incubate the cells with 10 μM DAPI/TBST for 10 min at room temperature.
9. After brief washing with TBST, cover the cells by a coverslip with VECTASHIELD Mounting Medium.
10. Analyze the cells with fluorescence microscope or confocal laser microscope (see Fig. 3).

3.6. Analysis of Sequestration of Transcriptional Factors by Polyglutamine Aggregates in HD Model Mouse Brain

Even though you could find the interaction of transcriptional factor with polyglutamine aggregates in transfected cultured cells, you have to confirm the interaction in vivo to show the pathological significance of this interaction. Currently, mouse models for HD and some other polyglutamine diseases have been established and used as in vivo models. For the HD model mouse, R6/2 mouse, a heterozygous transgenic mouse carrying htt exon 1 with 150 CAG repeats, is widely used and can be obtained from

Jackson Laboratory. We introduce two methods to analyze the sequestration of transcriptional factors by mutant Htt aggregates in R6/2 mouse brain.

3.7. Immunofluorescence Microscopy

Immunofluorescence microscopy represents one way to analyze the interaction in vivo. For this assay, you need an antibody highly specific for the transcriptional factor of your interest. It is recommended to use several antibodies which recognize different epitopes of the transcriptional factor to avoid false positives. If your antibody's titer is very high, you can also try to use brain lysates for Western blot analysis and filter trap assay for confirmation of the interaction. In the following experiment, sections prepared from frozen brain are used because the antigen is preserved in these sections compared with those from paraffin-embedded brain. Because mutant Htt inclusions contain ubiquitinated proteins which can be detected with antibody against ubiquitin, anti-ubiquitin antibody is used in the following method (see Note 9).

1. Isolate the mouse brain from wild-type or R6/2 mouse brain and freeze them in a mounting reagent with dry ice or liquid nitrogen.
2. Cut the frozen block into 10 μm sections with a freezing microtome, put them on APS coated slide glasses and dry with a dryer or electric fan for more than 30 min at room temperature (see Note 10).
3. Fix the sections with 4% PFA/PBS for 10 min at room temperature.
4. Wash the sections with TBST three times for 5 min each at room temperature.
5. Permeabilize the cells with 0.1% TritonX-100/TBST for 10 min at room temperature.
6. Wash the sections with TBST three times for 5 min each at room temperature.
7. Incubate the sections with blocking solution (5% skim/TBST) for 60 min at room temperature.
8. Wash the sections with TBST twice, and incubate them with 0.1% BSA/TBST containing antibodies against ubiquitin and transcriptional factor for overnight at 4°C.
9. Wash the sections with TBST three times for 5 min each.
10. Incubate the sections with 0.1% BSA/TBST containing secondary antibodies conjugated with Alexa 488 or Alexa 546 for 1 h at room temperature.
11. Wash the sections with TBST three times for 5 min each.
12. Incubate the cells with 10 μM DAPI in TBST for 10 min at room temperature.

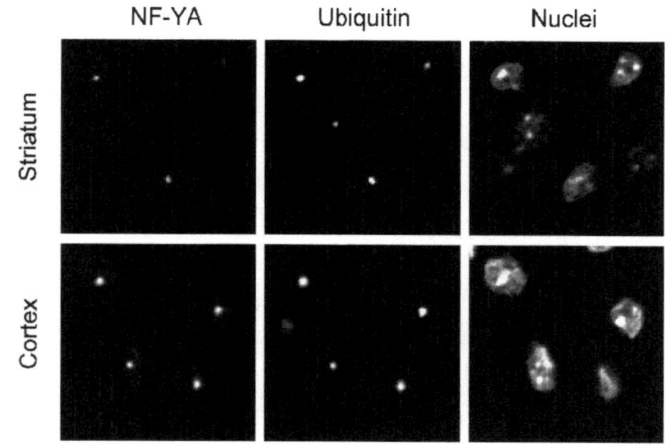

Fig. 4. Example data of immunofluorescence microscopy showing co-localization of transcriptional factor with mutant Htt inclusion in R6/2 mouse brain. 10 μm coronal sections prepared from frozen brain of a 12-week-old R6/2 mouse were fixed and stained with anti-NF-YA (sc-10779; SantaCruz) rabbit antibody together with anti-ubiquitin mouse antibody. Nuclei were detected with DAPI. Nuclear inclusions positive for ubiquitin were stained by NF-YA antibody in cortex and striatum.

13. Wash the sections with TBST for 10 min.
14. Cover the sections by coverslip with VECTASHIELD Mounting Medium.
15. Analyze the sections with confocal microscope system or fluorescence microscope (see Fig. 4).

3.8. Electrophoretic Mobility Shift Assay

EMSA is one way to examine the DNA binding ability of transcriptional factors. By EMSA, you can examine whether the sequestration of transcriptional factor leads to reduction of its function in vivo. In the following experiment, isolated striatum and cortex from R6/2 mouse brain are used because these regions are severely affected in HD mouse models and HD patients. Before doing this experiment, you have to check the expression level of your transcriptional factor in HD model mouse brain by RT-PCR using standard methods. This is very important because it is possible that reduction of DNA binding of the transcriptional factor is just caused by its reduced expression but not by sequestration. The following methods are modifications of the ones established in Dr. Morimoto's laboratory in Northwestern University (Chicago, USA) (15). The original methods can be found on the lab's homepage (http://www.biochem.northwestern.edu/morimoto/research/protocols.html).

3.9. Probe Preparation

1. Mix 1 μl of sense oligonucleotide (10 pmol/μl) with 1 μl of $\gamma[^{32}P]$-ATP (6,000 Ci/nmol), 1 μl T4 DNA kinase, 5 μl of 10× kinase buffer and 42 μl of H_2O.
2. Incubate at 37°C for 0.5–2 h.

3. Purify the ^{32}P-labeled oligonucleotide with Sepharose G-50 spin column (Amersham).
4. Mix 20 μl of purified ^{32}P-labeled sense oligonucleotide with 4 μl of unlabeled antisense oligonucleotide (10 pmol/μl) in a total 40 μl of buffer containing 10 mM Tris–HCl , pH 7.5, 10 mM $MgCl_2$, and 50 mM NaCl.
5. Heat the mixture at 95°C for 5 min and turn off the heat block to allow the mixture to cool slowly. The probe can be stored at –20°C.
6. To monitor the amount of probe forming double strands, run ^{32}P -labeled oligonucleotide annealed to unlabeled antisense oligonucleotide and its single-stranded oligonucleotide on a 15% polyacrylamide gel in 1× TBE and check whether the band for annealed oligonucleotide is shifted up completely.

3.10. Tissue Isolation, Lysis and Western Blot Analysis

1. Isolate tissues (cortex and striatum) from wild-type and R6/2 mouse brains, and freeze them in liquid nitrogen, weigh them and store at –80°C.
2. Add 10 v/w (e.g. 1 ml for 100 mg tissue) of cold lysis buffer to a glass homogenizer.
3. Transfer frozen tissue to the glass homogenizer containing lysis buffer and homogenize it for ten strokes on ice.
4. Add NaCl to a final concentration of 420 mM and incubate the lysate on ice for 20 min (see Note 11).
5. Transfer the lysate to a 1.5 ml tube and centrifuge it at 800 × *g* for 5 min at 4°C.
6. Transfer supernatant to a new tube and centrifuge it at 20,000 × *g* for 30 min at 4°C to clarify the lysates further.
7. Measure protein concentration.
8. Add Buffer C to the lysate to make the protein concentration 20 μg/5 μl, dispense it to tubes and store at –80°C (see Note 12).
9. Check protein expression of the transcriptional factor in the lysates by Western blot analysis (see Note 13).

3.11. EMSA

1. Prepare reaction mix (20 μl/sample): 12.5 μl of 2× binding buffer, 1.0 μl of BSA (10 mg/ml), 0.1 μl of poly(dI-dC) (5 mg/ml), 0.1 μl of probe and 6.3 μl of H_2O.
2. Add 5 μl of tissue lysates (20 μg protein) to 1.5 ml tube and place it on ice (see Note 14).
3. Pre-run the 4% polyacrylamide gel at room temperature in 0.25× TBE at 120–130 V for ~30 min until step 6.

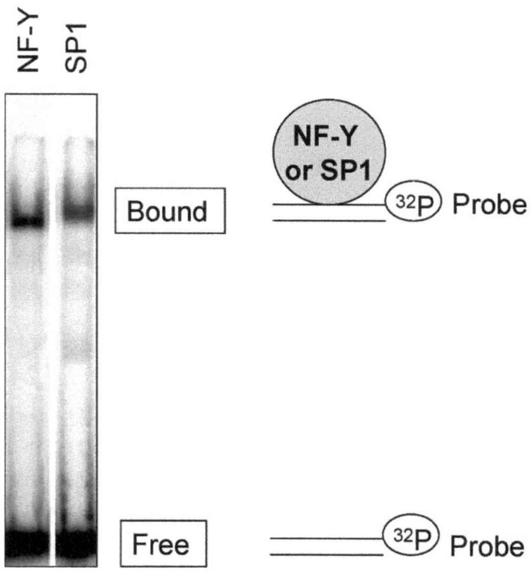

Fig. 5. Example data of EMSA showing binding of NF-Y and SP1 transcriptional factors in brain lysates to ^{32}P-labeld probes. Cortical lysates prepared from 12 week-old wild-type mouse were subjected to EMSA using probes containing NF-Y or SP1 binding site. In addition to bands for free labeled probes (Free), bands for NF-Y or SP1 bound probes (Bound) were observed in *upper region*, indicating that NF-Y and SP1 has DNA-binding activity in the brain cortex. By using R6/2 mouse brain lysates, you could examine whether the DNA binding activity of the transcriptional factor is affected by mutant Htt in vivo.

4. Add 20 μl of reaction mix to the tube containing tissue lysates (the total volume is 25 μl), and incubate the mixture at room temperature for 20 min.
5. Add 2.5 μl of loading dye and place the reactant on ice.
6. Stop the pre-run and load sample onto the well.
7. Run the gel at room temperature for ~2.5 h at 120–130 V in 0.25× TBE until the dye front reaches 1–2 cm from the gel bottom. Free probe will co-migrate with the blue dye.
8. Transfer the gel to paper filter and dry it by gel dryer.
9. Expose the dried gel to film or Imaging Plate. If the transcriptional factor binds to the probe, shifted band will be observed in addition to lower band for free probe (see Fig. 5).

4. Notes

1. Antibiotics (penicillin and streptomycin) are not included according to manufacturer's protocol because they may be toxic for cells during transfection with lipofectamine 2000 reagent.

2. If the expression level of the transcriptional factor seems to be much higher or lower than that of mutant Htt in transfected cells, it may be difficult to observe the interaction between these. In such a case, change the ratio of plasmid DNA for transcriptional factor to that for Nhttl50Q-EGFP-NLS. If this does not solve the problem, try other experimental conditions, such as the use of a different promoter for expression of transcriptional factor, or use of different type of cell for transfection, etc.
3. In the case of Nhttl50Q-EGFP-NLS, distinct nuclear inclusions are observed at 24 h after transfection (11). If you expressed other polyglutamine protein and could not observe inclusions at this time point, culture the cells one more day after changing the medium to new one. Alternatively, you can add dibutyryl cyclic AMP to medium to a final concentration of 5 mM in step 7. Because dibutyryl cyclic AMP suppresses cell proliferation to differentiate Neuro2a cells, the formation of polyglutamine aggregates is usually enhanced.
4. This overnight incubation seems to be critical for observation of aggregated proteins by Western blot analysis and filter trap assay. This may be due to low efficiency of interaction of antibody with the aggregates.
5. If the lysates are too sticky to go through the membrane, use lower amount of lysates. Brief sonication after step 2 may also avoid this problem.
6. This step seems to be necessary to fix the aggregates on the membrane well.
7. Because GFP is light sensitive, shield the slide chamber from light.
8. Alternatively, 5% skim milk/TBST can be used.
9. You can use mouse monoclonal antibody against htt (MAB5374: Chemicon) instead of anti-ubiquitin if the antibody against transcriptional factor is not derived from mouse.
10. You can store the sections in the freezer, but the frozen sections should be dried thoroughly before use. Without doing this, sections will be easily detached from the slide glass.
11. During this step, the transcriptional factors will be dissolved into the buffer.
12. If the protein concentration of the lysates is lower than 20 μg/5 μl, lower concentration (e.g. 10 μg/5 μl) is also possible. However, use of lysates with low concentration will reduce the signal of shifted band in EMSA.
13. If the transcriptional factor is sequestrated by mutant Htt aggregates, reduction of its protein level in R6/2 tissue lysates would be observed.

14. If you want to confirm that the binding protein to the probe is actually the transcriptional factor of your interest, add the antibody (around 0.5 μg) against the transcriptional factor to the lysates at this step and incubate for ~30 min on ice. Control IgG should be used as a negative control. The band shift will be further enhanced by binding of antibody to transcriptional factor-probe complex. Alternatively, shifted band is reduced or disappeared because of perturbation of interaction between transcriptional factor and probe by antibody. This assay is called Super Shift Assay.

References

1. The Huntington's Disease Collaborative Research and Group (1993) A novel gene containing a trinucleotide repeat that is expanded and unstable on Huntington's disease chromosomes. Cell 72:971–983
2. Landles C, Bates GP (2004) Huntingtin and the molecular pathogenesis of Huntington's disease. Fourth in molecular medicine review series. EMBO Rep 5:958–963
3. Li SH, Li XJ (2004) Huntingtin-protein interactions and the pathogenesis of Huntington's disease. Trends Genet 20:146–154
4. Harjes P, Wanker EE (2003) The hunt for huntingtin function: interaction partners tell many different stories. Trends Biochem Sci 28:425–433
5. Luthi-Carter R, Hanson SA, Strand AD, Bergstrom DA, Chun W, Peters NL, Woods AM, Chan EY, Kooperberg C, Krainc D, Young AB, Tapscott SJ, Olson JM (2002) Dysregulation of gene expression in the R6/2 model of polyglutamine disease: parallel changes in muscle and brain. Hum Mol Genet 11:1911–1926
6. Chan EY, Luthi-Carter R, Strand A, Solano SM, Hanson SA, DeJohn MM, Kooperberg C, Chase KO, DiFiglia M, Young AB, Leavitt BR, Cha JH, Aronin N, Hayden MR, Olson JM (2002) Increased huntingtin protein length reduces the number of polyglutamine-induced gene expression changes in mouse models of Huntington's disease. Hum Mol Genet 11:1939–1951
7. Kotliarova S, Jana NR, Sakamoto N, Kurosawa M, Miyazaki H, Nekooki M, Doi H, Machida Y, Wong HK, Suzuki T, Uchikawa C, Kotliarov Y, Uchida K, Nagao Y, Nagaoka U, Tamaoka A, Oyanagi K, Oyama F, Nukina N (2005) Decreased expression of hypothalamic neuropeptides in Huntington disease transgenic mice with expanded polyglutamine-EGFP fluorescent aggregates. J Neurochem 93:641–653
8. Sugars KL, Rubinsztein DC (2003) Transcriptional abnormalities in Huntington disease. Trends Genet 19:233–238
9. Riley BE, Orr HT (2006) Polyglutamine neurodegenerative diseases and regulation of transcription: assembling the puzzle. Genes Dev 20:2183–2192
10. Cha JH (2000) Transcriptional dysregulation in Huntington's disease. Trends Neurosci 23:387–392
11. Yamanaka T, Miyazaki H, Oyama F, Kurosawa M, Washizu C, Doi H, Nukina N (2008) Mutant Huntingtin reduces HSP70 expression through the sequestration of NF-Y transcription factor. EMBO J 27:827–839
12. Doi H, Mitsui K, Kurosawa M, Machida Y, Kuroiwa Y, Nukina N (2004) Identification of ubiquitin-interacting proteins in purified polyglutamine aggregates. FEBS Lett 571: 171–176
13. Busch A, Engemann S, Lurz R, Okazawa H, Lehrach H, Wanker EE (2003) Mutant huntingtin promotes the fibrillogenesis of wild-type huntingtin: a potential mechanism for loss of huntingtin function in Huntington's disease. J Biol Chem 278:41452–41461
14. Scherzinger E, Lurz R, Turmaine M, Mangiarini L, Hollenbach B, Hasenbank R, Bates GP, Davies SW, Lehrach H, Wanker EE (1997) Huntingtin-encoded polyglutamine expansions form amyloid-like protein aggregates in vitro and in vivo. Cell 90:549–558
15. Mosser DD, Theodorakis NG, Morimoto RI (1988) Coordinate changes in heat shock element-binding activity and HSP70 gene transcription rates in human cells. Mol Cell Biol 8:4736–4744

Chapter 15

Biological Membranes as Protein Aggregation Matrices and Targets of Amyloid Toxicity

Monica Bucciantini and Cristina Cecchi

Abstract

Aberrantly folded proteins and peptides are hallmarks of amyloid diseases. A deeper knowledge of the pathways leading to the formation of amyloid protein aggregates and of the mechanisms of their cytotoxicity is fundamental for a better understanding of several human diseases with amyloid deposition. Increasing evidence indicates that amyloids arising from different peptides and proteins behave similarly as for their cytotoxic effects. In general, different cell susceptibility to toxic protein aggregates depends on the efficiency of different cell types to accumulate amyloid precursors at their plasma membrane with subsequent growth of pre-fibrillar and fibrillar entities, resulting in membrane perturbation and cell damage. Actually, protein–lipid interaction displays a twofold aspect: on the one hand, the presence of a lipid membrane may influence protein unfolding and the aggregation process; on the other hand, protein aggregates may modify membrane structure and permeability. Understanding the molecular basis of the membrane–protein interaction (but, more extensively, of the surface–protein interaction) may help elucidating some of the factors affecting protein misfolding and aggregation *in vivo*. This topic has been investigated by a variety of techniques such as atomic force microscopy, transmission electron microscopy, confocal laser microscopy and flow cytometric analysis. In this overview, such techniques will be reviewed with special emphasis to their use in protein aggregation studies.

Key words: Plasma membranes, Amyloid aggregates, Loss of membrane integrity, Atomic force microscopy, Electron microscopy, Confocal laser microscopy, Flow cytometry

1. Introduction

An increasing body of evidence raised in the last few years has focussed the attention of the researchers on the high cytotoxic potential of small, prefibrillar protein aggregates arising initially in the path of protein fibrillization, even when they are formed *in vitro* from peptides or proteins not associated with any protein deposition disease (1). These data have led to propose that such prefibrillar assemblies share basic structural features that, at least

Peter Bross and Niels Gregersen (eds.), *Protein Misfolding and Cellular Stress in Disease and Aging: Concepts and Protocols*, Methods in Molecular Biology, vol. 648, DOI 10.1007/978-1-60761-756-3_15, in most cases, seem to underlie common biochemical mechanisms of cytotoxicity (2–4). One of the leading hypotheses of the molecular basis of amyloid toxicity claims that a sub-population of prefibrillar aggregates, assembled into a pore-like fashion, interacts with cell membranes in the form of toxic annular structures, similar to those arising from pore-forming toxins, responsible of membrane permeabilization (5, 6). Therefore, the protein-lipid environment seems to be a physico-chemical feature playing a crucial role on both aggregate growth kinetics and cytotoxicity.

2. Materials

2.1. Atomic Force Microscopy

1. De-ionized water.
2. Mica substrates.
3. Phosphate-buffered solution, (10 mM KH_2PO_4, 150 mM KCl, pH 6) or another suitable buffer.
4. TappingMode Fluid Cell.
5. Cantilevers (oxide-sharpened silicon nitride tips, Model DNP-S, works well).
6. Source of filtered (0.2 μm), compressed air or dry N_2.
7. Optional for cantilever cleaning: UV lamp, high intensity; Oriel Mod. 6035 pencil-style spectral calibration lamp or equivalent.
8. Optional: Fluid cell liquid lines (silicone tubing and fittings), o-ring, clamping devices (for liquid lines), syringes: 1 cc; 5 cc.

2.2. Transmission Electron Microscopy

1. Uranyl acetate staining: 2.0% (w/v) uranyl acetate (Sigma) in water (see Notes 1, 2, 4, and 5). The uranyl acetate is superior to other negative stains such as uranyl formate and phosphotungstate due to its higher penetrability and minimal grain size. The stained grids can be stored at room temperature without adverse effects.
2. Potassium phosphotungstate staining: a 1–3% solution of phosphotungstic acid (Sigma) in water; pH is adjusted to 7–7.3 using sodium hydroxide (see Notes 1, 3, and 4).
3. Dilution buffer: 20 mM Tris–HCl, pH 7.5, 200 mM KCl, 1.0 mM DTT or another suitable buffer.
4. Formvar/carbon-coated 400 mesh nickel grids (Agar Scientific, Stansted, UK).

2.3. Confocal Immunofluorescence and Flow Cytometric Analyses

1. Culture medium: Dulbecco's Modified Eagle's Medium (DMEM)/F-12 Ham (in a 1:1 ratio), 25 mM N-2-hydroxyethylpiperazine-N-2-ethanesulfonic acid (HEPES) and $NaHCO_3$ supplemented with 10% fetal bovine serum

(FBS, Sigma), 1.0% glutamine, and 1.0% antibiotics (60 μg/ml penicillin and 100 μg/ml streptomycin).

2. Solutions of trypsin (0.25%) and phosphate-buffered saline (PBS) are added to the culture dishes when required.
3. Microscope cover slips (20×20×0.25 mm) and microscope slides (25×75×1 mm).
4. Buffered paraformaldehyde (Sigma): 2.0% solution in PBS and store at 4°C.
5. Fluorescein-conjugated wheat germ agglutinin (Molecular Probes, Eugene, OR): 5.0 mg/ml solution in PBS; store at −20°C (see Note 6).
6. Glycerol solution: 3.0% (v/v) glycerol solution in PBS with 0.5% BSA (prepare fresh for each experiment).
7. Primary antibody: Mouse monoclonal 6E10 anti-Aβ antibody (Signet, Dedham, MA, USA); store at −20°C.
8. PBS with 1% fetal bovine serum (FBS).
9. Secondary antibody: Anti-mouse IgG conjugated to Texas Red (Vector Laboratories, DBA, Italy); store at 4°C (see Note 6).
10. Calcein-AM (Molecular Probes, Eugene, OR): 2.0 μM solution in DMSO; store at −20°C (see Note 6).
11. Vectashield gel mounting medium (Vector Laboratories, Inc., Burlingame, CA).
12. Confocal scanning microscope equipped with an argon laser source for fluorescence measurements (for example, Leica TCS SP5, Mannheim, Germany).
13. FACS tubes.
14. Fluorescent-activated cell sorting FACS equipment and appropriate software (for example: FACSCanto (BD, San Jose, CA) using the FACSDiva Software (BD, Milan, Italy)).

2.3.1. Preparation of Amyloid β-Derived Diffusible Ligands

1. Lyophilized Aβ42, (trifluoroacetate salt, Sigma) and Aβ42 amine-reactive succinimidyl esters of carboxyfluorescein (Aβ42-FAM) (see Note 6) (AnaSpec, San Jose, CA) is dissolved and incubated in hexafluoro-2-isopropanol (HFIP) (Sigma) to a final 1.0 mM concentration for 1.0 h at room temperature to allow complete peptide monomerization. The Aβ42 and Aβ42-FAM solutions are dried under N_2 and stored in aliquots at −20°C.
2. Aliquots of Aβ42 are dissolved in DMSO to a final concentration of 5.0 mM and incubated to a concentration of 100 μM in ice-cold F-12 Ham nutrient medium (Sigma) for 24 h at 4°C.
3. Aβ42-FAM aggregates are obtained using a mixture of Aβ42-FAM peptide with 2 molar equivalents of unlabeled Aβ42 peptide (at ratio of 1:2) (see Note 7).

4. Aliquots of Aβ42 aggregates in F-12 Ham medium are centrifuged at 14,000×*g* for 10 min to remove insoluble structures.
5. The supernatant, defined as the amyloid β-derived diffusible ligand (ADDL) preparation, consists of a fibril-free solution of globular structures, as assessed by tapping mode atomic force microscopy (AFM) (7).

3. Methods

3.1. Atomic Force Microscopy

The AFM is one of the most powerful tools for determining the surface topography of biomolecules in their native state at sub-nanometer resolution (8). Unlike X-ray crystallography and electron microscopy (EM), AFM allows biomolecules to be imaged not only under physiological conditions, but also while biological processes are at work. Because of the high signal-to-noise (S/N) ratio, the detailed topological information is not restricted to crystalline specimens. Hence, single biomolecules lacking inherent symmetry can be directly monitored in their native environment (9–11). The AFM works in the same way as our fingers in the dark which touch and probe the environment we cannot see. The use of a finger to "visualize" an object leads our brain to deduce its topography while touching it. The resolution we can get by AFM is determined by the radius of the fingertip. To achieve atomic scale resolution, a sharp stylus (radius ~1–2 nm) attached to a cantilever is used in the AFM to scan an object point by point and contouring it while a constant small force is applied to the stylus. The forces (e.g., Van der Waals, electrostatic, magnetic, etc.) generated between the AFM tip and the sample are transmitted to an attached flexible cantilever, causing it to bend. The bending of the cantilever is monitored by the deflection of a laser beam reflected by the cantilever. In the AFM, the role of the brain is taken over by a computer, while scanning the stylus is accomplished by a piezoelectric tube. The basic characteristics of AFM are as follows: (1) nearly no limits to samples; (2) high-resolution sample morphology in three dimensions; (3) working in various environments, such as vacuum, air, fluid and low temperature; (4) enabling dynamic observation in physiological environment. In the fluid, biological samples can maintain their native state avoiding any damage. According to the different ways of interaction between the AFM tip and the sample, the operation modes are classified as contact-mode, non-contact mode and intermittent contact mode.

3.1.1. Operation Modes

3.1.1.1. Contact Mode

The first and foremost mode of operation of AFM, contact mode, is widely used. As the tip is raster-scanned across the surface, it is deflected as it moves over the surface corrugation. In constant force mode, the tip is constantly adjusted to maintain a constant deflection and, therefore, constant height above the surface. Such an adjustment is displayed as data. However, the ability to track the surface in this manner is limited by the feedback circuit. Sometimes the tip is allowed to scan without this adjustment, and one measures only the deflection. This is useful for small, high-speed atomic resolution scans, and is known as variable-deflection mode. Because the tip is in hard contact with the surface, the stiffness of the lever needs to be less than the effective spring constant holding atoms together, which is on the order of 1–10 nN/nm. Most contact mode levers have a spring constant of <1 N/m.

3.1.1.2. Lateral Force Microscopy

LFM measures frictional forces on a surface. By measuring the "twist" of the cantilever, rather than merely its deflection, one can qualitatively determine areas of higher and lower friction.

3.1.1.3. Noncontact mode

Noncontact mode refers to the use of an oscillating cantilever. A stiff cantilever is oscillated in the attractive regime, meaning that the tip is quite close to the sample, but not touching it (hence, "noncontact"). The forces between the tip and sample are quite low, on the order of pN (10^{-12} N). The detection scheme is based on measuring changes to the resonant frequency or amplitude of the cantilever. Such a scanning mode minimizes mechanical stress to the sample.

3.1.1.4. Dynamic Force/Intermittent-Contact/"Tapping Mode" AFM

Commonly referred to as "tapping mode," it is also referred to as intermittent-contact or the more general term Dynamic Force Mode (DFM) AFM. A stiff cantilever is oscillated closer to the sample than in non-contact mode. Part of the oscillation extends into the repulsive regime, so the tip intermittently touches or "taps" the surface. Very stiff cantilevers are typically used, as tips can get "stuck" in the water contamination layer. The main advantage of tapping the surface is improved lateral resolution on soft samples. Lateral forces such as drag, common in contact mode, are virtually eliminated. For specimens poorly adsorbed on a substrate surface, the advantage is clearly seen.

3.1.1.5. Force Modulation

Force modulation refers to a method used to probe properties of materials through sample–tip interactions. The tip (or sample) is oscillated at a high frequency and pushed into the repulsive regime. The slope of the force–distance curve is measured and correlated to the sample's elasticity. The data can be acquired along with topography, which allows comparison of both height and material properties.

3.1.1.6. Phase Imaging

In Phase-mode imaging, the phase shift of the oscillating cantilever relative to the driving signal is measured. This phase shift can be correlated with specific material properties affecting tip–sample interaction. The phase shift can be used to differentiate areas on a sample with such differing properties as friction, adhesion, and viscoelasticity. The technique is used simultaneously with DFM mode, so that topography can be measured as well.

The samples for atomic force microscopy imaging should be immobilized on a rigid support. Biological specimens like cells, molecules (proteins), and Langmuir–Blodgett films are usually bound to a flat substrate like mica or glass. Mica is a very common support because the surface is:

- atomically flat
- clean after cleavage
- easy to cut to desired sizes
- relatively inexpensive
- negatively charged but it can be modified to make the surface positive

3.1.2. Sample Preparation

3.1.2.1. Sample Preparation: Protein

Sample preparation varies slightly depending on the protein itself. Typically, protein concentration is in the 1–100 μg/ml range. The incubation time also varies from as little as 15 min to up to 24 h. After proper incubation period, rinse the sample with fresh buffer (see Note 8). For best results, use the narrow silicon nitride tips for Tapping Mode AFM. It may also be necessary to adjust the electrolyte solution and pH of the buffer in order to minimize the force applied to the sample.

1. Dissolve protein
   - Dissolve the protein in suitable buffer and at proper concentration.
2. Prepare fluid cell
   - Prepare the fluid cell for Tapping Mode in fluid operation. Clean the fluid cell and load a probe. For best results, clean the probe with UV light.
3. Prepare mica
   - Cleave a fresh mica surface by first pressing some adhesive tape against the top mica surface, then peeling off the tape. Glue mica to a small puck (e.g. using epoxy).
4. Deposit protein solution on mica
   - Deposit 50 μL of protein solution on the freshly cleaved mica.
5. Allow protein to bind to substrate
   - Allow the protein solution to bind to the mica substrate for 20–30 min. Binding time may vary with different samples (it can be up to 24 h). For longer binding times,

put the mica in a covered dish with a wet piece of filter paper to keep the liquid from evaporating.

6. Rinse unbound proteins from fluid cell
   - Rinse the sample with a large quantity of buffer to remove unbound proteins. Leave a drop of buffer onto the mica.
7. Assemble fluid cell
   - Mount the sample on the scanner end cap.

3.1.2.2. Sample Preparation: Living Cells

Living cells in culture can be studied using both Contact and Tapping Mode AFM. In Contact Mode, images mainly show the cytoskeleton beneath the membrane. However, in Tapping Mode, more cellular details, including the membrane, the nucleus, and vesicles beneath/on the membrane, may be observed. The cells can be immobilized by the air-dry methods and imaged in air. In this case, dehydration may affect cell morphology. Alternatively, the cells can be immobilized in fluid for AFM imaging: For optimal imaging, cells grown in 60 mm petri dishes should be ~70–80% confluent. In most cases, no additional preparation is necessary (glutaraldehyde fixation, etc.); the cells should attach to the bottom surface of the petri dish, especially if they are adherent cell lines. If the cells are not firmly attaching to the petri dish, it can be necessary to use such adhesives as poly-l-lysine, collagen, laminin, entactin, CellTak™, or polyethylene glycol (PEG) derivatives. The immobilization methods by using poly-l-lysine, that are positively charged at physiological pH and can hold the cells by electrostatic interaction, are simple and reproducible. However, the disadvantage is that poly-l-Lysine can induce structural alterations in cells. The narrow, long (200 μm length) V-shaped cantilever (DNP) with a nominal spring constant of 0.06 N/m is used to minimize applied forces. Use a non-sharpened (or a tipless cantilever) to avoid cell membrane destruction. To minimize damage to the cells, select a small RMS free amplitude (~0.5 V) (TappingMode) or a small deflection signal of ~–0.5 V (Contact Mode). For both imaging techniques, we recommend a slow scan rate of 0.25–0.50 Hz to start.

3.1.2.3. Sample Preparation: Lipids

Two techniques commonly used to attach lipids to bare mica are the Langmuir–Blodgett (LB) technique and the Vesicle Fusion Method. Lipid films usually adsorb spontaneously on mica and can be imaged in air or in liquid, in both Contact Mode and Tapping Mode, sometimes with molecular resolution.

(a) Langmuir–Boldgett (LB) technique

For the LB technique, replace the usual glass surface used as a support with a piece of mica of the same size. After deposition of the lipid film, the mica can be cut and glued to a sample puck using epoxy for 5 min when necessary. The challenge at this point is to keep the sample under liquid to preserve the molecular structure. This technique is reproducible and enables observing large sheets of organized molecules.

(b) Vesicle fusion method

The Vesicle Fusion Method requires to get first a vesicle solution in buffer. The latter is deposited onto mica or another suitable substrate. During the incubation period, the vesicles will settle, spread, and fuse to the surface at room temperature in a relatively short time (e.g. about 15 min for phosphatidylcholine vesicles on mica). Usually the lipid films are not as large as with the LB technique. They form patches on the surface or include "defects", where the substrate is not covered but exposed. This protocol is useful to integrate proteins inside the film to more closely simulate cell membranes and to study lipid–protein interactions.

3.2. Transmission Electron Microscopy

The electron microscope (EM) extracts structural information carried by the scattered electrons. In the fixed-beam transmission, EM electrons scattered by the irradiated sample are collected over a narrow solid angle and focused by the objective lens onto the image plane. The elastically scattered electrons, that is, those which have changed their direction but not lost any energy on interaction with the specimen, interfere with the unscattered electrons to produce a phase contrast image. The inelastically scattered electrons, that is, those which have changed both direction and energy on interaction with the specimen, generate a diffuse background image that is in some EMs eliminated by an energy filter. With a resolution of about 2 nm, the electron microscopy offers a useful technique for the ultrastructural characterization of the oligomeric aggregates and the more complex prefibrillar structures eventually growing into protofibrils and mature amyloid fibrils The negative staining technique for specimen preparation has been widely used for transmission electron microscopy (TEM), although alternative techniques such as high resolution metal shadowing and cryo-TEM of unstained or cryo-negatively stained fibrils can be used as well. Negative fibril staining is performed using uranyl acetate, lead citrate, platinum palladium, and phosphotungstic acid. Negative staining produces quite flat images of fibrils characterized by two dark edges suggesting a possible fibril structure of a ribbon formed by protofilaments running side by side. By using this technique, Tashima et al. investigated the interaction of amyloid beta peptide (Abeta)1–40 and Abeta1–42 with model membranes and the influence of cholesterol and monosialganglioside on their deposition and aggregation (12). By TEM immunogold labelling of calcitonin, Diociaiuti et al. observed that the peptide interacts with model membranes and the binding is favoured by the presence of rafts. In addition, they found by TEM that oligomers of calcitonin inserted into the membrane bilayer with the characteristic pore-like morphology of the amyloid proteins (13). By electron microscopy analysis Engel et al. showed that hIAPP fibrils line the surface of distorted phospholipid vesicles, in agreement with the notion that hIAPP fibril growth at the membrane and membrane damage are physically associated (14). In our lab, we documented the interaction between native-like Ure2p aggregates and synthetic phospholipid unilamellar vesicles or biological membranes to provide novel information on the molecular events that induce plasma membrane permeabilization, a key event underlying amyloid and non-amyloid aggregate cytotoxicity (15).

1. 2-Dioleoyl-*sn*-Glycero-3-[Phospho-l-Serine] (DOPS) and 1,2-Dioleoyl-*sn*-Glycero-3-Phosphocholine (DOPC) were used to prepare small unilamellar vesicles (SUV).
2. DOPS or DOPC small unilamellar vesicles were diluted in dilution buffer at a final concentration of ~600 μM in the presence or in the absence of 10 μM of the different types of Ure2p assemblies.
3. To investigate the mutual interaction between lipid vesicles and Ure2p during protein assembly, Ure2p was incubated for 1.0 h or 48 h at 50 μM concentration in the absence or in the presence of 1,2 mM DOPS or DOPC vesicles at 4°C without shaking.
4. The lipid-protein suspensions were adsorbed onto formvar/carbon-coated 400 mesh nickel grids by floating the grids onto 10 μl drops of the lipid-protein sample for 1 min.
5. The excess of sample was removed by touching the grid with a filter paper and by air-dry.
6. Then a small drop of potassium phosphotungstate or uranyl acetate (prepared as above) was applied to the grid.
7. After 10 s the excess stain was removed by touching the edge with a filter paper.
8. The grids were allowed to air-dry and examined using a JEM 1010 transmission electron microscope at 80 kV excitation voltage.

3.3. Confocal Immunofluorescence and Flow Cytometric Analyses

The amyloid cascade hypothesis causally links the clinico-pathological process of Alzheimer's disease (AD) and neuronal cell death with the aggregation and deposition of Aβ (16). Nevertheless, a recently modified amyloid cascade hypothesis for AD indicates that prefibrillar Aβ40 and Aβ42 oligomers, also known as ADDLs, are the main factors responsible of cytotoxicity and synapse failure (17–22). Confocal immunofluorescence detection and flow cytometric analysis can easily supply evidence supporting the rising consensus on the role of membranes as initial triggers of the biochemical modifications culminating with cell death. A time- and dose-dependent increase of amyloid pick up at the plasma membranes, in cells facing increased protein assembly at the cell surface, eventually results in a specific perturbation of membrane physicochemical features and loss of membrane integrity. This data identifies plasma membranes as key targets of amyloid damage as a result of their ability to recruit protein aggregates to the cell surface.

3.3.1. Confocal Immunofluorescence for ADDLs Binding to the Cell Membrane

1. Human SH-SY5Y neuroblastoma cells (A.T.C.C., Manassas, VA) were cultured in plastic Petri dishes (PBI International, Milan, Italy) at a seeding density of $2 \times 10^4/\text{cm}^2$ in DMEM/F-12 Ham with a 5.0% CO_2 humidified atmosphere at 37°C.
2. The culture medium was replaced once every 3 days and the cells were grown until 80% confluence.
3. The cells were washed with PBS, then incubated with trypsin for 2–3 min. Then the culture medium was added to block trypsin activity and the cells were centrifuged for 6 min at $700 \times g$. The supernatant was discarded and the pellet resuspended in culture medium.
4. Human SH-SY5Y neuroblastoma cells were plated at a density of 5×10^4 per glass cover slip and incubated in culture medium for 24 h in a six-well plate in a 5.0% CO_2 humidified atmosphere at 37°C.
5. The cells were exposed to 1.0 μM soluble Aβ42 oligomers (ADDLs) in culture medium for different periods of time (0, 10, 30, 60 min) in a 5.0% CO_2 humidified atmosphere at 37°C.
6. The cells were extensively washed with PBS (see Note 9) and then counterstained with 5 μg/mL fluorescein-conjugated wheat germ agglutinin (650 μL per cover slip in a six-well plate) for 10 min to detect plasma membrane profiles.
7. Then the cells were fixed in 2% buffered paraformaldehyde (1.5 mL per each cover slip) for 10 min at room temperature.
8. After washing three times with PBS, plasma membrane permeabilization was achieved with a 3% glycerol solution for 5 min at room temperature.
9. After washing three times with PBS, the cover slips were incubated with mouse monoclonal 6E10 anti-Aβ antibodies diluted 1:800 in PBS with 1.0% FBS (650 μL per each cover slip in a six-well plate) for 60 min at 37°C (see Note 10).
10. The immunoreaction was revealed by incubation with Texas Red-conjugated anti-mouse secondary antibodies diluted 1:1,000 in PBS with 1.0% FBS for 90 min at 37°C.

11. The cover slips were subsequently washed three times with PBS, dried for 30 min at room temperature and mounted up-side-down on microscope slides with a drop of Vectashield gel mounting medium. Then the cover slips were enameled around the edges to prevent a shift during confocal analysis.
12. Cell fluorescence was analyzed by a confocal Leica TCS SP5 scanning microscope using excitation lines at 488 nm and 568 nm for fluorescein and Texas Red, respectively. A series of optical sections (1024 × 1024 pixels), 1.0 μm in thickness, were taken through the cell depth for each examined sample using a Leica Plan Apo 63× oil immersion objective and projected as a single composite image by superimposition. A variable number of cells (ranging from 10 to 20) were estimated on regions of interest in each experiment.

3.3.2. Quantitation of Cell Membrane-Bound ADDLs by Flow Cytometric Analysis

1. 5×10^5 human SH-SY5Y neuroblastoma cells were resuspended in 300 μL culture medium and transferred to a FACS tube.
2. Cells were exposed to 3.0 μM Aβ42-FAM ADDLs in culture medium for different periods of time (0, 10, 30, 60 min) in a 5.0% CO_2 humidified atmosphere at 37°C under moderate agitation (see Note 7).
3. Cells were analyzed by a FACSCanto (BD, San Jose, CA) using the FACSDiva Software (BD, Milan, Italy).

3.3.3. Confocal Microscopy Analysis of Plasma Membrane Permeabilisation

1. Human SH-SY5Y neuroblastoma cells were plated at a density of 5×10^4 per glass cover slip and incubated in culture medium for 24 h in a six-well plate in a 5.0% CO_2 humidified atmosphere at 37°C.
2. The cells were loaded with 2.0 μM calcein-AM (650 μL per each cover slip in a six-well plate) for 20 min in a 5.0% CO_2 humidified atmosphere at 37°C.
3. Following several washes with PBS (see Note 9), the cells were exposed to 1.0 μM ADDLs added to the culture medium for different periods of time (0, 10, 30, 60 min) in a 5.0% CO_2 humidified atmosphere at 37°C.
4. After several washes with PBS, the cells were fixed in 2% buffered paraformaldehyde for 10 min at room temperature.
5. After washing three times with PBS, the cover slips were dried for 30 min at room temperature and mounted up-side-down on microscope slides with a drop of Vectashield gel mounting medium. Then the cover slips were enameled around the edges to prevent a cover slip shift during confocal analysis.
6. The decay of calcein fluorescence was analyzed by a confocal Leica TCS SP5 scanning microscope equipped with an argon laser source for fluorescence measurements at an excitation

wavelength of 488 nm. A series of optical sections, (1024 × 1024 pixels) 1.0 μm in thickness, were taken through the cell depth for each examined sample using a Leica Plan Apo 63× oil immersion objective and projected as a single composite image by superimposition. A variable number of cells (ranging from 10 to 20) were estimated on regions of interest in each experiment. The fluorescence intensity of cells was quantified using ImageJ and data statistically analysed by ANOVA and the Bonferroni *t*-test where appropriate.

4. Notes

1. Keeps for approximately 3 months – store in a brown glass bottle.
2. The stain has a low pH, so this solution is not recommended for particles that are unstable in acid conditions.
3. The stain produces less contrast than the uranyl acetate.
4. All staining solutions should either be filtered through polycarbonate 0.2-μm pore filters (Millipore) or centrifuged before use.
5. The uranyl salts are toxic and radioactive and should be handled with care.
6. Protect from light and avoid repeated freezing and thawing.
7. Aβ42-FAM peptide was mixed with 2 molar equivalents of unlabeled Aβ42 peptide to minimize possible interference of the fluorophore with the aggregation process, while retaining sufficient fluorescent signal.
8. It is also possible to prepare samples inside the fluid cell by flowing the protein solution through the fluid cell. In this case, it may be helpful to engage the tip in Contact Mode with a zero SCAN SIZE to protect proteins from binding to the tip.
9. To ensure that the cells do not detach from cover slips, avoid applying excessive pressure during the washing process.
10. Negative controls were obtained by substituting PBS with 1.0% FBS solution for the mouse monoclonal 6E10 anti-Aβ antibody.

References

1. Stefani M, Dobson CM (2003) Protein aggregation and aggregate toxicity: new insights into protein folding, misfolding diseases and biological evolution. J Mol Med 81:678–699
2. Bucciantini M, Giannoni E, Chiti F, Baroni F, Formigli L, Zurdo J, Taddei N, Ramponi G, Dobson CM, Stefani M (2002) Inherent toxicity of aggregates implies a common mechanism for protein misfolding diseases. Nature 416:507–511
3. Kayed R, Head E, Thompson JL, McIntire TM, Milton SC, Cotman CW, Glabe CG

(2003) Common structure of soluble amyloid oligomers implies common mechanism of pathogenesis. Science 300:486–489
4. Cecchi C, Baglioni S, Fiorillo C, Pensalfini A, Liguri G, Nosi D, Rigacci S, Bucciantini M, Stefani M (2005) Insights into the molecular basis of the differing susceptibility of varying cell types to the toxicity of amyloid aggregates. J Cell Sci 118:3459–3470
5. Kourie JI, Henry CL (2002) Ion channel formation and membrane-linked pathologies of misfolded hydrophobic proteins: the role of dangerous unchaperoned molecules. Clin Exp Pharmacol Physiol 29:741–753
6. Kayed R, Sokolov Y, Edmonds B, McIntire TM, Milton SC, Hall JE, Glabe CG (2004) Permeabilization of lipid bilayers is a common conformation-dependent activity of soluble amyloid oligomers in protein misfolding diseases. J Biol Chem 279:46363–46366
7. Cecchi C, Nichino D, Zampagni M, Bernacchioni C, Evangelisti E, Liguri G, Pensalfini A, Gliozzi A, Stefani M, Relini A (2009) A protective role for lipid raft cholesterol against amyloid-induced membrane damage in human neuroblastoma cells. Biochim Biophys Acta 1788:2204–2216
8. Gosal WS, Meyers SL, Radford SE, Thomson NH (2006) Amyloid under the atomic force microscope. Protein Pept Lett 13:261–270
9. Santos NC, Castanho MA (2004) An overwiew of the biophysical applications of atomic force microscopy. Biophys Chem 2:133–149
10. Möller C, Allen M, Elings V, Engel A, Müller DJ (1999) Tapping-mode atomic microscopy produces faithful high-resolution images of protein surfaces. Biophys J 77:1150–1158
11. Müller DJ, Fotiadis D, Scheurig S, Müller SA, Engel A (1999) Electrostatically balanced subnanometer imaging of biological specimens by atomic force microscopy. Biophys J 76:1101–1111
12. Tashima Y, Oe R, Lee S, Sugihara G (2004) The effect of cholesterol and monosialoganglioside (GM1) on the release and aggregation of amyloid beta-peptide from liposomes prepared from brain membrane-like lipids. J Biol Chem 279:17587–17595
13. Diociaiuti M, Polzi LZ, Valvo L, Malchiodi-Albedi F, Bombelli C, Gaudiano MC (2006) Calcitonin forms oligomeric pore-like structures in lipid membranes. Biophys J 91:2275–2281
14. Engel MFM, Khemtemourian L, Kleijer CC, Meeldijk HJD, Jacobs J, Verkleij AJ, de Kruijff B, Killian JA, Hoppener JWM (2008) Membrane damage by human islet amyloid polypeptide through fibril growth at the membrane. PNAS 105:6033–6038
15. Pieri L, Bucciantini M, Guasti P, Savistchenk J, Melki R, Stefani M (2009) Synthetic lipid vesicles recruit native-like aggregates and affect the aggregation process of the prion Ure2p: insights on vesicles permeabilization and charge selectivity. Biophis J 96:3319–3330
16. Selkoe DJ (2001) Alzheimer's disease: genes, proteins, and therapy. PhysiolRev 81:741–766
17. Dahlgren KN, Manelli AM, Stine WB, Baker LK, Krafft GA, LaDu MJ (2002) Oligomeric and fibrillar species of amyloid-β peptides differentially affect neuronal viability. J Biol Chem 277:32046–32053
18. Klein WL (2002) Aβ toxicity in Alzheimer's disease: globular oligomers (ADDLs) as new vaccine and drug targets. Neurochem Intl 41:345–352
19. Lacor PN, Buniel MC, Chang L, Fernandez SJ, Gong Y, Viola KL, Lambert MP, Velasco PT, Bigio EH, Finch CE, Krafft GA, Klein WI (2004) Synaptic targeting by Alzheimer's-related amyloid β oligomers. J Neurosci 24:10191–10200
20. Wakabayashi M, Okada T, Kozutsumi Y, Matsuzaki K (2005) GM1 ganglioside-mediated accumulation of amyloid beta-protein on cell membranes. Biochim Biophys Res Com 328:1019–1023
21. Demuro A, Mina E, Kayed R, Milton S, Parker I, Glabe CG (2005) Calcium dysregulation and membrane disruption as a ubiquitous neurotoxic mechanism of soluble amyloid oligomers. J Biol Chem 280:17294–17300
22. Kayed R, Pensalfini A, Margol L, Sokolov Y, Sarsoza F, Head E, Hall J, Glabe CG (2009) Annular protofibrils are a structurally and functionally distinct type of amyloid oligomer. J Biol Chem 284:4230–4237

Chapter 16

Measurement of Mitochondrial ROS Production

Anatoly A. Starkov

Abstract

The significance of reactive oxygen species (ROS) as aggravating or primary factors in numerous pathologies is widely recognized, with mitochondria being considered the major intracellular source of ROS. It is not yet possible to routinely measure mitochondrial ROS in animals or cultured cells with a reasonable degree of certainty. However, at the level of isolated mitochondria, one can easily monitor and quantify the rate of ROS production, identify major sites of ROS production, and compare the rates of ROS production in mitochondria isolated from normal and diseased tissue. In this chapter, we describe in detail the most recent and reliable method to measure mitochondrial ROS as the rate of H_2O_2 emission. This method may be employed with minimal modifications to measure H_2O_2 production by mitochondria isolated from various tissues and under a wide variety of experimental conditions.

Key words: Reactive oxygen species, Mitochondria, Amplex, Oxidative stress

1. Introduction

The significance of reactive oxygen species (ROS) as aggravating or primary factors in numerous pathologies, ischemia, excitotoxicity, neurodegenerative diseases, and senescence is widely recognized and extensively reviewed elsewhere (1–11). More recent data strongly suggest that ROS, and specifically mitochondria-generated ROS, are involved in physiological signaling cascades regulating various cellular and organ functions (3, 12–14), with H_2O_2 being a chief messenger molecule (15).

Mitochondria are believed to be the major intracellular source of ROS. Several decades of research have firmly established that ROS production is inherent to mitochondrial oxidative metabolism. This literature has been extensively reviewed by us and others elsewhere (8, 12, 16–23).

Whereas the primary ROS species produced in mitochondria is thought to be superoxide (17, 24), by its chemical nature it is

Peter Bross and Niels Gregersen (eds.), *Protein Misfolding and Cellular Stress in Disease and Aging: Concepts and Protocols*, Methods in Molecular Biology, vol. 648, DOI 10.1007/978-1-60761-756-3_16, neither particularly reactive (21) nor capable of "long-distance" travel because it does not easily permeate cell and mitochondrial membranes due to its negative charge (17, 21). However, its electrically neutral dismutation product, H_2O_2, is more reactive toward critical cellular targets such as protein SH groups and it is known to be a precursor of even more aggressive ROS, hydroxyl radical. Measuring H_2O_2 emission is a preferred and well-established method of evaluating mitochondrial ROS production (17, 25–35). It is justified because (see for details (17)) most of the superoxide produced in mitochondria is almost immediately converted into H_2O_2 by the matrix-located superoxide dismutase (MnSOD) or dismutates spontaneously with a high rate into H_2O_2. Moreover, some mitochondrial sites such as flavins may produce H_2O_2 directly. Perhaps the most important reason to measure mitochondrial ROS as H_2O_2 emission is that methods to measure it are more quantitative and less labor-consuming and equipment-demanding than those to detect superoxide.

In this chapter, we describe in details the most recent and reliable method to measure low levels of H_2O_2 in vitro. It employs horseradish peroxidase to trap emitted H_2O_2 with high selectivity and affinity and Amplex Red Ultra (a derivative of 10-acetyl-3,7-dihydroxyphenoxazine) as a sensitive fluorescent probe for H_2O_2 (36).

2. Materials

1. Isolated mitochondria (see Note 1).
2. Incubation buffer: 125 mM KCl, 4 mM KH_2PO_4, 14 mM NaCl, 20 mM HEPES-NaOH, pH 7.2, 1 mM $MgCl_2$, 0.2% of fatty acids free bovine serum albumin, and 0.020 mM EGTA (see Note 2).
3. Amplex Red Ultra (Invitrogen) (see Notes 3, 5).
4. Horseradish peroxidase (e.g. Type VI-A, essentially salt-free, lyophilized powder, ~1,000 units/mg solid) (Sigma) (see Notes 4, 5).
5. H_2O_2 solution in water, 30–32 wt.%, semiconductor grade, 99.999% trace metals basis (Sigma).
6. Respiratory substrates and inhibitors (see Notes 5, 6).
7. (Optional) Catalase (e.g. from bovine liver, lyophilized powder, ~10,000 units/mg protein) (Sigma) (see Notes 5, 7).
8. (Optional) Sodium Azide (Sigma) (see Note 8).
9. (Optional) Cu, Zn Superoxide dismutase (e.g. from bovine erythrocytes, 2,500–7,000 units/mg protein, lyophilized powder) (Sigma) (see Note 9).

10. Small (5–10 ml) glass tubes.
11. Glass (preferable) or plastic disposable cuvettes with all walls transparent and flat bottom and matching stirring bars.
12. A quartz cuvette.
13. Fluorimeter equipped with a stirred thermostated cuvette holder such as "Hitachi "F2500, F4700, or F8000 models (see Note 10).
14. Spectrophotometer capable of quantitative absorbance detection at 240 nm wavelength.
15. Common labware such as pipettors and tips.

3. Methods

3.1. Prepare the H_2O_2 Calibration Solution

Fill two glass tubes with 5 ml of de-ionized water, each. To the first tube, add 5 μl of 30% (v/v) H_2O_2 solution (1:1,000 dilution), mix well. Determine the concentration of H_2O_2 spectrophotometrically by placing 2 ml of this solution in a quartz cuvette and reading its absorbance at 240 nm; calculate the concentration of H_2O_2 employing the extinction coefficient $E_{240}=43.6/M/cm$. Dilute H_2O_2 solution to ~0.1 mM by taking 50 μl of H_2O_2 solution from this tube and adding it to the second tube (1:100 dilution), mix well. Keep the second tube in ice and use within 1 h to calibrate the H_2O_2 assay (see Note 11).

3.2. Calibrate the Assay

Set the fluorimeter at 555 nm excitation and 581 nm emission wavelengths and turn on the cuvette holder's thermostat set at the desired temperature (25–37°C). Fill a cuvette with the incubation buffer (item 2, Subheading 2), add magnetic stirring bar, turn on the stirrer, and wait until the cuvette reaches the desired temperature (25–37°C). Add 4 U/ml of horseradish peroxidase, 10 μM Amplex Red Ultra, 40 U/ml superoxide dismutase (optional, see Note 12) and mitochondria (0.03–0.1 mg/ml) to the cuvette. Record the fluorescence changes of Amplex for ~100 s. Make 6–8 additions of ~0.1 mM H_2O_2 by ~100 nmol (~1 μl/ml) each, with about 30 s between the additions. You should obtain a "staircase"-like trace with about equal "steps" after each addition (see Fig. 1). Take the fluorescence numbers approximately at the points indicated by crosses (Fig. 1). Subtract the value of fluorescent signal right before the first addition of H_2O_2 from all other values and plot them vs. the amount of H_2O_2 added (Fig. 1). Calculate the slope coefficient of the graph. This coefficient is used to calculate the rate of H_2O_2 production by mitochondria, in nmols or pmols of H_2O_2 per mg mitochondrial protein per minute. The R square value should be at least 0.95.

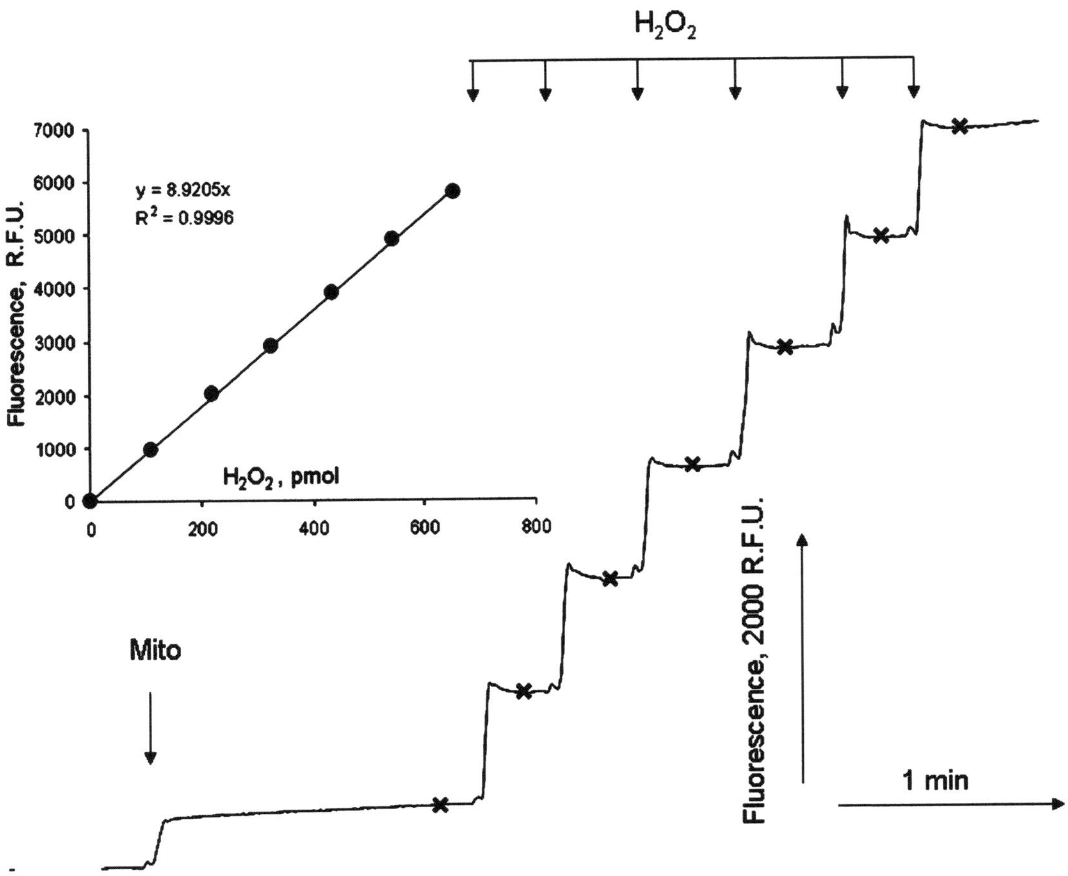

Fig. 1. Calibrating the H_2O_2 assay. See text for the details.

3.3. Measure Mitochondrial H_2O_2 Emission

Fill a cuvette with the incubation buffer (item 2, Subheading 2, see Note 13), add magnetic stirring bar, turn on the stirrer, and wait until the cuvette reaches the desired temperature (25–37°C). Add 4 U/ml of horseradish peroxidase, 10 μM Amplex Red Ultra, 40 U/ml superoxide dismutase (optional, see Note 12) and the same amount of mitochondria as used in step 3.3 to build the calibration curve. Record the fluorescence for ~150 s. Add respiratory substrates (see Note 6) and record H_2O_2 emission.

To illustrate a typical experimental protocol, Fig. 2 presents recordings of H_2O_2 production by isolated mouse brain mitochondria oxidizing NAD^+-dependent substrates or succinate. The H_2O_2 generation is triggered by the addition of a respiratory substrate (succinate, Fig. 2a or pyruvate and malate, Fig. 2b, see Note 14). With NAD^+-dependent substrates, H_2O_2 production was stimulated by rotenone, which inhibits NADH oxidation at Complex I (Fig. 2b). With succinate, rotenone inhibited H_2O_2 production indicating that it was fueled by reverse electron transfer from succinate to a site in Complex I (24). With either substrate, H_2O_2 production was stimulated by an inhibitor of Complex III (Antimycin A) (24, 37).

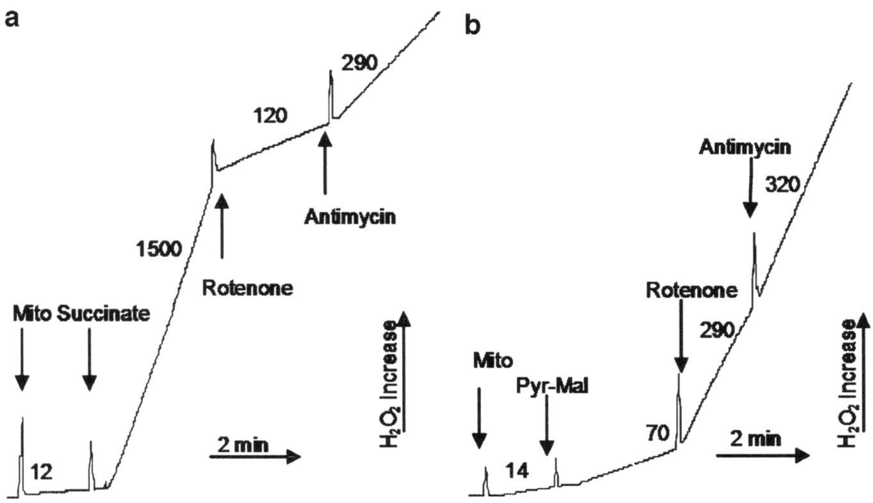

Fig. 2. H_2O_2 emission by mouse brain mitochondria. Incubation medium (37°C) (Subheading 4.2) is supplemented with 4 U/ml horseradish peroxidase, 40 U/ml superoxide dismutase, 0.010 mM Amplex Red, and respiratory substrates as indicated. (**a**) Mouse brain mitochondria oxidizing succinate. (**b**) Mouse brain mitochondria oxidizing pyruvate and malate. Additions: Mito, 0.05 mg/ml mouse brain mitochondria; Succinate, 10 mM; Pyruvate:malate, 5:1 mM; Rotenone, 0.5 μM; Antimycin, 1 μg/ml. Numbers near the tracings are the rates of H_2O_2 production expressed in pmol/min/mg of mitochondria protein.

3.4. Quantifying the Rates of H_2O_2 Emission

Calculate the rate of H_2O_2 emission using the slope coefficient obtained in step 3.3 by measuring the slope of the fluorescence traces in relative fluorescence units per minute and dividing it by the slope coefficient and by the amount of mitochondrial protein present in the cuvette. Present the rates of H_2O_2 emission in pmol/min/mg mitochondria.

3.5. Control Experiments (Optional)

If desired, two simple control experiments can be performed to ensure that what is measured is H_2O_2 and that the reaction is catalyzed by horseradish peroxidase (as opposed to some contaminants in your mitochondrial preparations). First, the incubation buffer can be supplemented with 200–400 U/ml of catalase. This should greatly suppress the observed rate of H_2O_2 emission, although not necessarily completely. Second, 1 mM azide (a strong inhibitor of horseradish peroxidase) can be included in the incubation medium; this should suppress the observed rate of H_2O_2 emission completely.

3.6. Interpreting the Obtained Data

Performing the assay as shown on Fig. 2 allows one to estimate the ROS production from several mitochondrial sites and can be interpreted in a reasonably straightforward way as extensively discussed elsewhere (17, 24–27, 35, 37, 38). This protocol yields the rates of ROS production from Complex I of the respiratory chain in both forward and reversed electron flux conditions, the rate of ROS production by the Complex III, and that by the matrix dehydrogenases (see (17) for discussion). To note, ROS production capacity of mitochondria is controlled by the factors affecting and reflecting the metabolic state of intact mitochondria. It has been found that most important factors controlling the ROS production in mitochondria are the chemical nature of the respiratory substrates, the amplitude of the membrane potential, the pH of the mitochondrial matrix, and the oxygen tension in their surrounding (17, 39). Perhaps the most important, irrespective of the nature of the respiratory substrate, is that the rates of ROS emission are directly related to the magnitude of the membrane potential of mitochondria (26, 27, 30, 31, 38); with NAD^+-dependent substrates, the rate of ROS emission is modulated by the level of NAD(P)H reduction in mitochondria. Experimentally, this necessitates measuring the membrane potential of mitochondria in addition to their ROS emission rates when the latter are measured in the absence of respiratory chain inhibitors. However, further discussion of these issues is beyond the scope of this publication.

4. Notes

1. Quality of and purity of mitochondria is very important. Contamination of mitochondrial preparation with cytosolic structures such as peroxisomes, synaptic terminals, etc. and with fragments of broken mitochondria can significantly interfere with ROS measurements. In essence, a meaningful and reproducible ROS assay cannot be performed with crude mitochondrial pellets and requires at least some further purification of the mitochondrial preparation, for example, by an isopycnic density gradient.
   (a) A discussion of mitochondria isolation and purification procedures is beyond the scope of this manuscript. However, we feel compelled to briefly describe the most common procedure for isolation and purification of mouse brain mitochondria which was employed in experiments presented on Fig 2.
   (b) Non-synaptic mouse brain mitochondria were isolated by a modified isopycnic centrifugation procedure (40) employing Percoll density gradient. Briefly, cortex brain tissue was homogenized in the MSEGTA buffer comprising 225 mM mannitol, 75 mM sucrose, 20 mM HEPES, adjusted to pH 7.4, 1 mM EGTA, 0.2% (w/v) fatty acid free bovine serum albumin (BSA) and centrifuged at $2{,}000 \times g \times 4$ min. The supernatant was collected and centrifuged at $12{,}000 \times g \times 10$ min. The pellet was resuspended in MSEGTA and layered over 25% (v/v) Percoll /MSEGTA (100% Percoll solution contains 225 mM nannitol, 75 mM sucrose, 20 mM HEPES, 1 mM EGTA dissolved in 100%

Percoll and adjustesd to pH 7.4. This Percoll solution is diluted with MSEGTA mixture described above to 25% v/v) and centrifuged at 30,000 × *g* × 10 min. Purified mitochondria fraction was collected at the bottom of the tube, resuspended in MSEGTA without added BSA and washed 2 times by centrifuging at 12,000 g × 10 min. Final mitochondrial pellet was resuspended in MS buffer comprising 225 mM mannitol, 75 mM sucrose, 20 mM HEPES-NaOH, pH 7.4 and stored on ice. Protein content was estimated by a commercial BCA assay ("Pierce Biotechnology"/"Thermo Scientific," USA). Typically this procedure yields essentially pure, primarily non-synaptic, well-coupled mitochondria with respiratory control index 7–9 on glutamate plus malate.

2. Prepare the base buffer comprising 125 mM KCl, 4 mM KH_2PO_4, 14 mM NaCl, 20 mM HEPES-NaOH, pH 7.2, 1 mM $MgCl_2$, and 0.020 mM EGTA, in de-ionized water (Milli-Q, "Millipore"), adjust pH with KOH and sterilize the buffer by filtering it through 0.22 μm membrane. Stable for at least 3 months if kept sterile. Store at room temperature (or at +4°C) in an air-tight glass container. DO NOT USE ANY PLASTIC VESSELS TO STORE THE BUFFER! Avoid freezing the buffer. Prepare 1 ml of 100 mg/ml solution of bovine serum albumin, essentially fatty acids free, dissolved in de-ionized water (Milli-Q, "Millipore"); store frozen at −20°C. This can be frozen and thawed many times. Reconstitute the complete incubation buffer mixture right before the experiments; discard unused buffer after the experiment.

   (a) Note that the composition of our incubation buffer mimics the ionic composition of cell cytosol and allows one to study various and many mitochondrial functional activities under standard, directly comparable experimental conditions. Bovine serum albumin and EGTA are added to increase the degree of coupling of mitochondria by removing loosely bound long chain fatty acids and residual calcium contaminating other chemicals, respectively. Although the composition of incubation buffer can affect the measurements of H_2O_2, we have had good experience in using various other buffers that are commonly used in experiments with isolated mitochondria. For example, simplest iso-osmotic buffer suitable for H_2O_2 measurement can be composed of 225 mM mannitol, 75 mM sucrose, and 20 mM HEPES-NaOH (pH 7.4). Basically, the choice of a buffer should suit the experimental needs of the study. It is, however, strongly advised to avoid including reducing sugars such as glucose, bicarbonate, and compounds that are substrates or inhibitors of horseradish peroxidase. An optimal pH range for these measurements is 7.2–7.6.

3. Prepare 10 mM stock solution by adding 330 μl of anhydrous DMSO to the content of 1 mg package. Wrap in foil to protect from light. Aliquot by 10–20 μl and keep frozen at –80°C; an aliquot can normally be thawed-re-frozen once or twice. Discard when a faint pink color appears.
4. Prepare the stock solution by dissolving 1,000 U of horseradish peroxidase in 1 ml of de-ionized water (Milli-Q, "Millipore"). Aliquot by 20–50 μl and keep frozen at -80°C; discard unused solution after the experiment (do not re-use).
5. For all aliquoted reagents, use only non-colored, clear plastic tubes (e.g. PCR tubes or Eppendorf™ type 1.7 ml tubes), or – whenever possible – glass tubes.
6. To note, de-energized mitochondria do not routinely produce ROS. It is therefore suggested to have ready at least some common mitochondrial respiratory substrates and respiratory chain inhibitors, although the exact set depends on the purpose and the design of the ROS assay. In either case, some standard substrates and inhibitors can come up handy, such as (stock solutions) 1 M sodium glutamate, 1 M sodium malate, 1 M sodium succinate, all prepared in de-ionized water; also 1 mg/ml Antimycin A and 1 mM rotenone, prepared in absolute ethanol. These reagents can be frozen-thawed multiple times.
7. Prepare small amount (~250 μl) of ~40,000 U/ml stock solution in de-ionized water (Milli-Q, "Millipore") right before the experiment. Do not freeze; keep on ice protected from light. Discard unused solution after the experiment.
8. Prepare 1 M solution in de-ionized water (Milli-Q, "Millipore"). Store frozen.
9. Prepare the stock solution by dissolving 10,000 U of superoxide dismutase in 1 ml of de-ionized water (Milli-Q, "Millipore"). Aliquot by 20–40 μl and keep frozen at –80°C; discard unused solution after the experiment (do not re-use).
10. In theory, a spectrophotometer can be used to detect ROS, as the compound produced in peroxidase catalyzed oxidation of Amplex Red Ultra by H_2O_2 has large absorption coefficient (please contact "Invitrogen" company for details). To the best of our knowledge, there were no publications where mitochondrial ROS were detected with Amplex Red Ultra spectrophotometrically. In either case, all the steps, reagents, and other information presented in this manuscript remain relevant irrespectively of whether a fluorimeter or a spectrophotometer is used to detect ROS.
11. DO NOT USE METAL-CONTAINING DEVICES (e.g. Hamilton syringes with steel needles) to handle H_2O_2 solutions, substrate solutions, inhibitors, mitochondria or other liquids.

12. Superoxide dismutase is generally not required with experimental protocols resembling the one presented on Fig. 2. However, supplementing the incubation buffer with superoxide dismutase is recommended when it contains NAD(P)H or other compounds capable of superoxide-initiated, peroxidase-dependent redox cycling. In such conditions, superoxide dismutase will greatly improve the reproducibility and sensitivity of H_2O_2 assay. Considering that superoxide dismutase is a relatively cheap reagent, and that it does not negatively affect the assay sensitivity and reproducibility, we recommend routinely adding it to the assay buffer.
13. Do not keep the tube containing the incubation buffer anywhere close to the ventilation exhaust of a fluorimeter's light source compartment. It will increase the residual concentration of H_2O_2 in the buffer and the baseline fluorescence.
14. Although pyruvate and alpha-ketoglutarate have recently gained some attention as "scavengers" of H_2O_2 in cellular systems (41–45), these substrates do not normally interfere with H_2O_2 measurements. Chemical decarboxylation of alpha-keto acids by H_2O_2 was reported more than a century ago (46); however, according to our experience and other published reports (45) the reaction of pyruvate and alpha-ketoglutarate with H_2O_2 is very slow particularly at low concentrations of H_2O_2: ~15 μM H_2O_2 was not decomposed by 2 mM pyruvate in over 20 min at 37°C (45). Therefore, alpha-keto acids cannot compete for H_2O_2 with other targets such as our horseradish peroxidase-based H_2O_2 detection system or other intramitochondrial dehydrogenases.

Acknowledgement

This work was supported by NIH/NIA grant AG014930.

References

1. Orrenius S (2007) Reactive oxygen species in mitochondria-mediated cell death. Drug Metab Rev 39:443–455
2. Droge W, Schipper HM (2007) Oxidative stress and aberrant signaling in aging and cognitive decline. Aging Cell 6:361–370
3. Valko M, Leibfritz D, Moncol J, Cronin MT, Mazur M, Telser J (2007) Free radicals and antioxidants in normal physiological functions and human disease. Int J Biochem Cell Biol 39:44–84
4. Giordano FJ (2005) Oxygen, oxidative stress, hypoxia, and heart failure. J Clin Invest 115:500–508
5. Busija DW, Gaspar T, Domoki F, Katakam PV, Bari F (2008) Mitochondrial-mediated suppression of ROS production upon exposure of neurons to lethal stress: mitochondrial targeted preconditioning. Adv Drug Deliv Rev 60:1471–1477
6. Chinopoulos C, Adam-Vizi V (2006) Calcium, mitochondria and oxidative stress in neuronal

pathology. Novel aspects of an enduring theme. FEBS J 273:433–450
7. Christophe M, Nicolas S (2006) Mitochondria: a target for neuroprotective interventions in cerebral ischemia-reperfusion. Curr Pharm Des 12:739–757
8. Starkov AA, Chinopoulos C, Fiskum G (2004) Mitochondrial calcium and oxidative stress as mediators of ischemic brain injury. Cell Calcium 36:257–264
9. Kuroda S, Siesjo BK (1997) Reperfusion damage following focal ischemia: pathophysiology and therapeutic windows. Clin Neurosci 4:199–212
10. Perez-Pinzon MA, Dave KR, Raval AP (2005) Role of reactive oxygen species and protein kinase C in ischemic tolerance in the brain. Antioxid Redox Signal 7:1150–1157
11. Siesjo BK, Elmer E, Janelidze S, Keep M, Kristian T, Ouyang YB, Uchino H (1999) Role and mechanisms of secondary mitochondrial failure. Acta Neurochir Suppl 73:7–13
12. Droge W (2002) Free radicals in the physiological control of cell function. Physiol Rev 82:47–95
13. Giorgio M, Trinei M, Migliaccio E, Pelicci PG (2007) Hydrogen peroxide: a metabolic by-product or a common mediator of ageing signals? Nat Rev Mol Cell Biol 8:722–728
14. Afanas'ev IB (2007) Signaling functions of free radicals superoxide & nitric oxide under physiological & pathological conditions. Mol Biotechnol 37:2–4
15. Stone JR, Yang S (2006) Hydrogen peroxide: a signaling messenger. Antioxid Redox Signal 8:243–270
16. Adam-Vizi V, Chinopoulos C (2006) Bioenergetics and the formation of mitochondrial reactive oxygen species. Trends Pharmacol Sci 27:639–645
17. Andreyev AY, Kushnareva YE, Starkov AA (2005) Mitochondrial metabolism of reactive oxygen species. Biochemistry (Mosc) 70:200–214
18. Boveris A, Cadenas E (1997) In: Clerch LB, Massaro DJ (ed) Oxygen, gene expression and cellular function, Marcel Dekker, New York, pp 1–25
19. Boveris A, Chance B (1973) The mitochondrial generation of hydrogen peroxide. General properties and effect of hyperbaric oxygen. Biochem J 134:707–716
20. Chance B, Sies H, Boveris A (1979) Hydroperoxide metabolism in mammalian organs. Physiol Rev 59:527–605
21. Sawyer DT, Valentine JS (1981) How super is superoxide? Acc Chem Res 14:393–400
22. Starkov AA (2006) Protein-mediated energy-dissipating pathways in mitochondria. Chem Biol Interact 163:133–144
23. Turrens JF (2003) Mitochondrial formation of reactive oxygen species. J Physiol 552:335–344
24. Turrens JF (1997) Superoxide production by the mitochondrial respiratory chain. Biosci Rep 17:3–8
25. Starkov A, Fiskum G (2002) Generation of reactive oxygen species by brain mitochondria mediated by ketoglutarate dehydrogenase. Abstract Viewer/Itinerary Planner Online, Program No. 194.17. Society for Neuroscience, Washington, DC
26. Starkov AA, Fiskum G (2003) Regulation of brain mitochondrial H_2O_2 production by membrane potential and NAD(P)H redox state. J Neurochem 86:1101–1107
27. Starkov AA, Polster BM, Fiskum G (2002) Regulation of hydrogen peroxide production by brain mitochondria by calcium and Bax. J Neurochem 83:220–228
28. Patole MS, Swaroop A, Ramasarma T (1986) Generation of H_2O_2 in brain mitochondria. J Neurochem 47:1–8
29. Cino M, Del Maestro RF (1989) Generation of hydrogen peroxide by brain mitochondria: the effect of reoxygenation following postdecapitative ischemia. Arch Biochem Biophys 269:623–638
30. Hansford RG, Hogue BA, Mildaziene V (1997) Dependence of H_2O_2 formation by rat heart mitochondria on substrate availability and donor age. J Bioenerg Biomembr 29:89–95
31. Korshunov SS, Skulachev VP, Starkov AA (1997) High protonic potential actuates a mechanism of production of reactive oxygen species in mitochondria. FEBS Lett 416:15–18
32. Kushnareva Y, Murphy AN, Andreyev A (2002) Complex I-mediated reactive oxygen species generation: modulation by cytochrome c and NAD(P)+oxidation–reduction state. Biochem J 368:545–553
33. Popov VN, Simonian RA, Skulachev VP, Starkov AA (1997) Inhibition of the alternative oxidase stimulates H_2O_2 production in plant mitochondria. FEBS Lett 415:87–90
34. Sorgato MC, Sartorelli L, Loschen G, Azzi A (1974) Oxygen radicals and hydrogen peroxide in rat brain mitochondria. FEBS Lett 45:92–95
35. Tretter L, Adam-Vizi V (2004) Generation of reactive oxygen species in the reaction catalyzed by alpha-ketoglutarate dehydrogenase. J Neurosci 24:7771–7778

36. Zhou M, Diwu Z, Panchuk-Voloshina N, Haugland RP (1997) A stable nonfluorescent derivative of resorufin for the fluorometric determination of trace hydrogen peroxide: applications in detecting the activity of phagocyte NADPH oxidase and other oxidases. Anal Biochem 253:162–168
37. Starkov AA, Fiskum G (2001) Myxothiazol induces H(2)O(2) production from mitochondrial respiratory chain. Biochem Biophys Res Commun 281:645–650
38. Votyakova TV, Reynolds IJ (2001) DeltaPsi(m)-Dependent and -independent production of reactive oxygen species by rat brain mitochondria. J Neurochem 79:266–277
39. Starkov AA (2008) The role of mitochondria in reactive oxygen species metabolism and signaling. Ann NY Acad Sci 1147:37–52
40. Sims NR (1990) Rapid isolation of metabolically active mitochondria from rat brain and subregions using Percoll density gradient centrifugation. J Neurochem 55:698–707
41. O'Donnell-Tormey J, Nathan CF, Lanks K, DeBoer CJ, de la Harpe J (1987) Secretion of pyruvate. An antioxidant defense of mammalian cells. J Exp Med 165:500–514
42. Varma SD, Devamanoharan PS, Morris SM (1990) Photoinduction of cataracts in rat lens in vitro. Preventive effect of pyruvate. Exp Eye Res 50:805–812
43. Nath KA, Ngo EO, Hebbel RP, Croatt AJ, Zhou B, Nutter LM (1995) Alpha-ketoacids scavenge H_2O_2 in vitro and in vivo and reduce menadione-induced DNA injury and cytotoxicity. Am J Physiol 268: C227–C236
44. Nath KA, Enright H, Nutter L, Fischereder M, Zou JN, Hebbel RP (1994) Effect of pyruvate on oxidant injury to isolated and cellular DNA. Kidney Int 45:166–176
45. Desagher S, Glowinski J, Premont J (1997) Pyruvate protects neurons against hydrogen peroxide-induced toxicity. J Neurosci 17:9060–9067
46. Holleman MAF (1904) Notice sur l'action de l'eau oxygenee sur les acides a-cetoniques et sur les dicetones 1.2. Recl Trav Chim Pays-Bas Belg 23:169–172

Chapter 17

In Vivo Detection of Oxidized Proteins: A Practical Approach to Tissue-Derived Mitochondria

Francesca Maltecca and Giorgio Casari

Abstract

Mitochondria are the major producers of free radical oxygen species (ROS) as well as the major target of oxidative damage. Defects in the mitochondrial respiratory chain complexes can increase ROS production and reduce ROS removal, leading to oxidative modification of proteins, lipids, and DNA. AAA proteases of the inner mitochondrial membrane, paraplegin and AFG3L2, participate in the biogenesis and maintenance of respiratory chain complexes. These proteins form hetero-oligomeric paraplegin/AFG3L2 and homo-oligomeric AFG3L2 complexes named *m*-AAA proteases. Inactivation of *m*-AAA proteases causes respiratory defects and altered mitochondrial morphology both in yeast and in mammals. In fact, mouse models defective for *Afg3l2* display a very severe neurological syndrome and die within two weeks after birth. They display widespread morphological alterations of mitochondria in the central and peripheral nervous system and deficiencies in respiratory chain complex I and in complex III, which are major producers of ROS in physiological and especially in pathological conditions. Therefore, an efficient and reliable methodology to monitor the effect of increased ROS production is useful for accurately phenotyping cellular and animal models mutants in *m*-AAA. By measuring carbonyl formation as marker of protein oxidation, we have shown that respiratory defects cause oxidative damage in *Afg3l2* mutants, indicating that oxidative stress is crucial in the pathogenesis of *m*-AAA deficiency.

Key words: Reactive oxygen species, Carbonylated proteins, Mitochondrial insults, Mitochondrial AAA-protease, Neurological disease, Protein quality control

1. Introduction

Mitochondria are important physiologic generators of ROS for cell signaling, but they can also be important contributors to oxidative damage. They contain multiple electron carriers capable of producing ROS, as well as an extensive network of antioxidant defence. Mitochondrial insults, including oxidative damage itself, can cause an imbalance between ROS production and removal,

Peter Bross and Niels Gregersen (eds.), *Protein Misfolding and Cellular Stress in Disease and Aging: Concepts and Protocols*, Methods in Molecular Biology, vol. 648, DOI 10.1007/978-1-60761-756-3_17, © Springer Science+Business Media, LLC 2010

resulting in net ROS production and oxidative modification of proteins, lipids and DNA (1, 2). Mitochondrial dysfunction and oxidative damage are known to be joint contributors to neurodegeneration and aging (3).

The electron transport chain is a major generator of ROS in cells: high-throughput electron transport down the respiratory chain inevitably results in the escape of unpaired electrons and generation of superoxide anion ($O_2^{\bullet -}$), largely at complex I and complex III (2, 4). Pharmacological inhibition of these complexes is known to diminish ATP production, increase leakage of electrons, and increase superoxide production as well as reduce ROS removal (2, 5).

Key components in the control of assembly and functionality of respiratory chain complexes in yeast and in mammals are ATP-dependent metalloproteases of the AAA-superfamily (ATPases associated with a variety of cellular activities): paraplegin and AFG3L2 (6–9). They are ubiquitous nuclear-encoded mitochondrial proteins that form hetero-oligomeric paraplegin/AFG3L2 and homo-oligomeric AFG3L2 complexes in the inner mitochondrial membrane, named *m*-AAA proteases (10). Dysfunction of *m*-AAA protease indeed results in impaired respiration both in yeast and in mammals (6, 8, 9).

Mutations of paraplegin cause hereditary spastic paraplegia (HSP), characterized by axonal degeneration of the longest motor and sensory axons of the central nervous system (11). We previously demonstrated that HSP fibroblasts lacking the *m*-AAA complex have a reduced respiratory complex I activity and increased sensitivity to oxidant stress (8).

We have developed and characterized mutant mouse models defective in *Afg3l2*: a spontaneous mutant carrier of a missense mutation in the core AAA domain of the protein ($Afg3l2^{par/par}$) and a complete knockout ($Afg3l2^{Emv66/Emv66}$). Contrasting with the mild axonal degeneration of paraplegin-deficient mouse (12), both mouse models defective in *Afg3l2* display a severe defect in axonal development, which results in early lethality at P16. The lack of functional AFG3L2 has deep effects on mitochondrial morphology and metabolism. Massive accumulation of swollen mitochondria with damaged cristae has been observed in neuronal cell bodies of the central and peripheral nervous system. Moreover, biochemical studies have revealed that enzymatic activities of respiratory chain complex I and complex III are strikingly impaired in *Afg3l2* models and result in a highly reduced ATP production (9). We thus evaluated whether these respiratory defects cause oxidative damage in *Afg3l2* mutants measuring carbonyl formation, an easy detectable marker of protein oxidation (13). Carbonyl levels after reaction with dinitrophenyl hydrazine (DNPH) are measured by Western blot on brain mitochondrial-enriched preparations from both *Afg3l2* mutant mice and controls.

We found that the level of protein carbonylation is increased in *Afg3l2* mutants suggesting that oxidative stress can play a role in determining *Afg3l2* mice phenotype.

2. Materials

2.1. Preparation of Mitochondria from Mouse Tissue

1. 3-N-morpholino propanesulfonic acid (MOPS, Sigma Aldrich, St. Louis, Missouri, USA) is dissolved in sterile water at 0.5 M; NaOH is used to adjust pH to 7.2.
2. Solutions of sucrose (1 M) and ethylenediamine tetraacetic acid (EDTA) (0.5 M) are prepared by dissolving the powder in sterile water.
3. 1,4-Dithio-DL-threitol (DTT): is dissolved at 1 M in sterile water and stored in aliquots at –20°C.
4. Phenylmethylsulphonyl fluoride (PMSF, Sigma Aldrich): is dissolved at 1 M in ethanol and stored at –20°C.
5. Triton X-100.
6. Tissue lysis buffer: 20 mM MOPS, pH 7.2, 0.44 M Sucrose, 1 mM EDTA, 0.2 mM PMSF, 50 mM DTT.
7. Mitochondria lysis buffer: 20 mM MOPS, pH 7.2, 0.44 M Sucrose, 1 mM EDTA, 0.2 mM PMSF, 50 mM DTT, 0.5% Triton X-100 (v/v).
8. Dounce tissue grinder (Wheaton, Millville NJ, USA).

2.2. Derivatization Reaction

1. Dinitrophenyl hydrazine (DNPH, Fluka-Sigma Aldrich): 10 mM, solution is prepared by carefully dissolving the powder at in HCl 2 N under a chemical hood.
2. Trichloroacetic acid (TCA): 20% (w/v).
3. Ethanol/ethyl acetate (Merck, Whitehouse Station, N.J. USA): ratio 1:1 (v/v).
4. Sample loading buffer 6X: 0.35 M Tris–HCl, pH 6.8, 30% (w/v) glycerol, 10% (w/v) SDS, 0.6 M DTT, 0.012% (w/v) bromophenol blue. Aliquots are stored at -20°C.

2.3. SDS-Polyacrylamide Gel Electrophoresis

1. Separating buffer: 1.5 M Tris–HCl, pH 8.8.
2. Stacking buffer: 0.5 M Tris–HCl, pH 6.8.
3. Acrylamide mix: 30% (w/v) acrylamide/bis solution: 37.5:1; Serva.
4. *N,N,N′,N′*-Tetramethylethylenediamine (TEMED).
5. Ammonium persulfate (APS): dissolve at 10% (w/v) in sterile water and immediately freeze in single use aliquots at –20°C.
6. Isopropanol.

7. Running buffer (10X): 250 mM Tris, 1.92 M glycine, 1% (w/v) SDS.
8. Prestained molecular weight marker: 7–175 kDa (New England Biolabs, Ipswich, MA, USA).

2.4. Western Blot of DNP-Derivatized Proteins

1. Methanol from Merck.
2. Phosphate Buffered Saline (PBS) 10× (Invitrogen).
3. Transfer buffer: 250 mM Tris, 1,92 M glycine, 20% Methanol (v/v). Store at 4°C.
4. Hybond-ECL nitrocellulose membrane (GE Healthcare, Uppsala, Sweden).
5. Filter papers (3MM, Whatman)
6. PBS with Tween (PBS-T): dilute 10× PBS to 1× in sterile water with the addition of 0.5% Tween-20 (v/v).
7. Red Ponceau S (Sigma Aldrich) solution: dissolve at 0.1% (w/v) in 1% (v/v) acetic acid.
8. Blocking buffer and antibody dilution buffer: 5% (w/v) non-fat dry milk in PBS-T.
9. Rabbit polyclonal anti-DNPH antibody (Molecular Probe, Invitrogen), mouse monoclonal antibody against respiratory chain complex I 39 kDa subunit (NDUFA9) and anti-porin antibody (Invitrogen).
10. Secondary antibodies: anti-mouse and anti-rabbit IgG, horseradish peroxidase (HRP)-linked species-specific whole antibodies (GE Healthcare).
11. Enhanced Chemiluminescence (ECL) HRP Substrate Reagent (GE Healthcare) and ECL hyperfilms (GE Healthcare).

3. Methods

Proteins are one of the major targets of oxygen-free radicals and other reactive species and the oxidative modification can modulate biochemical characteristics of proteins such as enzymatic activity, DNA-binding activities of transcription factors, and the susceptibility to proteolytic degradation. Metal catalyzed oxidation of proteins introduces carbonyl groups (aldehydes and ketones) at lysine, arginine, proline or threonine residues in a site-specific manner (13, 14). The experimental procedure described in this chapter allows simple and sensitive immunodetection of carbonyl groups, providing information on the oxidation status of proteins.

The carbonyl groups in the protein side chains are derivatized to 2,4 dinitrophenylhydrazone (DNP-hydrazone) by reaction

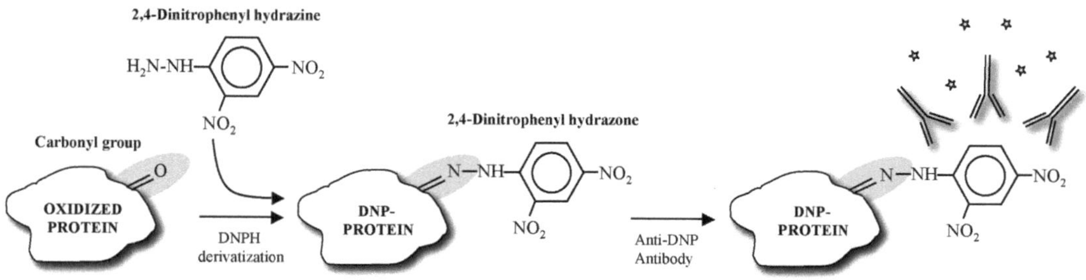

Fig. 1. Schematic representation of DNPH derivatization.

with 2,4-dinitrophenylhydrazine (DNPH). The DNP-derivatized protein samples are separated by SDS-polyacrylamide gel electrophoresis (SDS-PAGE) followed by immunoblot using a primary antibody specific to the DNP moiety of the proteins (Fig. 1).

The protocol is suited for in vivo evaluation of oxidized mitochondrial proteins derived from mouse tissue. In addition, the procedure allows the quantitative comparison of carbonyl levels in physiological and pathological conditions.

In order to avoid the oxidation of proteins that may occur after tissue lysis, we recommend adding a reducing agent to the lysis buffer (50 mM DTT), which is sufficient to prevent oxidation, but has no adverse effect on the derivatization reaction. For the same reason, we suggest to perform the derivatization reaction on fresh mitochondrial lysates, possibly avoiding frozen preparation.

3.1. Isolation of Mitochondria from Mouse Brain

1. *Afg3l2*$^{Emv66/Emv66}$, *Afg3l2*$^{par/par}$ mice and their respective syngenic controls are sacrificed by decapitation at P14 (see Note 1). Skin and ears are removed from skull, exposing the ear canals. The skull is then opened to expose the brain, which is rapidly picked up using a spatula and removed. One part (about 150 mg) is cut and rapidly submerged in 10 ml ice-cold PBS 1X in a small beaker to remove excess blood, while the rest of the brain is immediately frozen. Usually, two washes are sufficient to completely clarify PBS. The tissue is then minced into small pieces using scissors. This should be performed while keeping the beaker in an ice bath. PBS used during the mincing is discarded and replaced with 2 ml of ice-cold fresh tissue lysis buffer (see Notes 2 and 3). The suspension is then transferred to the glass potter and homogenized with 20 up and down strokes (see Notes 4 and 5). Tissue residues must be completely dissolved.
2. Brain homogenate is then transferred in a 2-ml eppendorf tube and centrifuged at 1,000 g for 10 min at 4°C to pellet cell debris and nuclei. The supernatant is transferred to 2-ml eppendorf tubes and centrifuged at 11,000 g for 20 min at 4°C.

The supernatant is discarded and the mitochondrial pellet is resuspended in the mitochondrial lysis buffer (see Notes 6 and 8). To avoid the formation of bubbles during the resuspension process, a 1-ml pipettor is used. After centrifugation at 11,000 g for 10 min, the supernatant is collected for derivatization. The mitochondrial protein suspension is stored on ice for the experiments (see Note 7).

3. Mitochondrial protein concentration is measured using a modified Bradford method (Bio-Rad Laboratories, Hercules, CA, USA according to the manufacturer's instruction) (see Notes 8 and 9).

3.2. Derivatization of Mitochondrial Proteins

1. Two aliquots of each sample to be analyzed are treated. One aliquot is subjected to the derivatization reaction, while the other aliquot will serve as negative control. Fifty micrograms of mitochondrial proteins from each sample are diluted with lysis buffer to 100 μl and incubated with an equal volume of DNPH (see Notes 10 and 11). Negative controls are incubated with 100 μl of 2 N HCl only. The reaction mixture is allowed to stand for 15 min at room temperature (see Notes 12 and 13).
2. Proteins are precipitated adding an equal volume of TCA 20% (10% final concentration). Each sample is left for 20 min on ice and then centrifuged at $8{,}000\times g$ for 10 min. Supernatant is carefully removed, leaving protein pellet intact.
3. Protein pellet is washed three times with 1 ml of ethanol/ethyl acetate (1:1) to remove excess DNPH. After the third step, ethanol/ethyl acetate must be carefully removed from the washed pellet evaporating under a hood for at least 1 h or using a rotary evaporator.
4. Dried pellet is resuspended in 50 μl of 20 mM Tris–HCl, pH 6.8 and SDS 0.2% and heated 30′ at 50°C in a hot block to allow complete solubilization (see Note 14).
5. A proper volume of sample loading buffer 6× is added to the resuspended sample (1× final concentration).

3.3. SDS-PAGE

1. The protocol is formatted for a Miniprotean III electrophoresis cell (Biorad). It is crucial that glass plates and combs are cleaned with a laboratory detergent (e.g., Alconox) and rinsed with distilled water after every use.
2. A 10% separating gel is prepared by mixing 2.5 ml of Tris–HCl pH 8.8 1.5 M with 3.3 ml of acrylamide mix, 0.1 ml of SDS10%, 4 ml water, 0.1 ml APS10%, and 4 μl TEMED (total volume: 10 ml) in a disposable 15 ml Falcon tube. The gel is poured between the glass plates placed on the casting stand leaving space for stacking gel, and overlayed with isopropanol.

The gel should polymerize in about 30 min (see Note 15). The isopropanol is then poured off and the top of the gel is rinsed twice with sterile water.

3. The stacking gel is prepared by mixing 0.38 ml of 0.5 M Tris–HCl, pH 6.8 with 0.5 ml of acrylamide mix, 30 μl of 10% SDS, 4 ml water, 30 μl of 10% APS and 3 μl of TEMED (total volume: 3 ml). After pouring the stacking up to the top of the gel, the comb is inserted. The stacking gel should polymerize in about 30 min.
4. The running buffer is prepared by diluting 100 ml of the 10× running buffer and 10 ml of 10% SDS in 890 ml of sterile water in a measuring cylinder. The cylinder is then covered with parafilm and inverted to allow solution mixing.
5. Once the stacking gel has set, the comb is removed by gently pulling it straight up. The wells have to be completely rinsed with distilled water or running buffer to remove residues of polymerized acrylamide.
6. After being removed from the casting stand, the gel cassette is placed into the electrophoresis module. The upper and lower chambers of the gel unit are filled with running buffer. 18 μl of each sample (corresponding to about 10 μg of derivatized proteins) are loaded. It is important to include one well for the untreated samples (treated with HCl only) and for prestained molecular weight marker.
7. After completing its assembly, the gel unit is connected to a power supply. The gel is run at 60 V through the stacking gel and at 80 V through the running gel. The run is stopped when the dye front reaches the bottom of the gel, in order to retain all the bands of the molecular weight marker.

3.4. Western Blotting of DNP-Proteins

1. Samples that have been separated by SDS-PAGE are then electrophoretically transferred to Hybond-ECL nitrocellulose membrane. The protocol is formatted for the use of a Mini Trans-Blot Electrophoretic Transfer Cell (Biorad).
2. The ice cooling unit is taken out from the laboratory freezer at −80°C ready to use. The Hybond membrane and the 3MM filter papers (see Note 16) are cut to the dimension of the gel and prewet, together with sponges in Transfer buffer 1× for 15 min. The gel is also equilibrated in Transfer buffer 1×. The stacking gel is removed and discarded.
3. The gel sandwich is then prepared, ensuring that no bubbles are trapped between the gel and the membrane. The transfer cassette, with the black side down, is placed on a clean surface. The pre-wetted sponges are placed on the black side of the cassette, followed by two sheets of filter paper, the equilibrated gel and the pre-wetted membrane. Sandwich preparation

is completed when placing two sheets of filter paper on the membrane, followed by the last fiber sponge. The sandwich is then placed in the transfer apparatus, containing the frozen Bio-Ice cooling unit. The tank is then completely filled with transfer buffer. It is important using chilled buffer to improve heat dissipation.

4. The lid is put on the tank and the power supply activated. Transfer can be accomplished at either 30 V overnight or 100 V for 1 h. A standard stir bar is also added to help maintain even buffer temperature and ion distribution in the tank. Upon completion of the run, the blotting sandwich is disassembled and the blotted membrane removed for development. The membrane is rinsed well with deionized water, immersed in a proper amount of Ponceau S Staining Solution and stained for 5 min to check transfer efficiency (see Notes 17 and 18). After staining, membrane desaturation is achieved by washing three times in water. Images are digitally recorded using a scanner.
5. The membrane is then washed in PBS-T until the red staining is completely disappeared. At this point, it is immersed in Blocking buffer and blocked for 1 h at room temperature with constant agitation.
6. The blocking buffer is discarded and the blotted membrane is incubated for 2 h at room temperature with anti-DNP primary antibody diluted 1:5,000 in 5% milk in PBS-T.
7. The primary antibody is then discarded and the membrane washed three times for 10 min each with PBS-T.
8. The secondary anti-rabbit antibody is freshly prepared for each experiment as 1:10,000-fold dilution in 5% milk in PBS-T and added to the membrane for 1 h at room temperature with constant agitation. The secondary antibody is then discarded and the membrane washed three times for 10 min each with PBS-T. After the final wash, equal volumes of each portion of the ECL reagent (0.1 ml/cm^2) are mixed together and equilibrated for 5 min. The membrane is incubated with detection solution for 1 min at room temperature in a dark room under safe light condition. The excess detection mix is then removed and the membrane is placed inside a plastic page protector, protein side up, in a film cassette.
9. In the dark room, an ECL film is carefully placed on top of the membrane. The film has to be exposed for the appropriate amount of time. Thirty seconds exposition should be sufficient to get signal with good intensity (Fig. 2).
10. The membrane is washed for 30 min in PBS-T and incubated for 2 h at room temperature with anti-respiratory chain complex I 39 kDa subunit (NDUFA9) or anti-porin antibody for

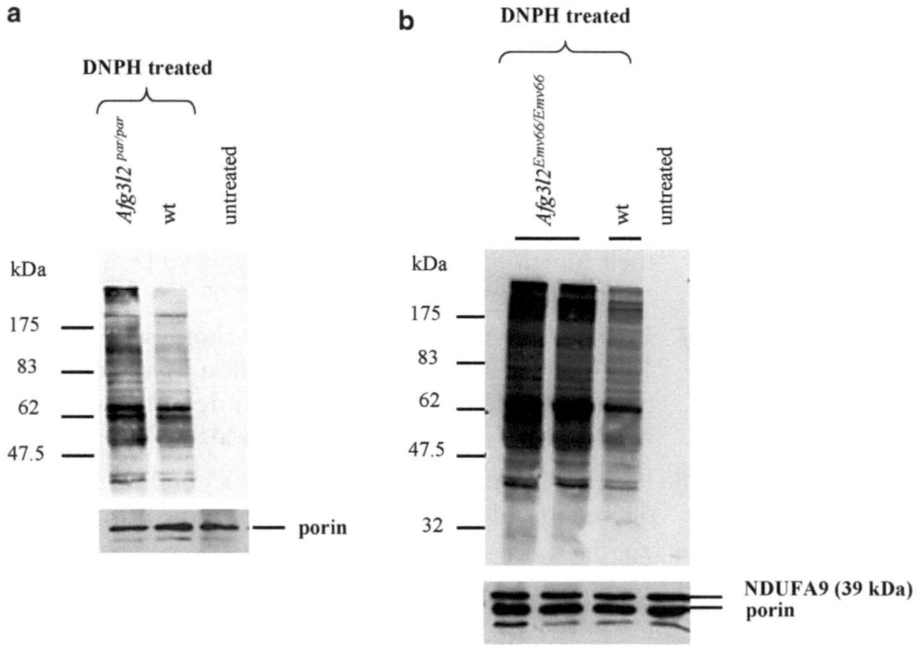

Fig. 2. Increased level of oxidized mitochondrial proteins in *Afg3l2* mutant mice. Carbonyl levels after reaction with DNPH are measured by Western blot on mitochondrial-enriched preparations from brain from both *Afg3l2* mutant mice and controls. Protein carbonylation is increased in *Afg3l2* mutants. Mitochondrial enrichments from brain tissue (*par/par* panel a, *Emv66/Emv66* panel b) are treated with DNPH. Derivatized proteins are analyzed by SDS-PAGE on a 10% gel. Anti-porin and anti-39 kDa antibodies are used to verify equal loading.

loading control. The procedure for secondary antibody incubation and ECL detection follows exactly the same instructions described above.

4. Notes

1. Local and national regulations on animal care and handling vary. It is important to check holding the appropriate authorization to perform animal experiments.
2. All buffers are prepared the same day of the experiment, to avoid bacterial/yeast growth in stored buffers. Since pH depends on temperature, measure the pH of all solutions at 25°C.
3. The optimal ratio between tissue and isolation buffer ranges between 1:5 and 1:10 (w:v).
4. All glassware and dounce have to be washed three times with bidistilled water to avoid Ca^{2+} contamination. Ca^{2+} overload is the most common cause for the dysfunction of isolated mitochondria.

5. Homogenization, as well as the following steps, must be performed at 4°C to minimize activation of damaging phospholipases and proteases. Dounce should be precooled in an ice-bath 5 min before starting the procedure.
6. The minimum volume of buffer should be used to resuspend mitochondria. The protein concentration should be between 5 and 10 mg/ml. Avoid diluting mitochondria as they retain their functionality for a longer time when kept concentrated, minimizing exposure to oxygen.
7. Perform the derivatization of mitochondrial proteins within 1–3 h after mitochondrial preparation. Avoid freezing mitochondrial preparation and perform derivatization later. This will help preventing sample oxidation.
8. For protein quantification, the mitochondrial lysate must be diluted in order to avoid saturation of the probe.
9. DNPH is very toxic and harmful if swallowed or inhaled; avoid contact with skin and eyes, wear gloves and eye protection, use under chemical hood.
10. Generally, 20–50 μg of protein is recommended per derivatization reaction.
11. Avoid high protein concentration (>10 μg/μl) to assure solubility during the derivatization reaction.
12. Do not allow the derivatization reaction to proceed more than 15 min, as side reactions other than hydrazone linkage may occur. This also prevents the effect of acid hydrolysis of proteins.
13. Some protocols suggest the addition of SDS (1–5% w/v) to protein samples to increase the efficiency of DNPH reaction. We have observed no differences in the efficiency of derivatization between protein samples treated with or without SDS.
14. Do not heat samples prior to loading into the gel. This precaution preserves the stability of the derivatized residues.
15. Check that remaining gel in Falcon tube is polymerized before proceeding.
16. It is important to wear gloves when handling membranes to prevent contamination.
17. Ponceau S staining is a reliable way of measuring total proteins transferred to the membrane and to demonstrate equal loading. Anyway, given the importance of the quantitative comparison in this experiment, more sensitive methods to check protein loading (immunoblot using antibodies against housekeeping proteins or Silver staining) are recommended.
18. Perform Ponceau S staining before blocking to avoid interference with the subsequent antibody reaction with derivatized proteins.

References

1. Jezek P, Hlavata L (2005) Mitochondria in homeostasis of reactive oxygen species in cell, tissues, and organism. Int J Biochem Cell Biol 37:2478–2503
2. Murphy MP (2009) How mitochondria produce reactive oxygen species. Biochem J 417: 1–13
3. Lin MT, Beal MF (2006) Mitochondrial dysfunction and oxidative stress in neurodegenerative diseases. Nature 443:787–795
4. St-Pierre J, Buckingham JA, Roebuck SJ, Brand MD (2002) Topology of superoxide production from different sites in the mitochondrial electron transport chain. J Biol Chem 277:44784–44790
5. Andreyev AY, Kushnareva YE, Starkov AA (2005) Mitochondrial metabolism of reactive oxygen species. Biochemistry (Mosc) 70:200–214
6. Arlt H, Tauer R, Feldmann H, Neupert W, Langer T (1996) The YTA10-12 complex, an AAA protease with chaperone-like activity in the inner membrane of mitochondria. Cell 85:875–885
7. Arlt H, Steglich G, Perryman R, Guiard B, Neupert W, Langer T (1998) The formation of respiratory chain complexes in mitochondria is under the proteolytic control of the m-AAA protease. EMBO J 17:4837–4847
8. Atorino L, Silvestri L, Koppen M, Cassina L, Ballabio A, Marconi R, Langer T, Casari G (2003) Loss of m-AAA protease in mitochondria causes complex I deficiency and increased sensitivity to oxidative stress in hereditary spastic paraplegia. J Cell Biol 163:777–787
9. Maltecca F, Aghaie A, Schroeder DG, Cassina L, Taylor BA, Phillips SJ, Malaguti M, Previtali S, Guenet JL, Quattrini A, Cox GA, Casari G (2008) The mitochondrial protease AFG3L2 is essential for axonal development. J Neurosci 28:2827–2836
10. Koppen M, Metodiev MD, Casari G, Rugarli EI, Langer T (2007) Variable and tissue-specific subunit composition of mitochondrial m-AAA protease complexes linked to hereditary spastic paraplegia. Mol Cell Biol 27:758–767
11. Casari G, De Fusco M, Ciarmatori S, Zeviani M, Mora M, Fernandez P, De Michele G, Filla A, Cocozza S, Marconi R, Dürr A, Fontaine B, Ballabio A (1998) Spastic paraplegia and OXPHOS impairment caused by mutations in paraplegin, a nuclear-encoded mitochondrial metalloprotease. Cell 93:973–983
12. Ferreirinha F, Quattrini A, Pirozzi M, Valsecchi V, Dina G, Broccoli V, Auricchio A, Piemonte F, Tozzi G, Gaeta L, Casari G, Ballabio A, Rugarli EI (2004) Axonal degeneration in paraplegin-deficient mice is associated with abnormal mitochondria and impairment of axonal transport. J Clin Invest 113:231–242
13. Requena JR, Levine RL, Stadtman ER (2003) Recent advances in the analysis of oxidized proteins. Amino Acids 25:221–226
14. Nystrom T (2005) Role of oxidative carbonylation in protein quality control and senescence. EMBO J 24:1311–1317

Chapter 18

Measurement of Oxidized/Reduced Glutathione Ratio

Joshua B. Owen and D. Allan Butterfield

Abstract

Glutathione (GSH) is the most abundant antioxidant in aerobic cells, present in micromolar (μM) concentrations in bodily fluids and in millimolar (mM) concentrations in tissue. GSH is critical for protecting the brain from oxidative stress, acting as a free radical scavenger and inhibitor of lipid peroxidation. GSH also participates in the detoxification of hydrogen peroxide by various glutathione peroxidases. The ratio of reduced GSH to oxidized GSH (GSSG) is an indicator of cellular health, with reduced GSH constituting up to 98% of cellular GSH under normal conditions. However, the GSH/GSSG ratio is reduced in neurodegenerative diseases, such as Parkinson's disease (PD) and Alzheimer's disease (AD). Measuring the GSH/GSSG ratio in pathological tissues and experimental models thereof in comparison to the results in controls is an excellent way to assess potential therapeutics efficacy in maintaining cellular redox potential. The availability of UV/Visible instruments equipped with 96-well plate readers as common laboratory equipment has made measuring the GSH/GSSG ratio on multiple samples a manageable procedure.

Key words: Glutathione, GSH, Glutathione disulfide, GSSG, GSH/GSSG ratio, Redox potential

1. Introduction

Glutathione (GSH) is a tri-peptide (γ-glutamylcysteinylglycine) that acts as an endogenous antioxidant, a xenobiotic detoxifier, and is involved in metabolic regulation. GSH is the most abundant antioxidant in aerobic cells, present in micromolar (μM) concentrations in bodily fluids and in millimolar (mM) concentrations in tissue (1). The central nervous system (CNS) has GSH concentrations ranging from 1 to 3 mM, depending on the region. The forebrain and cortex have the highest concentration of GSH, followed by the cerebellum, brain stem, and spinal cord (1). With high oxygen consumption and rich poly-unsaturated

Peter Bross and Niels Gregersen (eds.), *Protein Misfolding and Cellular Stress in Disease and Aging: Concepts and Protocols*, Methods in Molecular Biology, vol. 648, DOI 10.1007/978-1-60761-756-3_18, © Springer Science+Business Media, LLC 2010

fatty acid (PUFAs) content, the brain is particularly susceptible to oxidative stress. GSH is critical for protecting the brain from oxidative stress, acting as a free radical scavenger and inhibitor of lipid peroxidation (1).

GSH is synthesized from l-glutamate, l-cysteine, and l-glycine in two ATP requiring steps catalyzed by the enzymes γ-glutamyl-cysteine ligase and glutathione synthetase. The cysteine thiol moiety gives GSH its antioxidant properties. The thiol is oxidized by cellular pro-oxidants, such as free radicals and reactive aldehydes, to form oxidized GSH disulfide (GSSG). The reduction of GSSG back to GSH requires NADPH and is catalyzed by the enzyme glutathione reductase, thus regenerating GSH for cellular antioxidant defense. In addition to the reactions listed that produce and regenerate GSH, GSH is degraded by γ-glutamyl transpeptidase to form glutamate and cysteinyl glycine. The reactions described are summarized in Fig. 1.

Under normal conditions, reduced GSH is the most prevalent form of GSH, constituting up to 98% of the total GSH pool (2, 3). However, the GSH/GSSG ratio decreases during normal aging and is further depleted in neurodegenerative diseases, such as Parkinson's disease (PD) and Alzheimer's disease (AD) (1, 4–7). Using compounds to increase the GSH/GSSG ratio conceivably could be used to promote healthy aging and as therapeutics for

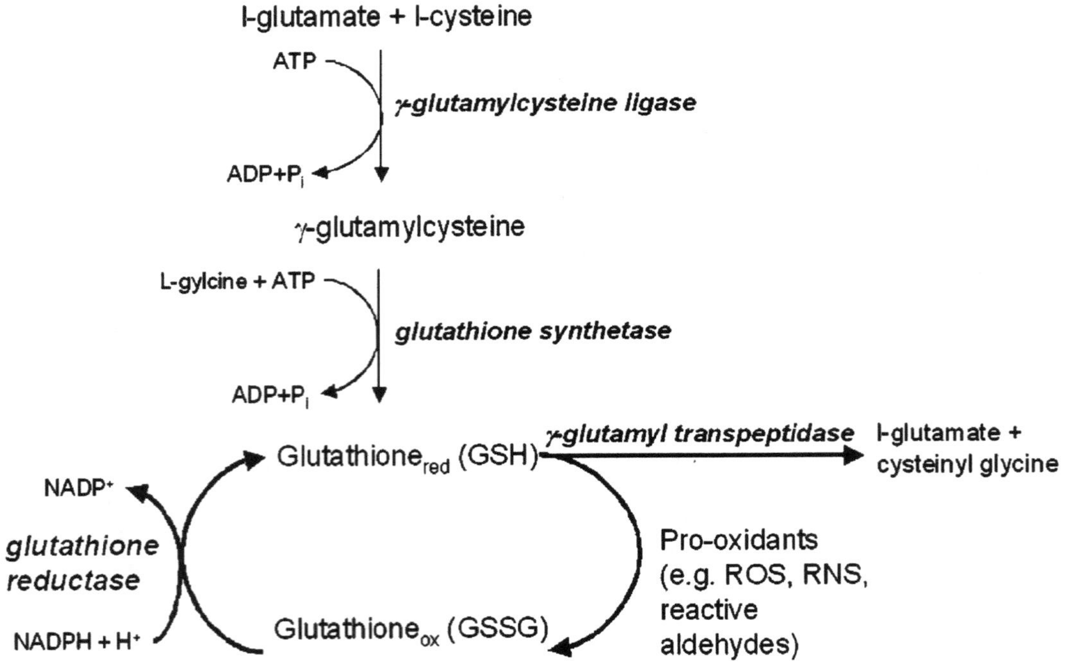

Fig. 1. Synthesis, degradation, and regeneration of glutathione: GSH is synthesized in 2 steps catalyzed by the enzymes γ-glutamyl cysteine ligase and glutathione synthetase. GSH is degraded by γ-glutamyl transpeptidase. GSSG is reduced to GSH by glutathione reductase at the expense of NADPH.

neurodegenerative diseases. Supporting this hypothesis, studies using γ-glutamyl cysteine ethyl ester (GCEE), a substrate that is converted to GSH, and GSH mimetics, such as tricyclodecan-9-yl-xanthogenate (D609), have proven protective in models of oxidative stress in neurodegenerative disease (7–12). Measuring the GSH/GSSG ratio in experimental models and comparing the results to control models is an excellent way of assessing the efficacy of potential therapeutics in maintaining cellular redox potential. The cellular redox potential is critical for normal cellular physiology, and the GSH/GSSG ratio as indicated, is the best indicator of this redox potential (13, 14).

2. Materials

2.1. Sample Preparation and Deproteination

1. Phosphate Buffer Solution (PBS): Prepare 176 mM dibasic sodium phosphate (Sigma, St. Louis, MO) and 2.57 M sodium chloride to make Solution A and prepare 230 mM monobasic sodium phosphate to make Solution B. The pH of Solution A is brought to 8 with Solution B. The resultant A/B solution is 10× concentration and can be stored at 4°C. Dilute to 1× concentration to desired volume.
2. 2-(N-morpholino)ethanesulphonic acid (MES) buffer (Caymen Chemical, Ann Arbor, MI) is composed of 0.4 M MES, 0.1 M phosphate, and 2 mM EDTA, at pH 6.
3. Triethanolamine (TEAM) (Sigma, St. Louis, MO) is diluted to 4 M.
4. 10% w/v Metaphosphoric Acid (MPA) solution (Sigma, St. Louis, MO): Prepare by dissolving 5 g of MPA in 50 mL of deionized (DI) water.
5. EDTA blood collection tubes (0.5 M, pH 7.00: 50–100 μL)
6. HPLC-grade water (Sigma, St. Louis, MO).

2.2. Experiment 1 Total GSH Determination

1. Ellman's reagent 5,5-Dithio-*bis*-(2-nitrobenzoic acid) (DTNB) (Caymen Chemical, Ann Arbor, MI)) comes in a lyophilized powder. Add 500 μL DI water to reconstitute the solution. Use the solution within 10 min.
2. GSSG standard (Caymen Chemical, Ann Arbor, MI), 25 μM and ready to use from the manufacturer.
3. GSH Co-Factor Mixture (Caymen Chemical, Ann Arbor, MI) contains lyophilized $NADP^+$ and glucose-6-phosphate. Reconstitute the compounds in 500 μL DI water and vortex. Solution is stable at 0–4°C for 2 weeks.
4. MES buffer (see Subheading 2.1 step 2).

5. GSH Enzyme Mixture (Caymen Chemical, Ann Arbor, MI) contains glutathione reductase and glucose-6-phosphate dehydrogenase in 200 μL of buffer. Add 2 mL of MES buffer (from step 2) to vial and vortex. Solution is stable at 0–4°C for 2 weeks.

2.3. Experiment 2 GSSG Determination

1. 2-Vinylpyridine (1 M) solution dissolved in ethanol.
2. All reagents from Subheading 2.2 (see Note 1).

3. Methods

Sample preparation for the GSH/GSSG ratio assay is critical and is dependent upon tissue and/or bodily fluid type (15). Tissues low in γ-glutamyl transpeptidase concentration, such as brain, can be frozen at −80°C and assayed for GSH/GSSG at a later date. However, tissues high in γ-glutamyl transpeptidase, such as the kidneys and pancreas, should be processed immediately upon sacrifice and never frozen. The thawing process accelerates γ-glutamyl transpeptidase activity, potentially leading to misleading results. Bodily fluids, such as plasma, must also be processed immediately after isolation because GSH oxidation occurs rapidly compared to tissue. However, once plasma has been processed, the aliquot can be frozen at −80°C (See Note 2).

Measuring the GSH/GSSG ratio can be accomplished using a UV-Visible spectrophotometer equipped with a 96-well plate reader. Such instrumentation is the best way to measure the GSH/GSSG ratio with multiple samples. Two separate experiments, that is, measuring total GSH levels followed by measuring GSSG levels, are required to determine the GSH/GSSG ratio.

Experiment 1: Measuring total GSH. Measuring total GSH utilizes the reaction of GSH and GSSG with Ellman's reagent (DTNB) to produce 2-nitrobenzoic acid anion (TNB^-), a compound containing a chromophore that absorbs at 412 nm (16). These reactions are catalyzed by glutathione reductase, which requires NADPH as a cofactor. (To supply the reactions with NADPH, glucose-6-phosphate (G6P) and glucose-6-phosphate dehydrogenase (G6PDH) are present in the enzyme mixture mentioned in Subheading 2.2.) Since both GSH and GSSG react with DTNB to form TNB^-, an intensely colored reagent that absorbs at 412 nm, the total level of GSH is measured in the cell.

Experiment 2: Measuring GSSG. To measure GSSG only, reduced GSH is derivatized with 2-vinylpyridine (17). This modification ensures that GSSG is the only form of glutathione that can react with the DTNB reagent. The GSSG level is subtracted

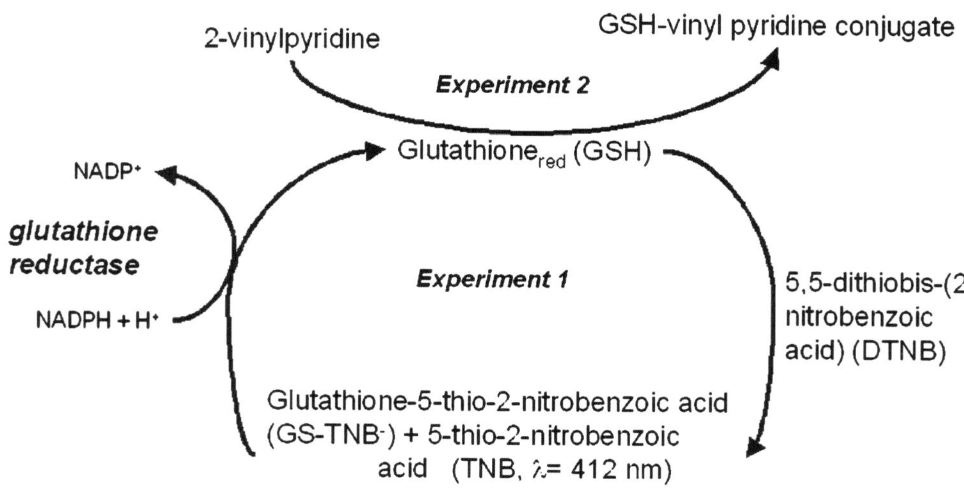

Fig. 2. Summary or experiments to measure total GSH and GSSG: Total GSH is measured in Experiment 1 by reacting GSH and GSSG with DTNB to yield 5-thio-2-nitrobenzoic acid (λ = 412 nm). Experiment 2 derivatizes GSH with 2-vinylpyridine and only GSSG undergoes the DTNB reaction, measuring GSSG levels.

from the results of Experiment 1 to yield the concentration of reduced GSH. The reactions of Experiment 1 and 2 are summarized in Fig. 2.

3.1. Sample Preparation for Tissue

1. Quickly and humanely sacrifice and exsanguinate all animals in accordance to all Institutional Animal Care and Use Committee (IACUC)-approved laboratory animal research protocols.
2. Remove tissue, beginning with those containing high levels of γ-glutamyl transpeptidase (e.g., kidneys, pancreas, and intestine). Low trans-peptidase tissues (e.g., liver, heart, and brain) can be washed and snap frozen in liquid nitrogen (−80°C) and stored for later use. Defrost tissue in cold water and perform steps 3–8. Never freeze tissue high in γ-glutamyl transpeptidase (see Note 3).
3. Trim excess fat from tissue.
4. Rinse tissue with PBS, blot dry, and weigh.
5. Place tissue in a tissue homogenizer, stored on ice, and add MES buffer solution (10% volume per gram of tissue). This ensures that each sample will have the same weight/volume concentration, independent of sample mass.
6. Homogenize tissue using a constant number of strokes for each sample.
7. Transfer the homogenate to a centrifuge tube and spin at 10,000 × *g* for 15 min at 4°C.
8. Use the supernatant for GSH assay.
9. Proceed to deproteination step (Subheading 3.3).

3.2. Sample Preparation for Plasma

1. Collect blood samples and add to EDTA blood collection tubes and swirl gently.
2. Transfer blood to a centrifuge tube with a pipette, leaving space at the top of the tube. Minimize air bubble formation during the transfer.
3. Centrifuge at 1,000 × *g* for 10 min at 4°C.
4. Remove the white layer (buffy coat-mostly leukocytes) from the top of the sample.
5. Add ice-cold HPLC grade water (4× the volume of the sample) to lyse the erythrocytes.
6. Centrifuge at 10,000 × *g* for 15 min at 4°C.
7. Collect supernatant and store on ice.
8. Proceed to deproteination step (Subheading 3.3). (Samples can be frozen before deproteination, but samples must be deproteinated before GSH/GSSG ratio assay.)

3.3. Sample Deproteination

1. Add an equal volume of 10% metaphosphoric acid (MPA) to volume of sample and vortex.
2. Let mixture stand at room temperature for 5 min and centrifuge at 2,000 × *g* for 2 min at 4°C.
3. Collect supernatant with caution, do not disturb the pellet. (The supernatant can be frozen at −20°C for 6 months.)
4. Add 50 μL of 4 M triethanolamine (TEAM) per mL of supernatant. Vortex and use the sample for GSH/GSSG ratio assay.

3.4. DTNB-GSSG Enzymatic Recycling Assay for Total GSH

1. Set UV-Visible spectrometer to 414 nm (see Note 4).
2. Program plate reader to read samples at 5 min intervals for 30 min (6 measurements total) (see Note 5).
3. Prepare GSH standards (0, 0.5, 1.0, 2.0, 4.0, 8.0, 12, and 16.0 μM) in tubes labeled A-H. (Example calculation: Add 5 μL GSSG standard and 495 μL MES buffer to obtain 0.25 μM GSSG/0.5 μM GSH.)
4. Add 50 μL of standard in triplicate to 96-well plate.
5. Add 50 μL of sample (see Subheadings 3.2 and 3.3) in triplicate to plate.
6. Cover plate.
7. Prepare assay cocktail composed of MES buffer (11.25 mL), reconstituted Cofactor mixture (0.45 mL), reconstituted Enzyme mixture (2.1 mL), DI water (2.3 mL), and DTNB (0.45 mL). This mixture provides enough cocktail for all 96 wells of the plate.

8. Add 150 μL of assay cocktail to each well, cover plate to ensure complete darkness, and incubate the plate on an orbital shaker to ensure thorough mixing.
9. Insert plate into plate reader and start measurements (see Note 5).
10. Measure the concentration of GSH using either the endpoint or kinetic method. Use the kinetic method if levels of cysteine and other free thiols are expected to be high (see Note 6).
11. The endpoint method requires the average absorbance value (do not forget to subtract the blank) of each standard at the 25 min interval to be plotted vs. standard concentration. Use the following equation to calculate GSH concentrations:

$$[\text{Total GSH}] = \frac{\{(\text{Absorbance at 414 nm}) - (Y\text{ - intercept})\}}{\text{slope}} \times 2 \times \text{Sample Dilution}$$

12. The kinetic method plots the resultant slope of each standard (*i*-slope) vs. GSH concentration. Take measurements every 5 min for 30 min to obtain the *i*-slope values (see Note 5). The standard slopes are plotted against the standard GSH concentrations and the plot slope (*f*-slope) is used in the equation below to calculate the sample GSH concentrations:

$$[\text{Total GSH}] = \frac{\{(\text{Sample}\, i\text{ - slope}) - (Y\text{ - intercept})\}}{f\text{ - slope}} \times 2 \times \text{Sample Dilution}$$

The *i*-slope value for each sample is calculated and inserted into the equation above (see Note 7).

3.5. DTNB-GSSG Enzymatic Recycling Assay for GSSG

1. Follow steps 1–4 in Subheading 3.4 Sample Deproteination.
2. Add 10 μL of 2-vinylpyridine solutions for every mL of sample.
3. Vortex sample and allow the solution to sit at room temperature for 1 h.
4. This procedure has a detection limit of 1 mM GSSG. If samples are suspected of concentrations greater than 1 mM, add TEAM to dilute sample.
5. Treat all standards as described in steps 1–4. 2-Vinylpyridine inhibits color development, and it is important that all samples and standards are incubated for the same amount of time and with identical 2-vinylpyridine treatments.
6. Follow steps 1–12 outlined in Subheading 3.4 'DTNB-GSSG Enzymatic Recycling Assay for Total GSH' to measure total GSSG level.

7. Once a concentration of GSSG is obtained, subtract the GSSG level value from the total GSH concentration obtained in Subheading 3.6 from the corresponding sample to obtain the value for reduced GSH.
8. Use the following equation to obtain the GSH/GSSG ratio:

$$\frac{\text{GSH}}{\text{GSSG}} = \frac{\text{Reduced GSH (Section 3.4 result} - \text{Section 3.5 result)}}{\text{GSSG (Section 3.5 result)}}$$

4. Notes

1. All the buffers and reagents described in Subheadings 2.2 and 2.3 can be purchased in a kit for easy preparation from Caymen Chemical.
2. Only the total GSH content should be measured in plasma using this assay. *The concentration of GSH in plasma is below the detection limit of this assay.* To circumvent this issue, plasma samples can be lyophilized and reconstituted with MES buffer to 1/3 the original volume. Proceed as described in Subheading 3.4.
3. The defrosting process of tissues high in levels of γ-glutamyl transpeptidase (e.g., kidneys, pancreas) will result in increased γ-glutamyl transpeptidase activity and poor experimental results.
4. The UV-Visible plate reader can be set in the range of 405–414 nm.
5. Always do a dry run with the plate reader to ensure that the GSH/GSSG ratio assay measurements are set up correctly. This is especially important using the kinetic measurement. Adding reagents and starting the reading only to find an error in the set up is wasteful of time, tissue, and resources. Checking the software will ensure time and materials are not wasted.
6. Use the kinetic assay if not much is known about the contents of the sample. The kinetic approach gives the most accurate results if the sample contains many free thiols.
7. The endpoint and kinetic method calculations are multiplied by two, in addition to sample dilutions, to account for protein deproteination steps.

References

1. Janaky R, Cruz-Aguado R, Oja SS, Shaw CA (2007) Glutathione in the nervous system: roles in neural function and health and implications for neurological disease. In: Lajtha A, Oja SS, Saransaari P, Schousboe A (eds) Handbook of neurochemistry and molecular neurobiology. Amino acids and peptides in the nervous system, 3rd edn. Springer, New York, pp 349–418
2. Slivka A, Spina MB, Cohen G (1987) Reduced and oxidized glutathione in human and monkey brain. Neurosci Lett 74:112–118
3. Kosower NS, Kosower EM (1978) The glutathione status of cells. Int Rev Cytol 54: 109–160
4. Lovell MA, Ehmann WD, Butler SM, Markesbery WR (1995) Elevated thiobarbituric acid-reactive substances and antioxidant enzyme activity in the brain in Alzheimer's disease. Neurology 45:1594–1601
5. Sofic E, Lange KW, Jellinger K, Riederer P (1992) Reduced and oxidized glutathione in the substantia nigra of patients with Parkinson's disease. Neurosci Lett 142:128–130
6. Adams JD Jr, Wang B, Klaidman LK, LeBel CP, Odunze IN, Shah D (1993) New aspects of brain oxidative stress induced by tert-butylhydroperoxide. Free Radic Biol Med 15: 195–202
7. Sultana R, Piroddi M, Galli F, Butterfield DA (2008) Protein levels and activity of some antioxidant enzymes in hippocampus of subjects with amnestic mild cognitive impairment. Neurochem Res 33:2540–2546
8. Ansari MA, Joshi G, Huang Q, Opii WO, Abdul HM, Sultana R, Butterfield DA (2006) In vivo administration of D609 leads to protection of subsequently isolated gerbil brain mitochondria subjected to in vitro oxidative stress induced by amyloid beta-peptide and other oxidative stressors: relevance to Alzheimer's disease and other oxidative stress-related neurodegenerative disorders. Free Radic Biol Med 41:1694–1703
9. Boyd-Kimball D, Sultana R, Abdul HM, Butterfield DA (2005) Gamma-glutamylcysteine ethyl ester-induced up-regulation of glutathione protects neurons against Abeta(1-42)-mediated oxidative stress and neurotoxicity: implications for Alzheimer's disease. J Neurosci Res 79:700–706
10. Joshi G, Hardas S, Sultana R, St Clair DK, Vore M, Butterfield DA (2007) Glutathione elevation by gamma-glutamyl cysteine ethyl ester as a potential therapeutic strategy for preventing oxidative stress in brain mediated by in vivo administration of adriamycin: implication for chemobrain. J Neurosci Res 85:497–503
11. Reed TT, Owen J, Pierce WM, Sebastian A, Sullivan PG, Butterfield DA (2009) Proteomic identification of nitrated brain proteins in traumatic brain-injured rats treated postinjury with gamma-glutamylcysteine ethyl ester: insights into the role of elevation of glutathione as a potential therapeutic strategy for traumatic brain injury. J Neurosci Res 87:408–417
12. Sultana R, Newman SF, Abdul HM, Cai J, Pierce WM, Klein JB, Merchant M, Butterfield DA (2006) Protective effect of D609 against amyloid-beta1-42-induced oxidative modification of neuronal proteins: redox proteomics study. J Neurosci Res 84:409–417
13. Butterfield DA, Pocernich CB, Drake J (2002) Elevated glutathione as a therapeutic strategy in Alzheimer's disease. Drug Dev Res 56:428–437
14. Kemp M, Go YM, Jones DP (2008) Nonequilibrium thermodynamics of thiol/disulfide redox systems: a perspective on redox systems biology. Free Radic Biol Med 44:921–937
15. Anderson, ME (1996) Glutathione in Free Radicals: a practical approach (Punchard, N, Kelly, FJ, ed.) Oxford University Press, New York, NY, pp 213–226
16. Baker MA, Cerniglia GJ, Zaman A (1990) Microtiter plate assay for the measurement of glutathione and glutathione disulfide in large numbers of biological samples. Anal Biochem 190:360–365
17. Griffith OW (1980) Determination of glutathione and glutathione disulfide using glutathione reductase and 2-vinylpyridine. Anal Biochem 106:207–212

Chapter 19

Determination of Altered Mitochondria Ultrastructure by Electron Microscopy

Shoichi Sasaki

Abstract

Mitochondria play a number of important roles, including production of ATP for the generation of energy, involvement in the regulation of excitotoxicity, involvement in the homeostasis of intracellular Ca^{2+}, production of reactive oxygen species, and the release of cytochrome c, a potent trigger of programmed cell death (apoptosis). Mitochondrial dysfunction has long been implicated in the pathogenesis of neurodegenerative disorders such as Parkinson's disease and amyotrophic lateral sclerosis (ALS), and in physiological conditions such as aging. Mitochondrial dysfunction has also been associated with endoplasmic reticulum (ER) stress caused by the accumulation of misfolded or unfolded proteins due to decreased ER-associated degradation (ERAD), by which misfolded or unfolded proteins are transported from the lumen of the ER to the cytoplasm, where they are ubiquitinated and eventually eliminated. The mitochondria may undergo various morphological alterations reflective of different pathological conditions and diseases. The transmission electron microscope (TEM) remains a powerful tool for the morphological examination of mitochondria and is expected to continue to enhance our understanding of cellular functions and dysfunction. Moreover, electron microscopic study has enabled us to confirm mitochondrial involvement in the pathomechanism of certain neurodegenerative diseases. Here, electron-microscopy procedures are outlined with particular emphasis on some practical aspects of this approach. Electron-micrographs of mitochondrial alterations under various normal and pathological conditions are provided, including images from human control individuals, patients with ALS, and familial ALS-related mutant SOD1 (G93A and H46R) transgenic mice.

Key words: Mitochondria, Outer membrane, Inner membrane intermembrane space, Cristae, Matrix, Ultrastructure, Electron microscopy

1. Introduction

Mitochondria are subcellular organelles that coordinate numerous metabolic reactions, including those of the respiratory complexes that produce ATP, by which cellular reactions are powered. Mitochondria assume a variety of shapes, including spheres, rods,

Peter Bross and Niels Gregersen (eds.), *Protein Misfolding and Cellular Stress in Disease and Aging: Concepts and Protocols*, Methods in Molecular Biology, vol. 648, DOI 10.1007/978-1-60761-756-3_19, © Springer Science+Business Media, LLC 2010

and a sausage-like morphology. The overall size and shape of mitochondria depend on their location within the cell. In the cytoplasm of neurons, they are generally sausage-shaped, about 0.1 μm wide and 1.0 μm long (1). In axons, the mitochondria tend to assume a highly elongated form, while they are more globoid in the synaptic terminals. Mitochondria are highly motile and are constantly moving in a directed manner along cytoskeletal tracks within cells. They are surrounded by an outer, smooth membrane and an inner, folded membrane forming the cristae which fill the interior of the organelle. The inner mitochondrial membrane encloses a space referred to as the matrix. The space between the inner and outer membranes is referred to as the intermembrane space. The outer membrane contains many voltage-dependent anion channels (also called mitochondrial porins). These channels are approximately 3 nm in diameter and are permeable to uncharged molecules up to 5,000 Da (2). Thus, although small molecules, ions, and metabolites can enter the intermembrane space, they cannot penetrate the inner membrane. Therefore, the environment of the intermembrane space is similar to that of the cytoplasm with respect to ions and small molecules. The inner membrane that forms the cristae contains proteins, the major functions of which include: (1) performing the oxidation reactions of the respiratory electron-transport chain, (2) synthesizing ATP, and (3) regulating transport of metabolites into and out of the matrix (2). The enzymes of the respiratory chain are attached to the inner membrane and project their heads into the matrix. The intermembrane space contains specific enzymes such as creatine kinase, adenylate kinase, and cytochrome c. The matrix contains the soluble enzymes of the citric acid cycle (Krebs cycle) and the enzymes involved in fatty acid beta-oxidation. When one mitochondrion merges with another one, not only the membranes of the two mitochondria but also their contents merge, including the matrix compartment that contains the mitochondrial DNA. Conversely, by a fission event, a single mitochondrion becomes two mitochondria. Because of these complementary processes, the identity of any individual mitochondrion is transient.

Mitochondria play a complex role in the pathogenesis of neurodegenerative diseases due to their involvement in a number of processes, including the regulation of excitotoxicity, the homeostasis of intracellular Ca^{2+}, the production of reactive oxygen species, and apoptotic processes. Although respiratory chain defects have been reported in patients with neurodegenerative diseases, clear pathogenic mutations in mtDNA or respiratory chain-related nuclear genes have been rarely observed (3, 4). On the other hand, numerous mutations in genes encoding mitochondrial proteins not related to or indirectly related to the respiratory chain have been identified (5) in the various different diseases. However, despite the abundance of studies dealing with the relationship between mtDNA and neurodegenerative diseases such as Parkinson's disease, Alzheimer's disease, and amyotrophic lateral sclerosis (ALS), no unequivocally convincing data are actually available. Accumulating evidence has indicated that mitochondrial damage is involved in motor neuron degeneration in sporadic ALS (6–8), familial ALS with Cu/Zn superoxide dismutase (SOD1) mutation (9, 10) and mutant SOD1 transgenic mice (11, 12). As regards currently available strategies for the morphological examination of mitochondria, the transmission electron microscope (TEM) remains a powerful tool for imaging the fine structure of cellular organelles and for locating particular structures. Thus, TEM is expected to continue to enhance our understanding of cellular functions and dysfunction. Moreover, electron microscopic study has enabled us to confirm mitochondrial involvement in the pathomechanism of certain neurodegenerative diseases.

2. Materials

2.1. Chemical Fixation

1. 2% Glutaraldehyde (pH 7.2) kept stable at 4°C for up to 1 month (see Note 1).
2. Fixative (2% glutaraldehyde [pH 7.2] and 0.1 M cacodylate buffer [pH 7.4]).
3. 0.1 M Cacodylate buffer (pH 7.4) kept stable at room temperature (~5–35°C) for up to approximately 6 months under unsealed conditions (see Note 2).
4. Osmium tetraoxide (OsO_4) ~1–2% (make fresh as required).
5. Fifty percent Ethanol stored at 4°C.
6. Sixty percent Ethanol stored at 4°C.
7. Seventy percent Ethanol stored at 4°C.
8. Eighty percent Ethanol stored at 4°C.
9. Ninety percent Ethanol stored at 4°C.
10. Ninety-five percent Ethanol stored at 4°C.
11. Hundred percent Ethanol kept at room temperature.
12. Propylene oxide or QY-1 (n-Butyl glycidyl ether) kept at room temperature.
13. Epoxy resin (TAAB Epon 812).
14. Uranyl acetate stored at 4°C.
15. Lead citrate stored at 4°C.

3. Methods

1. Removal of tissue samples.
   (a) Gentle handling while obtaining tissues of the nervous system is important, as rough handling of unfixed or incompletely fixed tissue can easily result in artifactitious deformities in both the brain and the spinal cord (see Note 3).
2. Tissue sectioning
   (a) Thin-sectioning of tissues soaked in fixative (see Note 4).
3. Prefixation
   (a) Two percent Glutaraldehyde (pH 7.2) and 0.1 M cacodylate buffer (pH 7.4) fixation at 4°C for more than 3 h (see Note 5).
4. Sectioning of prefixed tissues to proper size and thickness (see Note 6).
5. Rinsing
   (a) Rinsing of sectioned tissue in 0.1 M cacodylate buffer for 3 h at 4°C.
   (b) Rinsing of sectioned tissue in 0.1 M cacodylate buffer overnight at 4°C.
6. Postfixation
   (a) Fixation of tissues in 2% osmium tetroxide and 0.1 M cacodylate buffer for 2 h at 4°C (see Note 7).
7. Dehydration (see Note 8)
   (a) Fifty percent Ethanol for 20 min at 4°C.
   (b) Seventy percent Ethanol for 20 min at 4°C.
   (c) Ninety percent Ethanol for 30 min (or overnight) at 4°C (or at room temperature).
   (d) Hundred percent Ethanol for 30 min at room temperature (see Note 9).
   (e) Hundred percent Ethanol for 30 min at room temperature (see Note 9).
8. Infiltration (see Note 10)
   (a) Propylene oxide (or QY-1) for 30 min (see Note 11).
   (b) Propylene oxide for 30 min.
9. Infiltration into epoxy resin
   (a) Discard propylene oxide. Pour the mixed solution of propylene oxide/epoxy resin (1:1) onto the sample and let stand for more than 2 h.

10. Hundred percent Epoxy resin (Epon 812) applied overnight.
11. Hundred percent Epoxy resin (Epon 812) applied for more than 3 h (see Note 12).
12. Embedding (see Note 13).
13. Polymerization (see Note 14)
    (a) [Method 1] 37°C, 12 h; 37–60°C, 24 h; 60°C, 24 h.
    (b) [Method 2] 37°C, 12 h; 45°C, 12 h; 60°C, 48 h.
    (c) [Method 3] 60°C, 48 h.
14. Trimming of tissue blocks (see Note 15)
    (a) Trim tissue sample with a razor blade or a file, observing the tissue block with a substantial microscope all the while.
15. Semi-thin tissue sectioning
    (a) Tissue block is placed in an ultramicrotome.
    (b) Surface of the tissue block is trimmed with a glass knife (see Note 16).
    (c) Semi-thin sections (about 1 μm thick) are cut for light microscopy (see Note 17).
16. Semi-thin sections stained with toluidine blue and examined under a light microscope (see Note 18).
17. Tissue slicing into ultra-thin sections with a diamond knife (see Note 19).
18. Double-staining with uranyl acetate and lead citrate at room temperature for electron microscopy (see Note 20).
19. Electron microscopic examination
    (a) Mitochondrial alterations vary with pathological condition or disease (see Note 21).

4. Notes

1. The two most important factors in the storage of glutaraldehyde are temperature and pH value. Purified glutaraldehyde remains relatively stable for several months if the samples are stored at 4°C or below, provided the pH value is lowered to ~5.0. Purified glutaraldehyde can be stored for ~6 months at −14°C, and for ~1 month at 4°C without significant polymerization. Treatment with glutaraldehyde for 2 h at 0–4°C is generally preferred for routine fixation of animal tissues. However, glutaraldehyde can be used in many tissue types at temperatures ranging from 0 to 25°C, with little apparent difference in the appearance of the fine structure.

The temperature of fixation can occasionally be critical. For example, certain labile structures, such as certain microtubules, may be lost as a result of rearrangement when glutaraldehyde is used in a cold environment. The most effective means of minimizing the deterioration of purified glutaraldehyde is by storing as an unbuffered, 10–25% solution at subfreezing temperatures (~ −20°C). It is advisable not to mix the buffer and the fixative until immediately before use. Less polymerization occurs when glutaraldehyde is stored in a cacodylate buffer than when it is stored in a phosphate buffer at the same pH (7.4).

2. As cacodylate buffer contains toxic arsenic, careful attention should be paid to the management of the reagent and the treatment of discharge. Cacodylate buffer is quite effective within a pH range of ~6.4–7.4.
3. One well-known example is the so-called "tooth-paste" artifact of the spinal cord in which a portion of tissue is constricted leading to the abnormal displacement of tissue above and below the constriction. Regions of interest (e. g., infarcts, tumors) are especially prone to such changes.
4. Sections should be as thin as possible, i.e. <1 mm thick.
5. The nervous system is particularly vulnerable to postmortem delay. Postmortem autolysis is a common artifact resulting from delayed fixation. Speedy removal of the tissue samples of the nervous system in particular is important, as is the prompt fixation of specimens, in order to minimize the possibility of any delayed-fixation artifacts. In animal experiments, intracardial or intravenous perfusion with heparinized saline (pH 7.4) is performed to achieve suitable fixation, followed by perfusion with ice-cold 4% paraformaldehyde and 0.2% glutaraldehyde in 0.1 M phosphate buffer (pH 7.4).
6. Sectioned tissues should be as thin as possible (i.e. at least <1 mm thick) and when necessary, as large as possible (i.e. a 5 mm square from the cerebral cortex, or a large enough sample from the spinal cord to include the entire anterior horn to facilitate anatomical orientation). Sectioned tissue can then be trimmed to the appropriate size upon semi-thin tissue sectioning.
7. Higher concentrations of osmium tetroxide exhibit good penetration into tissues. The degree of tissue penetration is the same at temperature, ranging between at 4°C and room temperature, but with increasing temperature, tissue penetration inversely diminishes. Currently, the most commonly used concentrations of osmium tetroxide range from 1 to 2% in buffer. The procedure involving osmium tetroxide should be carried out in a draft chamber.

8. Dehydration procedures should be performed consistently and without long-time neglect in order to reduce the risk of one or more of the following: protein dissolution and dissolution of fixed cellular organelles at low concentrations of ethanol, lipid dissolution, and contraction of cells at high concentration.

   An alternative method of dehydration is given below.

   1. Fifty percent Ethanol for 10 min
   2. Sixty percent Ethanol for 10 min
   3. Seventy percent Ethanol for 10 min
   4. Eighty percent Ethanol for 10 min
   5. Ninety percent Ethanol for 10 min
   6. Ninety-five percent Ethanol for 10 min
   7. Ninety-nine and a half percent Ethanol for 10 min
9. The success of embedding a sample in epoxy resin is dependent on complete dehydration treatment with 100% ethanol. Thus, the final dehydration treatment with 100% ethanol should be consistently applied. Poor dehydration can lead to artifactitious pores within the tissues, heterogeneity of polymerization, and/or poor polymerization.
10. Infiltration should be complete as well, because any residual water or alcohol within the tissues will hinder the reaction between the epoxy resin and the hardening agent. This procedure should be carried out in a draft chamber.
11. Tilt the sample bottle and discard the 100% ethanol, then pour propylene oxide into the bottle. This step should be carried out twice within 30 min (during this period, a mixed solution of propylene oxide/epoxy resin [1:1] is prepared). As propylene oxide is very volatile, solution exchange should proceed quickly to avoid drying.
12. Preparation of the new epoxy resin is important, that is, progression of the polymerization of epoxy resin is enabled by preparation the day prior to use. This procedure should also be carried out in a draft chamber.
13. Place the tissue block in a gelatin or beem capsule containing the infused epoxy resin. There are various types of gelatin or beem capsules: choose a suitable one according to your purpose. Embedding of tissues (e.g. peripheral nerves and skeletal muscles) in a particular direction should be carried out in a flat-type silicon plate (TAAB) instead of in a gelatin or beem capsule. Largely sectioned tissue (ca. 5 mm square) should be placed in a cylindrical-shaped beem capsule, which in turn should be kept on adhesive tape on both sides of the

box in order to ensure stability. It is useful to attach a slip of paper bearing a name or identification number into capsules in order to differentiate between samples.

14. Use an exclusive polymerizer for polymerization. Always maintain low humidity, confirmed by a thermostat, with the use of a desiccant. Polymerized hard blocks should be preserved in a desiccator or in a storage bottle containing the desiccant, as such blocks are softened by humidification. The present author typically employs Method 1.
15. The angle between the superior and lateral aspects of the pyramid should measure about 120°, and should be sharp (i.e. lacking any roundness).
16. Glass knives outperform diamond knives for cutting semi-thin sections. It is important that both the angles of elevation and those of right and left sides remain at zero during the rough-cutting process. Several versatile ultramicrotomes are available (e.g. LKB Ultratomes I and III; Sorvall Porter-Blum MT-1, MT-2, and MT-7; and Reichert OMU2 and OMU3).
17. When the surface becomes even, sections (approximately 1-μm-thick) can be cut. After a satisfactory glass knife has been prepared, a trough is constructed to collect sections. A piece (~3 × 0.7 cm) of black electrical adhesive tape (Scotch Brand No. 3 or equivalent) is wrapped from one side to the other of the cutting edge of the glass knife. When the tape is wrapped correctly, its lower edge should be parallel to the lower edge of the knife; the upper edge of the tape should precisely reach the top of the cutting edge. To render the trough waterproof, the region of the heel of the trough should be sealed with paraffin or beeswax that has been melted in a dish on a hot plate or in any paraffin oven. To ensure a better seal, the paraffin or beeswax used for sealing the trough must be heated until it fumes. These fumes are considered carcinogenic; necessary precautions should be taken to avoid inhaling them. Alternatively, without making a trough, cut semi-thin sections and pick them up directly with sharp tweezers or a thin metallic rod, and place the sections onto a drop of water on the slide glass. In this case, tissue sectioned at a thickness of more than 1 μm (approximately 2.0–2.5 μm) is easier to handle. The present author prefers using a thin metallic rod without making a trough.
18. To create the toluidine blue staining solution, mix 100 ml distilled water with 0.5 g of sodium tetra-borate (borax) and then dissolve 0.5 g of toluidine blue completely in this solution.
    (a) Cut 1–2.5 μm-thick sections, place each section on the surface of a drop of distilled water on a clean slide glass, heat the sections on a heating unit, and fix the sections until they are dry.

(b) Stain the sections with 0.5% toluidine blue for several minutes at 150–200°C.

(c) Wash the slide glass in tap water or distilled water and then dry it on the heater, and if necessary, mount the sample.

19. The interference color of the section serves as an index of its approximate thickness: gray, 50–60 nm; silver, 70–90 nm; and gold, 90–150 nm. The author usually prefers silver–gold or gold coloring for tissues of the nervous system. The ultra-thin sections are collected onto grids. From among a variety of grids from coarse-meshed to fine-meshed, the present author prefers a grid of 150 meshes from the practical point of view.

20. Stain the sample with uranyl acetate for 5 min (Fig. 1a), followed by a 50% ethanol wash and a wash in distilled water (Fig. 1c). Then, stain the sample with lead citrate for 2 min (Fig. 1b), followed by a wash in distilled water (Fig. 1c) and subsequent drying. To create the uranyl acetate staining

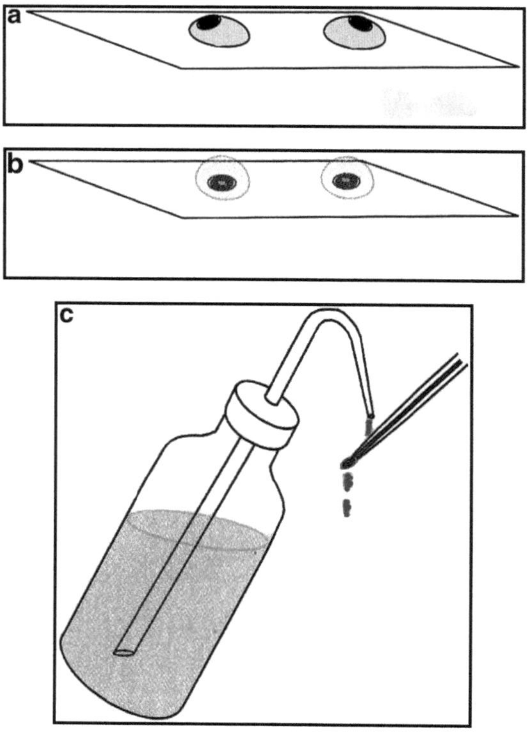

Fig. 1. Sample staining. For explanation see Note 20. (**a**) Place the side of the grid on which the tissue is attached on the surface of the uranyl acetate solution. (**b**) Sink the water washed grid with the tissue attached to the bottom of the lead citrate solution. (**c**) Prepare two bottles for washing the grid with the tissue attached. One of the bottles should contain a 50%-ethanol solution, and the other should contain distilled water. Pick up the uranyl acetate-stained grid with tweezers, and wash it directly with 50%-ethanol solution. Pick up the lead citrate-stained grid with tweezers and wash it directly with the distilled water.

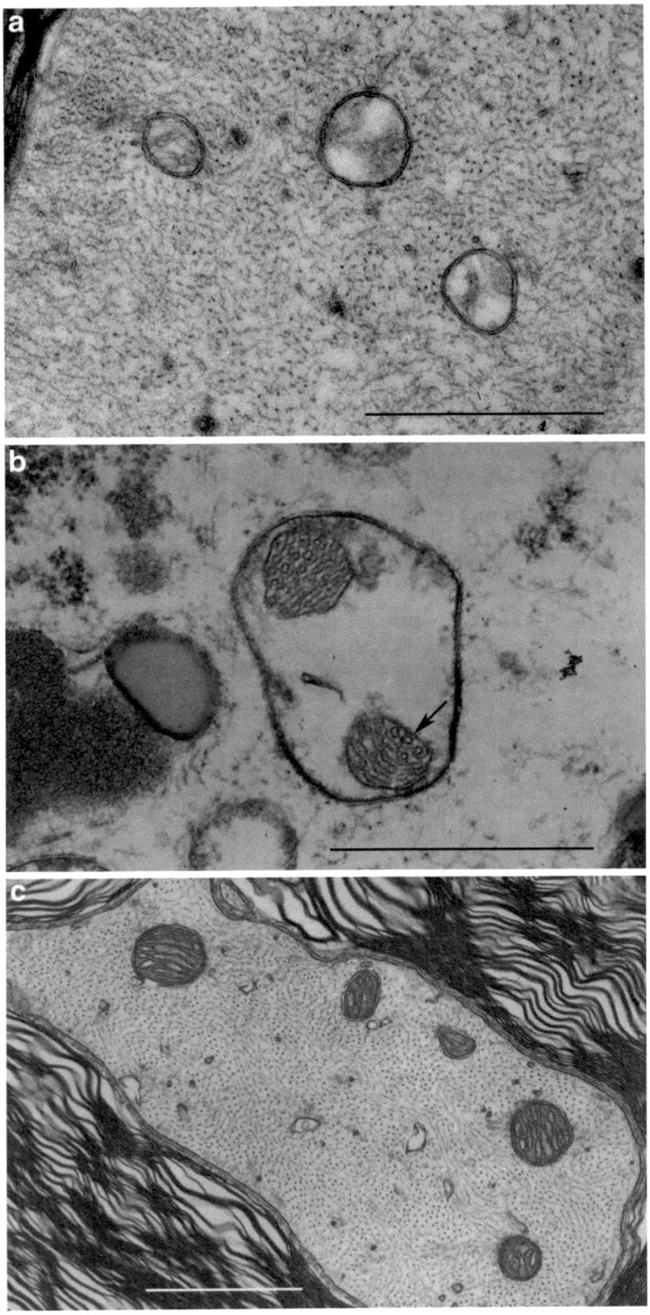

Fig. 2. Electron-micrographs of mitochondrial alterations under various conditions. For explanation see Note 21. Scale bars: 1 μm. (**a**) Relatively well-preserved mitochondria of the anterior horn neuron in a human control individual within 3 h of postmortem delay. Even with such a short postmortem delay, some artifacts, including disorganization of the inner compartment, will already have appeared in the mitochondria. (**b**) A mitochondrion of the dorsal root ganglion cell in a patient with ALS. A tubular structure (*arrow*) can be seen in the inner compartment on a transverse section. (**c**) Mitochondria in the axon of an anterior horn in a non-transgenic mouse littermate. Double membranes and the intermembrane space, cristae and matrix can be seen. (**d**) A mitochondrion in the axon in a mutant SOD1 (G93A) transgenic mouse, in which splitting of the intermembrane space (*arrow*) and formation of a large vacuole can be seen. (**e**) A mitochondrion in the axon in a mutant SOD1 (H46R) transgenic mouse, in which the electron-dense amorphous structure of the cristae (*arrow*) is seen.

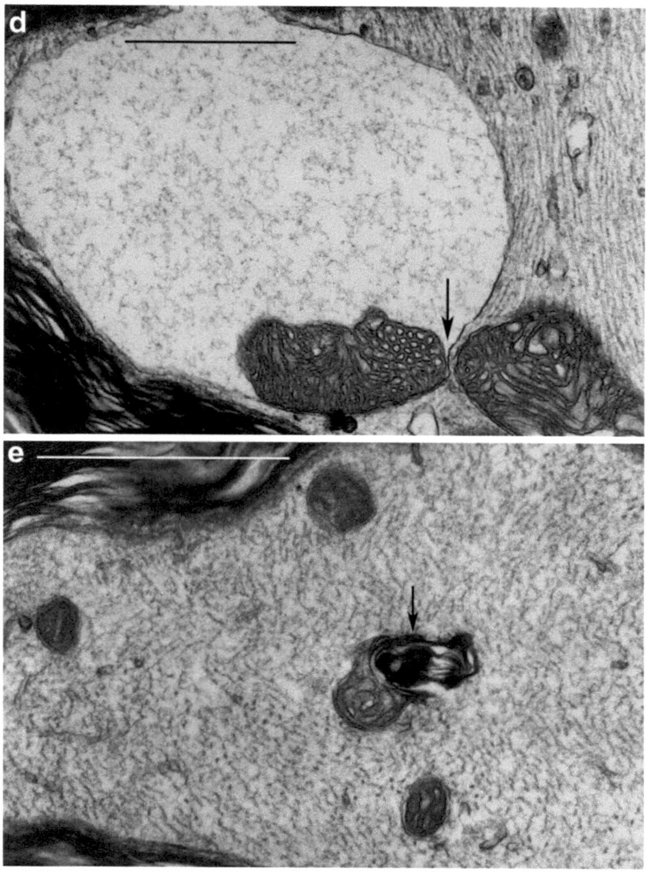

Fig. 2. (continued)

solution, dissolve uranyl acetate with 50% ethanol to yield a 5% concentration, that is, near saturation, and make sure to take the time necessary to dissolve uranyl acetate completely. This will yield a sufficient dyeing solution, provided that about 80% of the uranyl acetate is dissolved into 50% ethanol. Keep the solution in the refrigerator and in the shade, because it degenerates easily when exposed to ultraviolet rays. To create the lead citrate staining solution, boil 50 ml of ultra-purified water or distilled water until the CO_2 has been removed. Dissolve 0.2 g NaOH into water, and then filter out the water using filter paper (No. 1). The filtrate should be transparent. Subsequently, add the lead citrate (0.2 g) and dissolved completely. Again filter out the water with filter paper (No. 1). The final concentration should be 0.4% lead citrate/0.4% NaOH solution. If the final solution comes into contact with CO_2 in the air, it will turn milky and could be a contaminant. Therefore, keep the solution in a refrigerator in order to block access to the ambient air. The present author usually keeps this solution refrigerated in a 10 ml syringe with the outlet wrapped completely in wax paper.

21. Electron-micrographs of mitochondrial alterations under various conditions.

Acknowledgments

This work was supported by a Grant-in-Aid for General Scientific Research (C) from The Ministry of Education, Culture, Sports, Science and Technology of Japan.

References

1. Hirano A (ed) (1981) A guide to neuropathology. Igaku-Shoin, New York, USA, pp. 154–157
2. Ross, M. H. and Pawlina, W. (eds.) (2006) Mitochondria, 2 the cell cytoplasm. Membranous organelles, in *Histology.* A text and atlas. With correlated cell and molecular biology, 5th ed., Lippincott Williams & Wilkins, USA, pp. 52–54
3. Nikali K, Suomalainen A, Saharinen J, Kuokkanen M, Spelbrink JN, Lönnqvist T et al (2005) Infantile onset spinocerebellar ataxia is caused by recessive mutations in mitochondrial proteins Twinkle and Twinky. Hum Mol Genet 14:2981–2990
4. Comi GP, Bordoni A, Salani S, Franceschina L, Sciacco M, Prelle A et al (1998) Cytochrome c oxidase submit I microdeletion in a patient with motor neuron disease. Ann Neurol 43:110–116
5. Orth M, Schapira AH (2001) Mitochondria and degenerative disorders. Am J Med Genet 106:27–36
6. Sasaki S, Iwata M (1996) Ultrastructural study of synapses in the anterior neurons of patients with amyotrophic lateral sclerosis. Neurosci Lett 204:53–56
7. Siklos L, Engelhardt J, Harati Y, Smith RG, Joo F, Appel SH (1996) Ultrastructural evidence for altered calcium in motor nerve terminals in amyotrophic lateral sclerosis. Ann Neurol 39:203–216
8. Wiedemann FR, Manfredi G, Mawrin C, Beal MF, Schon EA (2002) Mitochondrial DNA and respiratory chain function in spinal cords of ALS patients. J Neurochem 80:616–625
9. Bowling AC, Schulz JB, Brown RH, Beal MF (1993) SOD activity, oxidative damage, and mitochondrial energy metabolism in familial and sporadic ALS. J Neurochem 61:2322–2325
10. Dhaliwal GK, Grewall RP (2000) Mitochondrial DNA deletion mutation levels are elevated in ALS brains. Neuroreport 11:2507–2509
11. Kong J, Xu Z (1998) Massive mitochondrial degeneration in motor neurons triggers the onset of amyotrophic lateral sclerosis in mice expressing a mutant SOD1. J Neurosci 18:3241–3250
12. Wong PC, Pardo CA, Borchelt DR, Lee MK, Copeland NG, Jenkins NA et al (1995) An adverse property of a familial ALS-linked SOD1 mutation causes motor neuron disease characterized by vacuolar degeneration of mitochondria. Neuron 14:1105–1116

Chapter 20

Assessing Bad Sub-cellular Localization Under Conditions Associated with Prevention or Promotion of Mitochondrial Permeability Transition-Dependent Toxicity

Liana Cerioni and Orazio Cantoni

Abstract

Cells belonging to the monocyte/macrophage lineage are in general highly resistant to peroxynitrite, a reactive nitrogen species extensively produced by these and other cell types under inflammatory conditions. Resistance is not dependent on the scavenging of peroxynitrite but is rather associated with the prompt activation of a survival signaling in response to various molecules largely available at the inflammatory sites, as arachidonic acid and products of the 5-lipoxygenase or cyclooxygenase pathways. We detected significant levels of Bad in the mitochondria of monocytes/macrophages and found that these signaling pathways converge in Bad phosphorylation, and thus in its cytosolic accumulation. Phosphorylation inhibits binding of Bad to Bcl-2, or $BclX_L$, and promotes its translocation to the cytosol, thereby enabling Bcl-2 and $BclX_L$ to exert effects leading to prevention of mitochondrial permeability transition (MPT). Upstream inhibition of the survival signaling indeed promotes the mitochondrial accumulation of Bad and the rapid onset of MPT-dependent toxicity. The above results contribute to the definition of the mechanism(s) whereby monocytes/macrophages survive to peroxynitrite in inflamed tissues.

Key words: Bad, Peroxynitrite, Mitochondria, Mitochondrial permeability transition, Inflammation, Sub-cellular localization, Western blot analysis

1. Introduction

The Bcl-2 family of proteins, a critical death regulator that resides immediately upstream of mitochondria, comprises anti-apoptotic/pro-survival members (Bcl-2, $BclX_L$, Bcl-w, Mcl-1 and A1) and two groups of pro-apoptotic/pro-death members, the "multidomain"

Peter Bross and Niels Gregersen (eds.), *Protein Misfolding and Cellular Stress in Disease and Aging: Concepts and Protocols*, Methods in Molecular Biology, vol. 648, DOI 10.1007/978-1-60761-756-3_20, © Springer Science+Business Media, LLC 2010

(Bax, Bak and Bok) and "BH3-domain only" members (Bim, Bad, Bid, Bik, Bmf, Puma, Noxa and Hrk). Pro- and anti-apoptotic family members may heterodimerize, thereby affecting their respective functions, and their activity is regulated either at the transcriptional or posttranslational levels.

Bad (Bcl-2 antagonist of cell death) is a positive regulator of cell death and, like other BH3-only proteins, selectively displaces Bax to heterodimerize with Bcl-2 or $BclX_L$. Thus, Bad may activate Bax simply by freeing it from the above pro-survival proteins. The pro-death effect of Bad is dependent on its ability to dimerise with pro-survival members of the Bcl-2 family (1, 2).

In the absence of survival stimuli, endogenous Bad is dephosphorylated and localized in the outer mitochondrial membrane. In the presence of survival factors, Bad is promptly phosphorylated on serine 136 by phosphatidylinositol 3-kinase/Akt (3, 4) or on serine 112 by either protein kinase A (PKA) or protein kinase C (PKC)/$p90^{RSK}$ (5, 6). Phosphorylation at serine 112, or 136, allows Bad sequestration in the cytoplasm, bound to 14–3–3 scaffold proteins. PKA-dependent phosphorylation at serine 155 has also been reported and shown to reduce the affinity of Bad for pro-survival Bcl-2 family members (7).

Our previous work has shown that pro-monocytic cell lines, and human monocytes/macrophages, survive to peroxynitrite because of their ability to respond to molecules largely available in the inflammatory milieu (e.g. arachidonic acid) with the triggering of events converging in Bad phosphorylation (8, 9). In particular, we showed that 5-hydroxyeicosatetraenoic acid (5-HETE), a product of 5-lipoxygenase (5-LO), promotes the mitochondrial translocation of PKCα and the ensuing phosphorylation of Bad (10). Inhibition of the survival signaling at the level of cytosolic phospholipase A_2, 5-LO or PKCα rapidly leads to the mitochondrial accumulation of Bad, associated with the rapid onset of mitochondrial permeability transition (MPT)-dependent toxicity. Cells, however, can nevertheless be rescued by supplementation of products downstream to the inhibited step or, in alternative, via EP_2 receptor-mediated, PKA-dependent Bad phosphorylation elicited by exogenous PGE_2 (11). A third Bad kinase, Akt, is severely inhibited by peroxynitrite in U937 cells.

In short, the survival strategy adopted by monocytes/macrophages to cope with peroxynitrite is simple, yet very effective: while committed to MPT-dependent toxicity, these cells nevertheless survive by promoting Bad phosphorylation via different pathways. It is interesting to note that these pathways are triggered by molecules extensively produced under inflammatory conditions (as peroxynitrite) by monocytes/macrophages themselves, as well as by other cell types.

2. Materials

2.1. Cell Culture and Treatments

1. RPMI 1640 Medium (Sigma-Aldrich) supplemented with 10% fetal bovine serum (HyClone Laboratories, Logan, UT), penicillin (100 U/ml), and streptomycin (100 μg/ml) (HyClone).
2. T-75 tissue culture flasks (Sarstedt, Nümbrecht, Germany).
3. Peroxynitrite was synthesized by the reaction of nitrite with acidified H_2O_2, as described previously (12), with minor modifications (13) (see Note 1).
4. Saline A: 8.182 g/l NaCl, 0.372 g/l KCl, 0.336 g/l $NaHCO_3$, and 0.9 g/l glucose, pH 7.4 at 37°C.

2.2. Cell Fractionation

1. Extraction buffer: 20 mM HEPES-KCl (pH 7.4), 10 mM KCl, 250 mM sucrose, 1.5 mM $MgCl_2$, 1 mM sodium EDTA, 1 mM sodium EGTA. Immediately before use add 1 mM dithiothreitol (DTT), 1 mM phenylmethylsulfonyl fluoride (PMSF), 10 μg/ml leupeptin, 10 μg/ml pepstatin, and 10 μg/ml aprotinin.
2. Lysis buffer: 20 mM Tris–HCl (pH 7.5), 150 mM NaCl, 1% NP40. Immediately before use add 1 mM DTT, 1 mM PMSF, 10 mM sodium orthovanadate (Na_3VO_4), 10 mM sodium fluoride (NaF), 10 μg/ml leupeptin, 10 μg/ml pepstatin, and 10 μg/ml aprotinin.

2.3. SDS-Polyacrylamide Gel Electrophoresis

1. Bio-Rad Dye Binding protein assay (Bio-Rad Laboratories, Hercules, CA).
2. Mini gel electrophoresis apparatus (e.g., BioRad Mini-PROTEAN 3 electrophoresis system).
3. Loading buffer: 125 mM Tris–HCl (see Note 2), 2% (w/v) sodium dodecyl sulphate (SDS), 10% glycerol, 0.002% bromophenol blue, pH 6.9. Prepare a 4× solution and store in aliquots at –20°C. Before use add 100 mM DTT.
4. Upper Tris (4×): 0.5 M Tris–HCl, pH 6.8. Store at 4°C (see Note 2).
5. Lower Tris (4×): 1.5 M Tris–HCl, pH 8.8. Store at 4°C (see Note 2).
6. Forty percent acrylamide/bisacrylamide solution: 39:1 w/w; this is a neurotoxin when unpolymerized and so care should be taken not to receive exposure (see Note 3).
7. Ammonium persulfate (APS): prepare a 10% solution in water (see Note 4).
8. Stacking gel solution (4%): prepare 100 ml of solution by mixing 64 ml of water, 10 ml of 40% acrylamide, 25 ml of

Upper Tris, 1 ml of 10% SDS. Store at 4°C. Add 4 µl of *N,N,N,N′*-Tetramethyl-ethylendiamide (TEMED, Sigma-Aldrich) (see Note 5) and 40 µl of APS to 4 ml of the stacking gel solution.

9. Running gel solution (12.5%): prepare 10 ml of solution by mixing 4.3 ml of water, 3.1 ml of 40% acrylamide, 2.5 ml of Lower Tris, 100 µl of 10% SDS, 14 µl of TEMED and 60 µl of APS.
10. Pre-stained molecular weight markers: Color Burst (Sigma-Aldrich).

2.4. Western Blotting for Mitochondrial and Cytosolic Fractions

1. Blotting apparatus (e.g., Bio-Rad Mini-PROTEAN 3 full immersion electroblotting system).
2. Electrophoresis Buffer (10×): 250 mM Tris (pH 8.3), 1.92 M glycine, 0.1% SDS. Store at room temperature.
3. Transfer Buffer (10×): 250 mM Tris (pH 8.3), 1.92 M glycine. Store at room temperature, dilute 1:10 in water and add 20% (v/v) methanol prior to use (see Note 6).
4. Polyvinyldiene (PVDF) membrane (GE Healthcare).
5. Tris-buffered saline with Tween (TBS-T): prepare a 10× solution with 1.40 M NaCl, 500 mM Tris–HCl, pH 7.2. Dilute 100 ml with 900 ml water for use and complete with 0.1% Tween-20 (see Note 7).
6. Blocking buffer: 5% (w/v) nonfat dry milk in TBS-T.
7. Primary antibody dilution buffer: dilute Bad primary antibody (BD Transduction Laboratories, Lexington, KY) 1:700 in 4 ml of blocking buffer (see Note 8). Dilute anti-actin and anti-HSP-60 antibodies (Santa Cruz, Santa Cruz, CA) 1:1,000 in 4 ml of blocking buffer.
8. Secondary antibody: dilute antimouse IgG conjugate to horse radish peroxidase (Santa Cruz) 1:2,000 in 4 ml of blocking buffer.
9. Enhanced chemiluminescent (ECL) reagent (Sigma-Aldrich): mix 10 ml of solution A (1 M Tris–HCl, pH 8.6, 1.25 mM luminol) with 3.2 µl of 30% H_2O_2 and 50 µl of solution B (6.7 mM p-Coumaric acid in DMSO).
10. X-ray film and exposure cassette.

3. Methods

A significant amount of Bad is normally detected in the mitochondria of untreated U937 cells. Exposure of these cells to a non-toxic, but nevertheless MPT-committing concentration of peroxynitrite, promotes the mitochondria to cytosol translocation of Bad, an event critical for survival, as MPT-dependent toxicity is readily observed under conditions in which upstream inhibition of the survival signaling enforces the mitochondrial accumulation of Bad. In these experiments, sub-cellular localization of Bad is conveniently measured with an anti-Bad antibody. As indicated in Fig. 1, peroxynitrite promotes the cytosolic accumulation of Bad via a mechanism sensitive to either AA861 (a 5-LO inhibitor) or Gö6850 (a PKC inhibitor). Fig. 1 also shows that loss of mitochondrial Bad, associated with the cytosolic accumulation of the protein, is promptly re-established upon addition of 5-HETE (a 5-LO product) to 5-LO inhibited cells. When this signaling is intercepted downstream to 5-LO, with a PKC inhibitor, accumulation of Bad can also be promoted via a PKA-dependent signaling triggered by PGE_2.

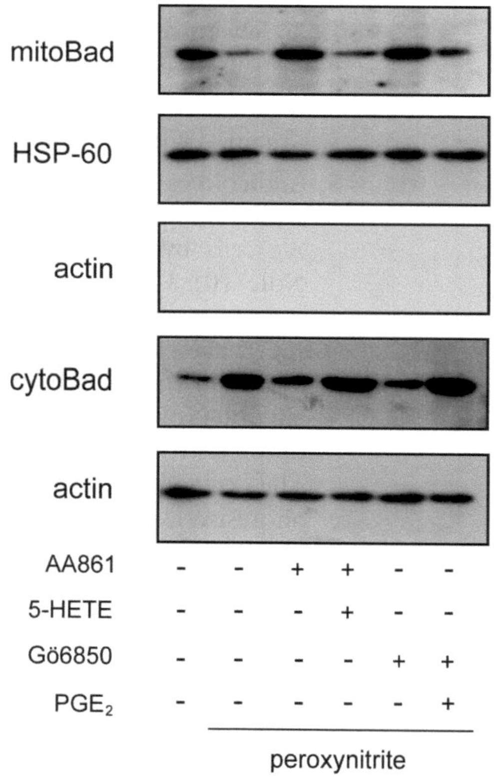

Fig. 1. Peroxynitrite promotes the mitochondrial loss of Bad and its cytosolic accumulation, via a 5-LO/PKC-dependent mechanism mimicked by exogenous PGE_2. U937 cells were treated for 3 min with 100 μM peroxynitrite and subsequently exposed for a further 7 min to 1 μM AA861 (5-LO inhibitor), 0.3 μM 5-HETE, 3 μM Gö6850 (PKC inhibitor) or 0.3 μM PGE_2, as detailed in the figure. After treatments, mitochondrial and cytosolic fractions were isolated and processed for Western blot analysis using an antibody against Bad. Blots were subsequently re-probed for actin and HSP-60 to assess the purity of mitochondrial fraction and the equal loading of the lanes. Blots are representative of three separate experiments.

3.1. Preparation of Samples for Detection of Bad by Western Blotting

1. U937 human myeloid leukaemia cells were cultured in suspension in RPMI 1640 Medium supplemented with 10% fetal bovine serum, penicillin, and streptomycin at 37° in T-75 tissue culture flasks gassed with an atmosphere of 95% air-5% CO_2. Isolation of cytosolic and mitochondria-enriched fractions requires at least 5×10^6 U937 cells.
2. All the materials required for treatments and cell fractionation are made ready: inhibitors of 5-LO and PKC, as well as 5-HETE and PGE_2, at appropriate stock concentrations for 1:1,000 dilution into the cultures; a centrifuge tube for each sample; a vacuum aspirator; cooled centrifuges, and cold extraction and lysis buffer.
3. Treatments were performed in pre-warmed saline A containing 2.5×10^5 cells/ml. The cell suspension (20 ml) was inoculated into 50 ml plastic tubes before addition of peroxynitrite. Peroxynitrite was added on the wall of these tubes and immediately mixed to equilibrate its concentration on the cell suspension (see Note 9).
4. After treatments (10 min), collect the cells by centrifugation (3 min, 4°C) at $1,300 \times g$.
5. Add cold extraction buffer to the cell pellet (35 µl extraction buffer/1×10^6 cells) and incubate for 10 min. Homogenize the cells by 40 passages through a 26-gauge needle (see Note 10). Centrifuge the cell lysate (5 min) at $1,000 \times g$ to remove nuclei, unbroken cells, and cell debris. The resulting supernatant should then be centrifuged (30 min) at $12,000 \times g$ to obtain the mitochondrial fraction (pellet). The cytosolic fraction is obtained from the supernatant upon further purification with an additional centrifugation (1 h) at $10,000 \times g$. The cytosolic fraction is maintained at 4°C prior to electrophoresis. The mitochondrial fraction is suspended in 55 µl of cold lysis buffer, incubated for 20 min on ice and finally centrifuged (5 min) at $15,000 \times g$ to obtain the solubilised enriched mitochondrial fraction.
6. Protein concentration is determined with the Bio-Rad Dye Binding protein assay (see Note 11).
7. Equal amounts (20 µg, see Note 12) of the mitochondrial and cytosolic fractions are diluted in loading buffer and incubated for 5–10 min at 95°C.

3.2. SDS-PAGE

1. Carefully clean and extensively rinse the glass plates of the gel cassette.
2. Prepare a 1.0 mm thick 12.5% gel by mixing 4.3 ml of water, 3.1 ml of 40% acrylamide, 2.5 ml of Lower Tris, 100 µl of 10% SDS, and 14 µl of TEMED. 500 µl of this solution should be mixed with an excess APS (50–60 µl, promoting a

rapid polymerization) to prepare a 2–3 mm thick border on the bottom of the gel cassette. Complete the running gel with the above gel solution (approximately 9.5 ml) and APS (60 μl) and immediately pour into the gel cassette (6 cm from the bottom). Overlay with isobutanol (about 700–800 μl) and allow polymerization for about 30 min.

3. Pour off the isobutanol layer, rinse twice the top of the gel with water and then carefully dry with a thin blotting-paper.
4. Add to the stacking gel solution (4 ml), prepared as detailed above (Subheading 2.3, point 8), 4 μl of TEMED and 40 μl of APS. Immediately pour this solution on top of the running gel. Insert the comb and allow polymerization for about 30 min.
5. Prepare the electrophoresis buffer by diluting 100 ml of the 10× electrophoresis buffer (Subheading 2.4, point 2) with 900 ml of water.
6. Electrophoresis was performed with a Bio-Rad Mini-PROTEAN 3 electrophoresis gel system. Place the gel cassette sandwich into the electrode assembly and then in the inner chamber of the electrophoresis module. The inner chamber is finally lowered into the mini tank.
7. Fill the inner chamber with the electrophoresis buffer and subsequently add the same solution in the mini tank (3 cm from the bottom). Carefully remove the comb and use a 50 μl syringe, fitted with a 22-gauge needle (see Note 13), to rinse the wells with the electrophoresis buffer.
8. Load each sample in a well with the same 50 μl syringe. Carefully rinse the syringe with electophoresis buffer when loading the different samples. Load one well with pre-stained molecular weight markers. The first and the last wells should be loaded with the loading buffer alone.
9. Complete the assembly of the gel unit and connect to a power supply. The gel is generally run at 55 V through the stacking gel and at 110–120 V through the running gel. In order to prevent loss of low molecular weight proteins, electrophoresis should be stopped immediately prior to or soon after the dye front runs off the gel. The dye front is blue because of the presence of bromophenol blue.

3.3. Western Blotting for Detection of Mitochondrial and Cytosolic Bad

1. After SDS-Polyacrylamide Gel Electrophoresis (SDS-PAGE), proteins are transferred to PVDF membranes using a Bio-Rad Mini-PROTEAN 3 full immersion electroblotting system.
2. Place the transfer cassette in a tray and add enough transfer buffer (Subheading 2.4, point 3) to submerge the lower side of the cassette, in which a piece of sponge sheet and three sheets of blotting-paper are sequentially layered.

3. The gel unit is disconnected from the power supply and disassembled. Remove the stacking gel and soak (30 s) the remaining portion of the gel in transfer buffer. Place the gel on top of the blotting-paper and make sure that it remains submerged in the transfer buffer.
4. Soak a PVDF membrane (approximately 6 × 9 cm, same size as the running gel) in 100% methanol (1 min). The membrane should then be rinsed with the transfer buffer (1–2 min) and equilibrated in the same buffer for at least 30 min. Finally, place the membrane on top of the gel in the transfer cassette.
5. Carefully lay three sheets of blotting-paper and a sponge (pre-wetted with the transfer buffer) on top of the PVDF membrane, ensuring that no bubbles are trapped in the resulting sandwich. Close the transfer cassette.
6. Place the cassette into the transfer tank. In order to prevent loss of the proteins into the transfer buffer, the PVDF membrane should be between the gel and the anode.
7. Insert the Bio-Ice cooling unit in the transfer tank. This is necessary to optimize protein transfer and to prevent damage to the electroblotting system due to heat generated during transfer.
8. Fill the transfer tank with the transfer buffer.
9. Complete the assembly of the electroblotting system and connect to a power supply (36 V for 75 min).
10. After protein transfer, the transfer cassette should be removed from the tank and immediately disassembled. Remove the upper sponge/blotting-paper sheets and lay the PVDF membrane on clean paper. The gel and blotting-paper can be discarded. Mark positions of molecular weight markers with a pen (see Note 14).
11. The membrane is then incubated for 1 h (at room temperature) in 20 ml of blocking buffer (Subheading 2.4, point 6) on a rocking platform to prevent nonspecific protein binding.
12. Discard the blocking buffer and quickly rinse the membrane prior to addition of the anti-Bad antibody (diluted as detailed in Subheading 2.4, point 7). The membrane is then incubated overnight (4°C) on a roller mixer.
13. After overnight incubation, wash the membrane with 20 ml TBS-T at room temperature (1 × quick rinse, 1 × 15 min and 2 × 10 min).
14. The secondary antibody should be freshly prepared (Subheading 2.4, point 8) in each experiment. Add the antibody-containing solution to the membrane and allow incubation for 90 min (at room temperature) on a rocking platform.

15. Discard the secondary antibody and wash the membrane three times (10 min each) with 20 ml of TBS-T.
16. Soak the membrane in ECL reagent under safe light conditions. Rotate the membrane by hand for 1 min (to ensure even coverage), remove the excess ECL reagent with blotting paper and immediately place the membrane between the leaves of an acetate sheet protector (cut to the size of an X-ray film cassette) (see Note 15).
17. The final step is performed in the dark room. Place the membrane-containing acetate sheet in an X-ray film cassette. Place the film on top of the membrane for a suitable exposure time, typically 10 min. This time can be decreased or increased, depending on the results obtained.

3.4. Re-probing Blots for Actin and HSP-60

1. After film exposure, strip the membrane of the Bad signal and re-probe with anti-actin or anti-HSP-60 antibodies to assess equal loading of the lanes and the purity of the mitochondrial fractions.
2. Since the above proteins display remarkably different molecular weights, the stripping procedure is simplified as detailed below.
3. Wash the membrane with TBS-T, for 1 h, on a rocking platform with occasional change of buffer. The membrane is now ready to be re-probed with anti-actin or anti-HSP-60 (diluted as detailed in Subheading 2.4, point 7) antibodies, using the same procedure described for the anti-Bad antibody.

4. Notes

1. Solutions should be prepared in distilled or deionized water.
2. Prepare the Tris buffer by dissolving the Tris base in water and adjusting the pH with HCl. Adjust to pH 6.8 for Upper Tris and to pH 8.8 for Lower Tris. Tris base cannot be replaced by Tris–Cl, or Trizma, as the salt concentration would be too high and, under these conditions, polypeptides migrate anomalously through the gel, yielding extremely diffuse bands.
3. In solution, acrylamide and bisacrylamide are slowly converted during storage to acrylic acid and bisacrylic acid. These reactions are catalyzed by light or alkali. Check that the pH of the solution is 7.0 or less. Store the solution in dark bottles. Fresh solutions should be prepared every few months.
4. APS provides the free radicals that drive polymerization of acrylamide and bisacrylamide. A small amount of a 10% (w/v)

stock solution should be prepared in water and stored at 4°C. APS decomposes slowly. Fresh solutions should be prepared weekly.

5. TEMED may decline in quality after opening (gels will take longer to polymerize).
6. Transfer buffer can be used for two transfers within 1 week.
7. Tween-20 is a viscous solution and causes problems during pipetting, partially avoided by cutting the pipette tip.
8. Once diluted, the primary antibody can be stored at –20°C for up to 3 weeks. In addition, the primary antibody can be used in three different blots with the only adjustment required for increasing length of exposure to film at the ECL step.
9. To avoid changes in pH due to the high alkalinity of the peroxynitrite stock solution, an appropriate amount of 1.5 N HCl, approximately the same amount of the peroxynitrite stock solution, normally made in 1.5 N NaOH, was also added to the wall of the tubes prior to peroxynitrite. Note that this procedure has to be fast since peroxynitrite rapidly decomposes at neutral pH.
10. Note that the numbers of passages necessary to lyse the cells may differ significantly among different cell types. The status of the cells can be conveniently monitored after addition of trypan blue under a microscope.
11. The fractionation procedure of U937 cells normally yields about 1.7 mg/ml proteins for the cytosolic fraction and 1.2 mg/ml proteins for the mitochondrial fraction.
12. Apply 5–20 μg proteins to each well of a 0.75–1.0 mm thick gel. For thicker gels (i.e. 1.5 mm), apply up to 25–40 μg in each well.
13. A syringe is normally used for 0.75–1.0 mm thick gels. Use a micropipette for 1.5 mm thick gels.
14. Interrupt the experiments if colored molecular weight markers are not clearly visible on the membrane. It indicates that something did not work during transfer.
15. Apply a luminescent sticker on the edge of the acetate sheet to provide an alignment mark for the film and the membrane.

Acknowledgments

This work was supported by grant from AIRC and PRIN (OC).

References

1. Adams J, Cory S (1998) The Bcl-2 protein family: arbiters of cell survival. Science 281:1322–1326
2. Zamzami BN, Susin SA, Marchetti P, Hirsch T, Gómez-Monterrey I, Castedo M, Kroemer G (1996) Mitochondrial control of nuclear apoptosis. J Exp Med 183:1533–1544
3. Datta SR, Dudek H, Tao X, Masters S, Fu H, Gotoh Y, Greenberg ME (1997) Akt phosphorylation of BAD couples survival signals to the cell-intrinsic death machinery. Cell 91:231–241
4. del Peso L, Gonzalez-Garcia M, Page C, Herrera R, Nunez G (1997) Interleukin-3- induced phosphorylation of BAD through the protein kinase Akt. Science 278:687–689
5. Tan Y, Ruan H, Demeter MR, Comb MJ (1999) p90(RSK) blocks Bad-mediated cell death via a protein kinase C-dependent pathway. J Biol Chem 274:34859–34867
6. Bertolotto C, Maulon L, Filippa N, Baier G, Auberger P (2000) Protein kinase C theta and epsilon promote T-cell survival by a rsk-dependent phosphorylation and inactivation of BAD. J Biol Chem 275:37246–37250
7. Datta SR, Katsov A, Hu L, Petros A, Fesik SW, Yaffem MB, Greenberg ME (2000) 14-3-3 proteins and survival kinases cooperate to inactivate BAD by BH3 domain phosphorylation. Mol Cell 6:41–51
8. Guidarelli A, Cerioni L, Tommasini I, Fiorani M, Brüne B, Cantoni O (2005) Role of Bcl-2 in the arachidonate-mediated survival signaling preventing mitochondrial permeability transition-dependent U937 cell necrosis induced by peroxynitrite. Free Rad Biol Med 39:1638–1649
9. Cerioni L, Palomba L, Brüne B, Cantoni O (2006) Peroxynitrite-induced mitochondrial translocation of PKCα causes U937 cell survival. Biochem Biophys Res Commun 339:126–131
10. Tommasini I, Palomba L, Guidarelli A, Cerioni L, Cantoni O (2006) 5-Hydroxyeicosatetraenoic acid is a key intermediate of the arachidonate-dependent protective signaling in monocytes/macrophages exposed to peroxynitrite. J Leukoc Biol 80:929–938
11. Tommasini I, Cerioni L, Palomba L, Cantoni O (2008) Prostaglandin E2 signals monocyte/macrophage survival to peroxynitrite via protein kinase A converging in Bad phosphorylation with the protein kinase Cα-dependent pathway driven by 5-hydroxyeicosatetraenoic acid. J Immunol 181:5637–5645
12. Radi R, Beckman JS, Bush KM, Freeman BA (1991) Peroxynitrite oxidation of sulfhydryls. The cytotoxic potential of superoxide and nitric oxide. J Biol Chem 266:4244–4250
13. Tommasini I, Sestili P, Guidarelli A, Cantoni O (2002) Peroxynitrite stimulates the activity of cytosolic phospholipase A_2 in U937 cells: the extent of arachidonic acid formation regulates the balance between cell survival or death. Cell Death Differ 9:1368–1376

Chapter 21

A Cellular Viability Assay to Monitor Drug Toxicity

Jakob Hansen and Peter Bross

Abstract

A central part of the research in protein misfolding and its associated disorders is the development of treatment strategies based on ensuring cellular protein homeostasis. This often includes testing chemical substances or drugs for their ability to counteract protein misfolding processes and to promote correct folding. Such investigations also include assessment of how the tested chemical substances affect cellular viability, that is, their cytotoxic effect. Investigations of cytotoxicity often require testing several different concentrations and drug exposure times using cells in culture. It is therefore attractive to use a viability test that permits the analysis of many samples with little handling time. This protocol describes a simple and fast methodology to analyze viability of lymphoblastoid cells and to test putative cytotoxic effects associated with exposure to a chemical substance, here exemplified by celastrol. The natural substance celastrol has been used for many years in traditional Chinese medicine and has subsequently been shown to induce transcription of genes encoding molecular chaperones (heat shock proteins) that are involved in promoting folding of cellular proteins. The well-described colorimetric tetrazolium salt (MTT) assay, which monitors metabolic activity of cultured cells, was adapted to analyze the viability of cells exposed to celastrol. After having established a suitable cell seeding density, the dose-dependence and time-course of viability reduction of lymphoblastoid cells treated with celastrol were determined. It was found that 4- and 24-h exposure to 0.8 µM celastrol reduced the viability of lymphoblastoid cells, with the most severe effect observed at 24 h with MTT reductions approaching 30% of non-exposed cells. For a series of incubations for 24 h, it was found that concentrations as low as 0.2 µM were sufficient to affect the viability, and celastrol concentrations of 0.5 µM reduced the MTT reduction rate to approximately half the level displayed by cells receiving vehicle alone.

Key words: Protein misfolding, Disease, Molecular chaperones, Cytotoxicity, MTT reduction, Cell viability

1. Introduction

Cellular protein homeostasis is required for the well-functioning of cells and disturbances to protein folding and degradation may cause human diseases. The "loss of function" of a specific cellular protein caused by inherited missense mutations that impair protein

Peter Bross and Niels Gregersen (eds.), *Protein Misfolding and Cellular Stress in Disease and Aging: Concepts and Protocols*, Methods in Molecular Biology, vol. 648, DOI 10.1007/978-1-60761-756-3_21, © Springer Science+Business Media, LLC 2010

folding is the underlying mechanism of several human disorders including enzyme deficiencies of mitochondrial fatty acid oxidation, cystic fibrosis, and phenylketonuria (1, 2). The folding-prone mutant protein may suffer from accelerated cellular degradation and impaired assembly into enzymatically active enzyme complexes. There is a growing interest in identifying strategies to fight loss of vital protein function that are based on enhancing the folding and assembly of folding-impaired proteins (3). One strategy has been to increase the cellular level of molecular chaperones, a group of protein that ensures cellular protein homeostasis by assisting and enhancing the protein folding process. Celastrol is a natural substance that has been used for many years in traditional Chinese medicine where it has been administered to individuals with inflammatory diseases. Subsequent biological studies have shown that celastrol is able to induce the transcription of molecular chaperones in several types of cells (4, 5) but that celastrol also may exert toxic effects in some cell types. A thorough investigation of the toxic and folding-enhancing effects of a drug requires that a relatively large number of different drug concentrations and incubation periods are analyzed in a cellular context. It is therefore vital to use an assay that can accurately evaluate the viability of cultured cells exposed to a drug in a fairly high sample throughput manner. Here, a protocol for testing the viability of cells cultured in 96-wells microplates is described that is based on the MTT cellular reduction assay originally developed by Mosmann (6). This colorimetric assay is based on quantification of dark-colored formazan produced by the reduction of the tetrazolium salt MTT (3-(4,5-dimethylthiazol-2-yl)-2,5-diphenyltetrazolium bromide) by metabolically active cells. Cells with decreased viability are considered to be less metabolically active and hence will reduce less MTT. The magnitude of a decrease in formazan production in cells exposed to certain drug is taken as a measurement of the cytotoxicity of the drug. The described protocol addresses a putative toxic effect of celastrol on the viability of lymphoblastoid cells. This cell type is chosen because it can readily be derived from a simple blood sample from, for example, patients with an inherited "loss of function" protein misfolding disorder. A patient-derived cell line expresses the disease-associated and mutated folding-prone protein in a genetically relevant context and can constitute a good cell model for the testing of putative folding-enhancing substances. However, the described protocol can be used for the evaluation of other chemical substances and with small adjustments for other cell types including adherent cells. The simplicity of the MTT protocol is reflected by the absence of washing steps and the fact that all steps following cell seeding are performed in a microplate. This combined with the broad availability of microplate readers have made MTT type assays widely used.

2. Materials

1. Human Epstein–Barr virus transformed lymphoblastoid cell lines from Coriell Cell Repositories (Camden, NJ).
2. Cell incubators providing a humidified atmosphere, 37°C, and 5% CO_2.
3. Flat-bottom 96-wells cell culturing microplates with lid.
4. Roswell Park Memorial Institute (RPMI 1640) medium (Gibco, Breda) with addition of 10% fetal calf serum, 2 mM glutamine, 100 units/ml penicillin, and 100 mg/ml streptomycin.
5. Automated cell counting system (Nucleocounter©, Chemometec, Alleroed) or any available manual or automated system for cell counting.
6. Celastrol (Gaia Chemical Corporation, Gaylordsville, CT), 10 mM stock solutions prepared in dimethyl sulfoxide (DMSO).
7. MTT Labeling Reagent (3-(4,5-dimethylthiazol-2-yl)-2, 5-diphenyltetrazolium bromide) dissolved in phosphate-buffered saline to 5 mg/ml (as, e.g., part of the cell proliferation kit MTT, Roche applied science, Mannheim).
8. Solubilization Solution (10% sodium dodecyl sulfate (SDS) in 0.01 M HCl) (see Note 1) (as, e.g., part of the cell proliferation kit MTT, Roche applied science, Mannheim).
9. Adhesive sheets for 96-wells microplates.
10. Microplate reader (as, e.g., iMark, Bio-Rad, Hercules, CA).

3. Methods

A huge number of different methodologies are available for the investigation of putative toxic effects associated with the exposure of cultured cells to a chemical substance including assays for cell viability and cell death. Here, measurements of viabilities are used to examine the cytotoxic effects of celastrol. Cellular viability is addressed by measuring the cell-specific reduction of the MTT reagent, and a typical assay includes, seeding cells at defined densities in 96-well plates, addition of MTT solution followed by a 4-h incubation period, termination of the reaction, and re-solubilization and quantification of reduced form of MTT. The MTT readily enters the cell interior and is then reduced by metabolic active cells but not by unviable cells. Originally, it was assumed that the cellular reduction of MTT was exclusively performed by mitochondrial enzymes; however, later studies showed that MTT reduction also takes place in microsomal and cytosolic fractions by oxido-reductase type enzymes transferring two electrons to MTT, most likely through reduced forms of NADH and NAPDH (7–9). The reduced form of MTT is an insoluble dark colored product, called formazan which upon re-solubilization can be quantified by measuring the absorbance at wavelengths in the range 550–600 nm. Before using the MTT assay for evaluation of a putative toxicity of a drug it is recommendable to perform a pilot study to test MTT reduction rates as a function of the cell seeding density. This will give an idea of the number of cells required to get a significant and detectable signal of a magnitude that later allows for detection of a decreased MTT reduction. The use of too many cells per well may lead to the depletion of nutrients such as glucose that has been shown to affect cellular MTT reduction rates (9). The magnitude of the MTT signal is often approaching a linear correlation with cell density (6), which is also evident from the data presented here using lymphoblastoid cells (Fig. 1).

Putative toxic effects are investigated by exposing a set of cells to celastrol in a period prior to addition of MTT and then taking

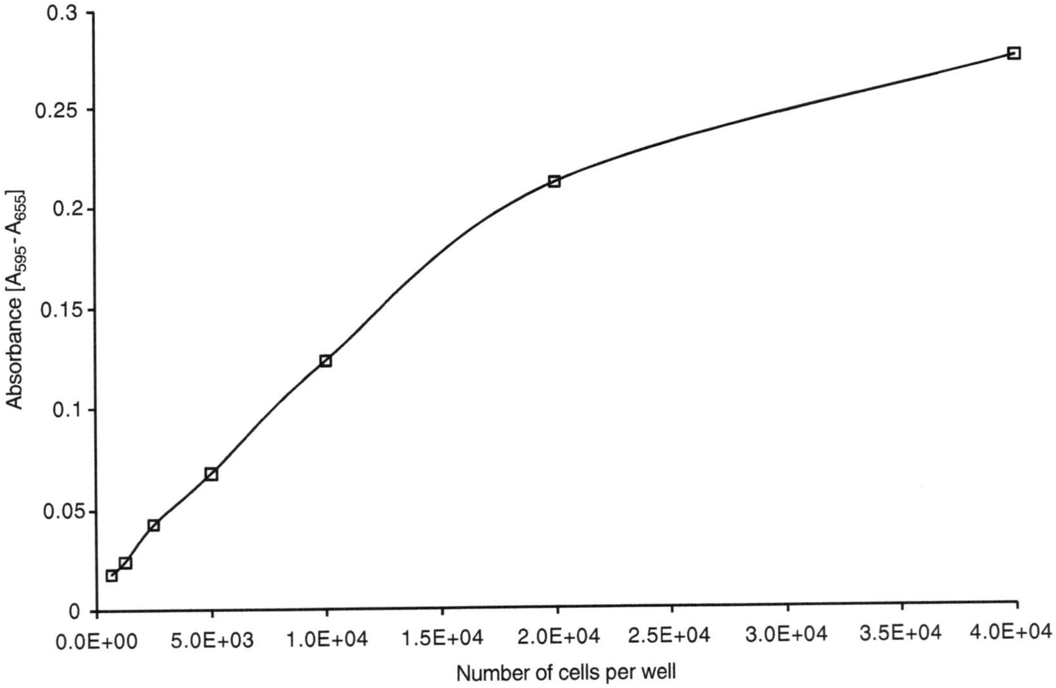

Fig. 1. Correlation between MTT reduction and cell seeding density. Lymphoblastoide cells were seeded at densities from 4.0E+04 cells per well to 6.2E+02 cells per well in 90 μl culture media in a 96-well microplate. Following overnight incubation, 10 μl MTT solution was added (final concentration 0.5 μg/μl). The MTT reduction was stopped after 4 h by addition of 100 μl solubilization buffer. Solubilized MTT was quantified by measuring the absorbance at 595 nm with reference at 655 nm ([A_{595}–A_{655}] and plotted as a function of cell density). Each value represents the mean of three experiments and the mean coefficient of variance for all cell densities was 16%.

a resulting decrease in MTT reduction in celastrol exposed cells as an indicator of a toxic effect. It should be emphasized that a decrease in MTT reduction rate following exposure to a putative cytotoxic agent cannot solely be ascribed to a reduction in the number of viable cells but more correctly to a decrease in the "mean" MTT-related metabolic activity of the cell population present in the well. We determined the dose-dependent effect of celastrol on the viability of cells treated for 24 h and demonstrated a clear cytotoxic effect that was evident at two different cell seeding densities (Fig. 2). We investigated the effect of 0.8 μM celastrol in greater detail by monitoring the MTT reduction in response to 1-, 4-, and 24-h exposures and found that a 4-h exposure was sufficient to reduce cell viability (Fig. 3).

3.1. Setting Up Cells and Determination of Appropriate Number of Cells per Well

1. Propagate a stock lymphoblastoid cell culture in RPMI 1640 culture medium, cells should be seeded at densities of no less than 2.00E+05 cells/ml and should be split and sub-cultured when the density is approximately 6.00E+05 cells/ml (see Note 2). The cells grow in suspension in culture flasks

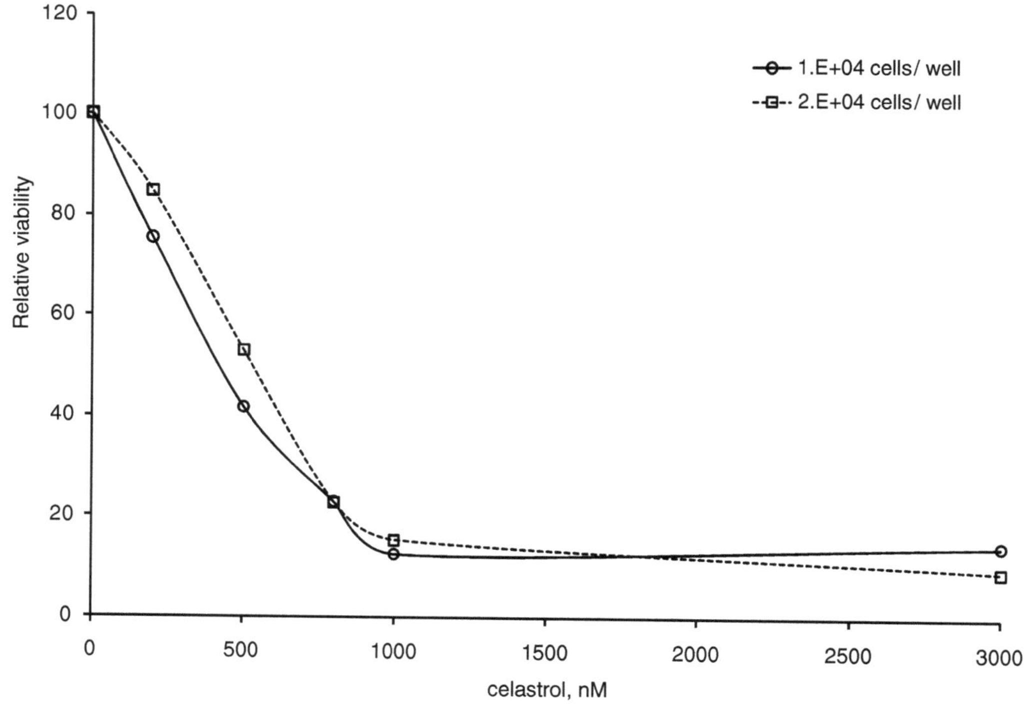

Fig. 2. Dose-dependent viability reduction in lymphoblastoid cells exposed to celastrol. Cells seeded at either 1.E+04 cells per well or 2.E+04 cells per well were incubated for 24 h in different celastrol concentrations or DMSO (controls). MTT reduction was measured as absorbance [A_{595}–A_{655}] following a 4-h incubation period in MTT. The MTT reduction of cells exposed to DMSO were set to 100. Each value represents the mean of three experiments.

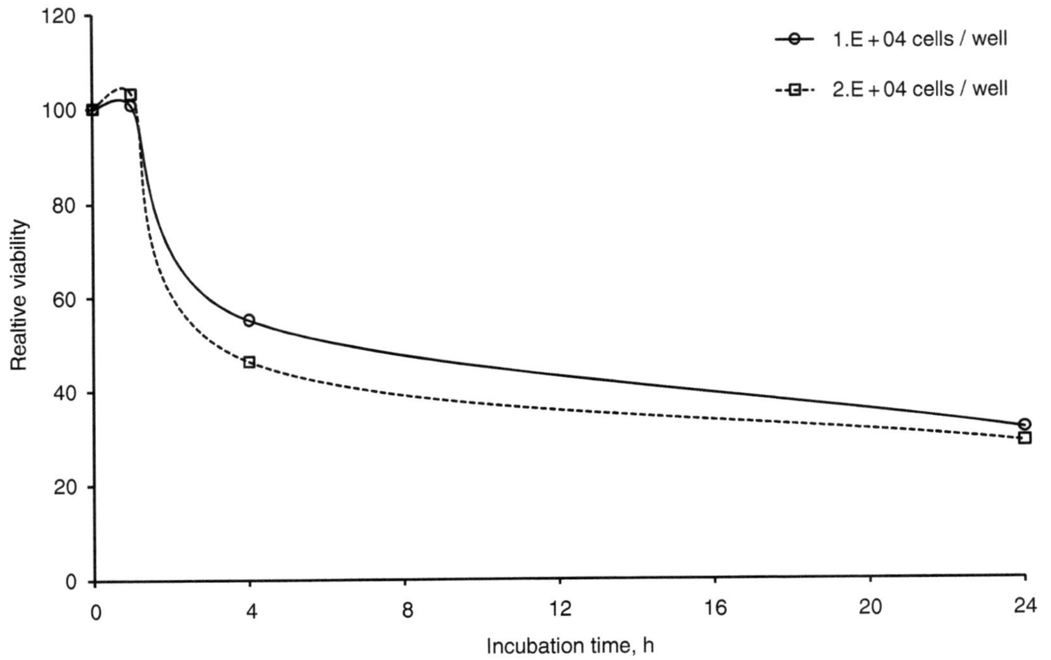

Fig. 3. Time-course of viability reduction in lymphoblastoid cells exposed to 0.8 μM celastrol. Cells seeded at either 1.E+04 cells per well or 2.E+04 cells per well were incubated in 0.8 μM celastrol. Separate wells were incubated for 1, 4, or 24 h in celastrol before addition of MTT for a 4-h period. MTT reduction was measured as absorbance [A_{595}–A_{655}] and the MTT reduction at $t=0$ (no addition of celastrol) was set to 100. Each value represents the mean of three experiments.

(in upright position) (see Note 3) in humidified atmosphere at 37°C and 5% CO_2.

2. Extract a small aliquot of homogeneous cell solution to determine cell density using a cell counting systems (e.g., chemometec nuclecounter system) (see Note 4).
3. Extract a volume of cells sufficient for performing an entire microplate MTT experiment, spin down cells, and re-dissolve in culture media to get a cell density of 4.44E+05 cells/ml corresponding to 4.00E+04 cells per 90 μl. Make a series of twofold dilutions (as e.g., down to 624 cells per well) in a suitable volume for analysis of the required number of replicates (see Note 5).
4. From each dilution transfer 90 μl to the appropriate number of wells of a flat-bottom microplate for cell culturing.
5. Apply lid to the microplate and incubate until the next day in a cell incubator (at 37°C and 5% CO_2 in humidified atmosphere). Continue in Subheading 3.3.

3.2. Cell Setup and Incubation with Celastrol

1. Propagate a stock cell culture as described in Subheading 3.1, steps 1 and 2.
2. Extract a cell volume sufficient for the entire experiment, spin down cells and re-dissolve to get the required cell density

(as previously determined in Subheading 3.1). Here, cell densities of 2×10^4 and 1.00E+04 cells/90 μl/well (corresponding to 2.22 and 1.11E+05 cells/ml, respectively) are used.

3. Transfer 90 μl cell solutions to each well in a microplate, apply the lid, and let cells incubate to the next day at 37°C and 5% CO_2 in a humidified atmosphere.
4. Just prior to use, prepare 10× working solutions of the preferred celastrol concentrations by serial dilutions of the 10 mM stock solution using culture media.
5. Add 10 μl celastrol working solution (10×) to the assay wells to obtain the desired final celastrol concentrations in 100 μl total volume per well. To another set of wells, add 10 μl "vehicle alone" (DMSO diluted in culture media). Allow the cells to incubate for the chosen amount of time.

3.3. MTT Reduction, Solubilization, and Quantification of Formazan

1. At the end of the incubation period, add 10 μl of MTT solution to each well.
2. Incubate for 4 h at 37°C and 5% CO_2 in humidified atmosphere.
3. Stop the reaction by adding 100 μl solubilization solution to each well and incubate cells at 37°C for at least 4 h (or overnight) to allow re-solubilization of formazan (see Note 6).
4. Mix the content of the wells by replacing the microplate lid with a tight adhesive sheet and turn the plate upside down three times followed by a brief centrifugation at 800 × *g* (see Note 7).
5. Remove adhesive sheet and read the plate in microplate reader at a wavelength in the interval 550–600 nm (in accordance with the filters available in the microplate reader) and using the absorbance at a wavelength >650 nm as reference (Many microplate readers can measure in a "dual mode" automatically giving the absorbance at the test wavelength minus the reference absorbance as, e.g., absorbance A_{590}–A_{655}).
6. Export readings to a spreadsheet for calculations and production of graphics.

4. Notes

1. The formazan solubilization buffer used in the original MTT assay (6) contained isopropanol that may cause precipitation of culture media constituents such as serum proteins that may interfere with absorbance reading. Several other solubilization buffers were subsequently developed (10), including the one used here containing SDS (11).

2. The proliferation rate of lymphoblastoid cell lines can differ; however, a cell population seeded at a density of 2×10^5 cells/ml generally reach a density of 6×10^5 in 2 days.
3. We use no more than 10 ml media in a T-25 culture flask in order to keep the surface to volume ratio in a range that allows sufficient gas exchange.
4. Lymphoblastoid cells grow in suspension but tend to form clusters of cells or cell aggregates. To ensure that a homogenous cell solution is used for cell counting, it is important to dissolve these aggregates by gently mixing with a Pasteur pipette (do not vortex). If it is desired to use an adherent type of cell line, a detaching step using a mild trypsin treatment is required before cell counting.
5. We use at least three wells for each condition/dilution.
6. Verify that formazan is completely solubilized, non-solubilized formazan will appear as dark colored crystals and if these are still present allow longer time for solubilization.
7. Not all laboratories perform this mixing step. If the skipping of this step is desired, it is recommended to test if mixing affects absorbance data, for example, by comparing the absorbance data obtained by reading a number of microplate assays both prior and after mixing.

Acknowledgments

We thank the Lundbeck Foundation and the Ludvig og Sara Elsass Fond for financial support.

References

1. Gregersen N, Bross P, Vang S, Christensen JH (2006) Protein misfolding and human disease. Annu Rev Genomics Hum Genet 7:103–124
2. Luheshi LM, Crowther DC, Dobson CM (2008) Protein misfolding and disease: From the test tube to the organism. Curr Opin Chem Biol 12:25–31
3. Balch WE, Morimoto RI, Dillin A, Kelly JW (2008) Adapting proteostasis for disease intervention. Science 319:916–919
4. Westerheide SD, Bosman JD, Mbadugha BN, Kawahara TL, Matsumoto G, Kim S, Gu W, Devlin JP, Silverman RB, Morimoto RI (2004) Celastrols as inducers of the heat shock response and cytoprotection. J Biol Chem 279:56053–56060
5. Chow AM, Brown IR (2007) Induction of heat shock proteins in differentiated human and rodent neurons by celastrol. Cell Stress Chaperones 12:237–244
6. Mosmann T (1983) Rapid colorimetric assay for cellular growth and survival: Application to proliferation and cytotoxicity assays. J Immunol Meth 65:55–63
7. Berridge MV, Tan AS (1993) Characterization of the cellular reduction of 3-(4, 5-dimethylthiazol-2-yl)-2, 5-diphenyltetrazolium bromide (MTT): Subcellular localization, substrate dependence, and involvement of mitochondrial electron transport in MTT reduction. Arch Biochem Biophys 303: 474–482

8. Berridge MV, Herst PM, Tan AS (2005) Tetrazolium dyes as tools in cell biology: New insights into their cellular reduction. Biotechnol Annu Rev 11:127–152
9. Vistica DT, Skehan P, Scudiero D, Monks A, Pittman A, Boyd MR (1991) Tetrazolium-based assays for cellular viability: A critical examination of selected parameters affecting formazan production. Cancer Res 51:2515–2520
10. Marshall NJ, Goodwin CJ, Holt SJ (1995) A critical assessment of the use of microculture tetrazolium assays to measure cell growth and function. Growth Regul 5:69–84
11. Tada H, Shiho O, Kuroshima K, Koyama M, Tsukamoto K (1986) An improved colorimetric assay for interleukin 2. J Immunol Methods 93:157–165

Chapter 22

Rescue of Misfolded Proteins and Stabilization by Small Molecules

Raymond C. Stevens, Javier Sancho, and Aurora Martinez

Abstract

Increasing stability of functional proteins by binding small compounds and ions has long been used to extend shelf-life of protein formulations in the pharmacological and biotechnological industry. Likewise, the therapeutic application of small molecules for in vivo recovery and maintenance of structure and function of proteins is steadily increasing. Compounds that can rescue misfolded proteins by stimulating their correct folding and/or the stabilization of native-like conformations in vivo are referred to as pharmacological chaperones. Here we present thermal-shift and isothermal methods for the high-throughput screening of stabilizing pharmacological chaperones for soluble and membrane proteins. The effect of selected hit compounds on the kinetics of protein synthesis is further evaluated by an in vitro transcription–translation rapid translation system. These procedures can be integrated in an interdisciplinary and translational approach for the search of personalized pharmacological chaperones in genetic misfolding diseases.

Key words: Cytoplasmatic and membrane proteins, High-throughput, Experimental screening, Protein stability, Misfolding diseases, Denaturation, Microscale fluorescence, Thermal-shift, Melting temperature, Compound libraries

Abbreviations

ANS	8-Anilino 1-naphthalene sulfonic acid
$C_{0.5}$	Concentration for half-maximal binding
CPM	*N*-[4-(7-diethylamino-4-methyl-3-coumarinyl)phenyl]maleimide
DDM	*n*-Dodecyl-β-D-maltoside
DMSO	Dimethyl sulfoxide
GPCR	G protein-coupled receptor
GnHCl	Guanidinium hydrochloride
HTP	High-throughput
λ_{em}	Emission wavelength
λ_{exc}	Excitation wavelength
PAH	Phenylalanine hydroxylase
PKU	Phenylketonuria
RTS	Rapid translation system

Peter Bross and Niels Gregersen (eds.), *Protein Misfolding and Cellular Stress in Disease and Aging: Concepts and Protocols*, Methods in Molecular Biology, vol. 648, DOI 10.1007/978-1-60761-756-3_22, © Springer Science+Business Media, LLC 2010

$T_{0.5}$ Half-denaturation temperature
$t_{1/2}$ Half-live

1. Introduction

Equilibrium ligands that bind to a specific site on the native state of a protein increase protein conformational stability depending on the concentration and affinity of the ligand (1, 2). Hence, hunting for small molecules with high affinity and stabilizing effect has been a constant when optimization of protein stability is the goal (3, 4). High-throughput (HTP) screening schemes have been developed to select appropriate stabilizing ligands. These schemes have often been based on thermal stability assays, using fluorescence dyes, for both soluble (5, 6) and membrane (7) proteins. Screening for stabilizing ligands is a requisite when searching for optimal formulation conditions to increase the operating shelf-life of biocatalysts and protein pharmaceuticals (8). Structural biology efforts applying both crystallization and nuclear magnetic resonance (NMR) also require stable and durable protein samples. Stabilizing additives and ligands have been shown to be helpful in improving protein formulations for crystallization (5). More recently, HTP assays, notably thermal-based fluorescence assays, have also been used in the first steps of the identification of compounds stabilizing misfolded proteins associated with conformational diseases. These stabilizing compounds have often been found to exert a therapeutic effect by aiding to a correct trafficking and function in vivo, and are referred to as pharmacological chaperones (9–13).

Protein instability, misfolding, and aggregation are established mechanisms in the pathology underlying several genetic diseases. Misfolding and early aggregate formation play also a central role in cytotoxicity associated to neurodegenerative and other conformational diseases. Pharmacological chaperones are small structure-directed molecules that specifically facilitate the folding of intermediate states toward the native state of the target protein and/or stabilize the final folded state (10, 13).

When searching for putative pharmacological chaperones with therapeutic potential, the preliminary selection of hit ligands should be followed by a careful characterization of their functional and physiological effects on the stability of wild-type and mutants of the target protein of interest. The characterization of the effect of the hit compounds and their preliminary evaluation as therapeutic drugs include, among other, (i) determination of the bind-ing affinity, (ii) analysis of protein function (activity or activation in the case of enzymes or receptors, respectively),

(iii) kinetics of synthesis, steady state level, and intracellular half-life of the proteins in in vitro transcription–translation systems and/or transiently or permanently expressing eukaryote cells, and, ideally (iv) testing in animal models (see Fig. 1 for a flow chart of the complete approach (13)). Knowledge of the structure of the targeted protein might accelerate and guide the search of hit compounds by virtual screening steps (14, 15). Further optimization of the drug will also largely benefit from knowledge of the structure of the ligand–protein complex.

Here we present the protocols associated with the HTP screening and in vitro transcription–translation for the selection of pharmacological chaperones, for both soluble and membrane proteins. HTP screening of stabilizing ligands of both types of proteins is performed by the fluorescence thermal shift assay (6, 7, 13). The positive hits from the HTP thermal assay can be further analyzed by the isothermal denaturation assay, which represents a medium throughput screening scheme, at least for stability assays of membrane proteins (16). A protocol for determining the binding affinity of positive hits from HTP screening by measuring the quenching of intrinsic fluorescence is also included. Finally, the effect of the hit compounds on the kinetics of protein synthesis

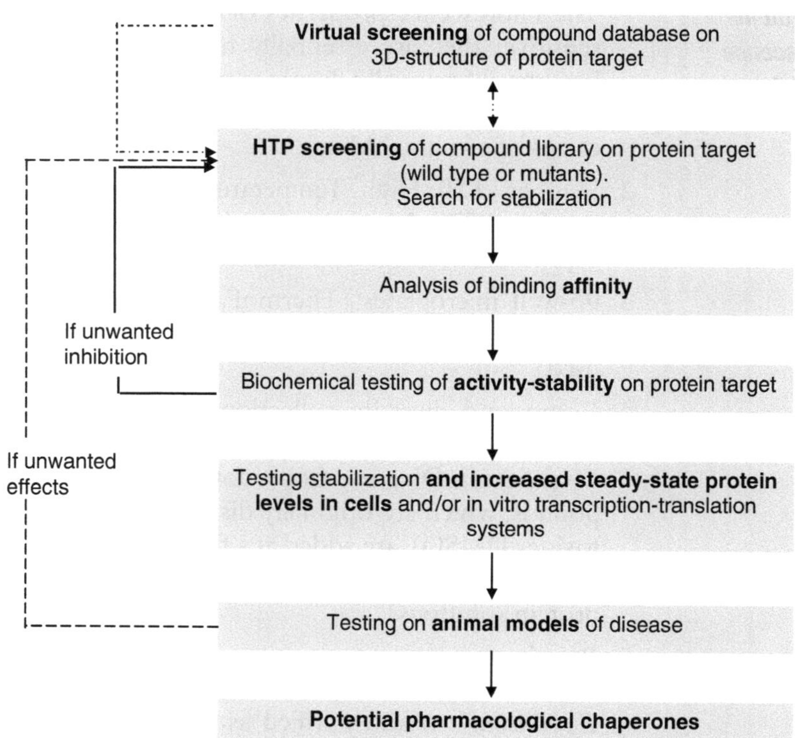

Fig. 1. Flowchart of translational approach for search and evaluation of small molecule stabilizing compounds. HTP, high throughput.

and/or degradation can be measured using in vitro transcription–translation systems (17). The protocols have been adapted for the search of pharmacological chaperones for a soluble protein: phenylalanine hydroxylase (PAH), and a membrane one: a G protein-coupled receptor (GPCR).

Over the last years, it has been shown that misfolding of mutated PAH – a cytosolic protein involved in hepatic catabolism of L-Phe – is the molecular basis of the genetic metabolic disease phenylketonuria (PKU; OMIM 261600) (18). Recent efforts have concentrated in the search of stabilizing compounds with therapeutic potential as pharmacological chaperones to treat PKU (13). On the other hand, GPCRs represent a large family of integral membrane proteins involved in signal transduction, and provide a very important target for therapeutic discovery (19). Microscale screening assays for selection of stabilizing ligands of GPCRs have also been recently developed (7).

2. Materials

2.1. HTP Screening of Stabilizing Ligands for Soluble Proteins by the Fluorescence Thermal Shift Assay

1. Fluorescence dyes such as 8-anilino 1-naphthalene sulfonic acid (ANS) or Sypro Orange (Molecular Probes, see Note 1). The fluorescence properties of these dyes are environment-sensitive; they are essentially non-fluorescent in water and become substantially fluorescent when bound to membranes or hydrophobic regions of proteins, thus being valuable indicators of protein folding.
2. FluoDia T70 High Temperature Fluorescence Microplate Reader (PTI). It can be substituted by an iCycler iQ Real Time PCR Detection System (Bio-Rad).
3. 96-well microplates (ThermoFast 96 skirted from Thermo Scientific or 96-well thin-wall PCR plates if the iCycler iQ is used).
4. Compound diversity collection (e.g., from the Maybridge HitFinder Collection database, Maybridge Ltd., or the MyriaScreen Diversity Collection, Sigma-Aldrich). The compounds, which are originally dissolved in 100% dimethyl sulfoxide (DMSO), are added at a final concentration of 100 μM and 2.5% DMSO to the 96-well microplates containing the protein solutions.
5. Protein sample to be analyzed, e.g., recombinant human PAH, wild-type or mutants associated with PKU, expressed in *Escherichia coli* and purified as described (20). The protein samples are prepared as 5–10 mg/ml in 20 mM Hepes-NaOH, pH 7.0, 200 mM NaCl, and added at a final concentration in the plates of 0.1 mg/ml PAH (~2 μM PAH subunit).

2.2. HTP Screening of Stabilizing Ligands for Membrane Proteins by the Fluorescence Thermal Shift Assay

1. Thiol-reactive *N*-[4-(7-diethylamino-4-methyl-3-coumarinyl) phenyl]maleimide (CPM) from Invitrogen (see Note 2). Prior to use the dye stock is diluted in dye dilution buffer (20 mM Hepes-NaOH, pH 7.5, 200 mM NaCl, 0.025% *n*-dodecyl-β-D-maltoside (DDM) (Sigma), incubated for 5 min at room temperature, and used immediately while protected from light to reduce photobleaching. Optimal dilution of the CPM dye should be determined empirically for each tested protein, but a 1:40 dilution is appropriate in most cases.
2. FluoDia T70 High Temperature Fluorescence Microplate Reader (PTI, Birmingham, NJ).
3. 96- or 384-well microplates.
4. Compound diversity collection (see 2.1, above).
5. Protein to be analyzed, for example, recombinant human β_2-adrenergic receptor (β_2AR), a well-characterized GPCR, expressed and purified from baculovirus-infected Sf9 cells as described (21).

2.3. Isothermal Denaturation Assay to Determine Ligand-Induced Stabilization of Membrane Proteins

1. CPM prepared as a 4 mg/ml stock in dimethyl formamide.
2. 1 M guanidinium hydrochloride (GnHCl). This chemical denaturant is included to speed the process of unfolding, thereby minimizing the effects of dehydration on data processing errors.
3. Membrane protein, for example, β_2AR.
4. 96- or 384-well microplates.
5. A Genios Pro fluorescent plate reader (Tecan Systems, Inc., San Jose, CA).
6. GraphPad Prism program (GraphPadPrism v.4.00 for Windows, Graphpad Sofware, San Diego, CA).

2.4. Determination of Binding Affinity of Positive Hits from HTP Screening by Fluorescence Quenching

1. PAH (1 μM subunit) in 20 mM Hepes-NaOH, pH 7.0, 200 mM NaCl.
2. 0–100 μM of the hit compounds in 2.5% DMSO.
3. Spectrofluorimeter, for example, Cary Eclipse spectrofluorimeter (Varian Inc., Mulgrave, Victoria, Australia).

2.5. Coupled In Vitro Transcription–Translation System

1. Rapid Translation System (RTS) (RTS ProteoMaster 500; Roche Molecular Biochemicals, Roche Diagnostics GmbH, Mannheim, Germany).
2. Enhanced *E. coli* lysate kit for RTS (RTS 100 E. coli HY Kit, Roche Molecular Biochemicals).
3. ^{35}S-L-Met (Amersham Pharmacia Biotech, Piscataway, NJ).
4. Expression vector with T7 promotor including the gene of the protein to be expressed (e.g., pcDNA3-wt-PAH).
5. Hit compounds at defined concentrations.

6. Unlabeled L-Met (6 mM), ribonuclease A (1.4 mg/ml), and DNase (1.4 mg/ml; final concentrations) from Sigma to stop the RTS assays.

2.6. Sodium Dodecyl Sulphate-Polyacrylamide Gel Electrophoresis and Autoradiography

1. SDS sample buffer (4×): 40% glycerol, 240 mM Tris–HCl, pH 6.8, 8% SDS, 8% β-mercaptoethanol, and 0.04% bromophenol blue.
2. Separating SDS buffer, stacking SDS buffer, 30% acrylamide/0.8% bisacrylamide solution (acrylamide is a neurotoxin when unpolymerized and so care should be taken not to receive exposure), *N,N,N,N′*-tetramethyl-ethylenediamine (TEMED), and ammonium persulfate.
3. Sodium Dodecyl Sulphate-Polyacrylamide Gel Electrophoresis (SDS-PAGE) equipment.
4. Autoradiography film.
5. Phosphorimager (e.g., Personal Molecular Imager from Bio-Rad).

3. Methods

3.1. HTP Screening of Stabilizing Ligands for Soluble Proteins by the Fluorescence Thermal Shift Assay (=Differential Scanning Fluorimetry)

1. The HTP screening, searching for stabilizing ligands, is performed by monitoring the thermal denaturation of proteins in the presence of the extrinsic fluorescent probe (e.g., ANS using a FluoDia T70).
2. The compounds from the diversity collection (e.g. from the Maybridge HitFinder Collection database) are manually dispensed into 96-well microplates with multichannel pipettes (100 ml of 100 mM compound in 2.5% DMSO), and overlaid with mineral oil. Robotic instrumentation can also be used to dispense the compounds. Protein solutions, e.g., recombinant human PAH (0.1 mg/ml, 2 μM PAH subunit) in 20 mM Hepes-NaOH, pH 7.0, 200 mM NaCl, and 100 μM ANS are then added. The final sample volume of ligand, protein, and ANS in each well is 200 μl.
3. The microplates are then incubated at 25°C for 30 min before loading onto the microplate reader.
4. Control samples without addition of DMSO and/or ligands are routinely included in each microplate.
5. Thermal denaturation is monitored by following the increase in ANS fluorescence intensity associated with protein unfolding ($\lambda_{exc} = 395$ and $\lambda_{em} = 500$ nm, where λ_{exc} and λ_{em} are the excitation and emission wavelengths, respectively) (Figs. 2a and 2b).

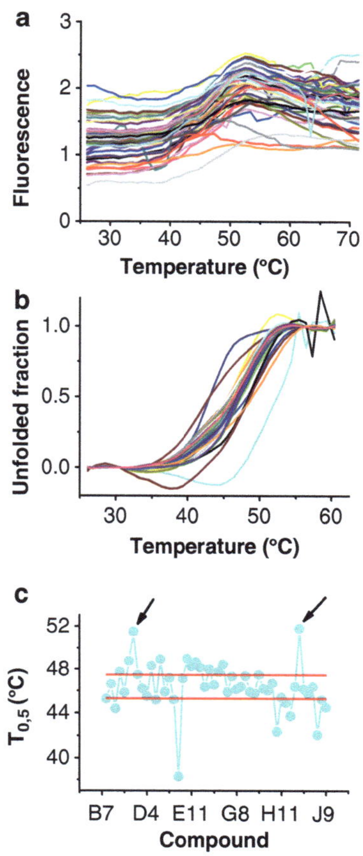

Fig. 2. HTP screening of stabilizing ligands by the fluorescence thermal shift assay. (**a**) Representative melting curves (raw data). (**b**) Smoothed and normalized curves. (**c**) Calculated half-denaturation temperature ($T_{0.5}$) for individual curves obtained in presence of different compounds. The red lines comprise the range $T_{0.5} \pm$ SD; compounds are considered to be potential hits (marked by *arrows*) if $T_{0.5}$(hit) > $T_{0.5}$(control) + 2 × SD.

6. Unfolding curves are registered from 25 to 75°C, at a 1°C/min scan rate; the system is allowed to equilibrate at each temperature for 1 min before fluorescence acquisition. In practice, this represents an operational heating rate of about 0.25°C/min.
7. For data analysis, the fraction of unfolded protein (χ_U) along each experimental unfolding curve is calculated as:

$$\chi_U = \left[F - (F_N + m_N \times T)\right] / \left[(F_U + m_U \times T) - (F_N + m_N \times T)\right] \qquad (1)$$

Where F is the experimental fluorescence, T is the temperature (in °C), F_N and F_U are the fluorescence values of the native and unfolded states, respectively, at a reference $T = 0$°C, and m_N and m_U are the slopes of their linear temperature dependencies.

8. For each curve, $T_{0.5}$ is calculated as the T value at which half of the protein molecules are in the unfolded state ($\chi_U = 0.5$). Compounds that increased the $T_{0.5}$ values of the

protein beyond 2× the standard error (SD) of the $T_{0.5}$ values determined in the corresponding control experiments (4–8 experiments with no added ligand performed in the same microplate), that is, $T_{0.5}(\text{hit}) > T_{0.5}(\text{control}) + 2 \times \text{SD}$, are considered as potential hits (see Fig. 2c for distribution of $T_{0.5}$- values and selection of hits).

9. The original curves of the potential positives are then individually fitted to a 2-state thermal unfolding equation (22) in order to discard false positives generated during the automated analysis.

3.2. HTP Screening of Stabilizing Ligands for Membrane Proteins by the Fluorescence Thermal Shift Assay

1. The thermal denaturation assay of membrane proteins was performed in a total volume of 130 μl (*7*). The tested protein (1–20 μg) is diluted in the appropriate buffer and added to the microplates where the compounds to be tested are added beforehand (see Subheading 3.1 step 2). After an incubation period (usually 5 min at room temperature), included to allow equilibration of the protein with the buffer components, 10 μl of diluted CPM is added and thoroughly mixed with the protein.
2. The microplates are heated in a controlled way with a ramp rate of 2°C/min with $\lambda_{exc} = 390$ and $\lambda_{em} = 465$ nm. Assays are performed over a temperature range starting from either 4 or 20°C and ending at 80 or 90°C.
3. In case the fluorescent signal shows considerable fluctuation at high temperature, the top plateau of the unfolding profile is constrained and made equal to the maximally attainable fluorescence reading immediately following the steep melting transition. The background fluorescence in the presence of buffer components is very low at temperatures below 60°C.
4. Data are processed with GraphPad Prism program. In order to determine the inflection point of the melting curves, which is assumed to equal the melting temperature (T_m), a Bolzmann sigmoidal equation is fitted to the raw data.
5. Raw data is used directly for the fitting algorithm because it was determined that fluorescent background correction has little effect on the fitted values of T_m.

3.3. Isothermal Denaturation Assay to Determine Ligand Induced Stabilization

1. In addition to the HTP-thermal shift assays for soluble and membrane proteins, that is, Subheadings 3.1 and 3.2, respectively, the effect of ligands on protein stability can also be studied in isothermal assays, which allow assessing protein stability and aggregation properties under physiological conditions (e.g., 37°C). The method has been recently described in detail for soluble proteins (*23*), and here we describe a medium throughput protocol adapted to membrane proteins, for example, β_2AR solubilized in the detergent DDM, for

which the extent of covalent coupling of receptor cysteines to the thiol-reactive dye CPM is monitored as a function of time at constant temperature (*16*).

2. Thermal stability data at 35°C are collected in the presence of CPM and of 1 M GnHCl (other concentrations of GnHCl may be evaluated).
3. The protein is prepared in a concentration of about 0.5–1 mg/ml in 50 mM Na-Hepes, pH 7.5, 150 mM NaCl, containing 2 mg of CPM dye (diluted from the 4 mg/ml stock in dimethyl formamide) and incubated at 4°C for 10 min.
4. Various combinations of ligands are then added to the mixture at 2× final concentration and allowed to equilibrate for 30 min at 4°C before the final dilution to working volume with the appropriate concentration of GnHCl in the assay buffer. Final volume of the assay is 60 μl.
5. Data are collected as soon as possible after the addition of the chemical denaturant to reduce the degree of unfolding prior to measurement.
6. Fluorescence of the CPM dye is measured every minute for 3 h on a fluorescent plate reader using 340-nm excitation filter with a 35-nm bandpass and a 475-nm emission filter with a 20-nm bandpass.
7. Data are fitted to a single exponential decay curve using GraphPad Prism Software so that first-order rate constants (k) for denaturation and half-lives ($t_{1/2}$) of protein denaturation ($t_{1/2} = -\ln(0.5)/k$) are obtained (see Fig. 3). The assay aims to select compounds that increase $t_{1/2}$ more than two-fold relative to the ligand-free protein.

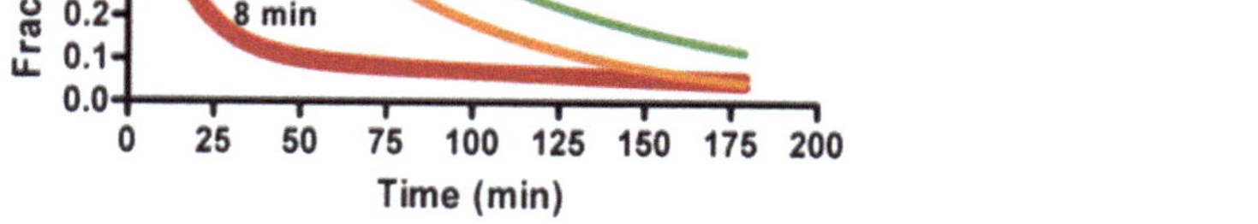

Fig. 3. Isothermal denaturation assay to determine ligand-induced stabilization of membrane proteins. Isothermal CPM determination of the half-life of denaturation of a recombinant form of human β_2-adrenergic receptor (β_2AR) in the presence of 1 M GnHCl at 35°C, with and without both cholesterol (CHS) and timolol (Tim). The thickness of the line represents the 95% confidence interval over three replicates, and the fitted half lives are indicated next to the respective curves. Both TIM and CHS cause an approximate fivefold increase in half-life under these conditions. In combination, the effect is almost 16-fold relative to the ligand-free apo protein. Copied with permission from (16).

3.4. Determination of Binding Affinity of Positive Hits from HTP Screening by Fluorescence Quenching

1. The protocol here presented is adapted to measure the apparent affinity of ligand compounds to soluble PAH (as concentrations for half-maximal binding: $C_{0.5}$). Prior to measurements, PAH samples (1 μM subunit) in 20 mM NaHepes, 200 mM NaCl, pH 7.0, 2.5% DMSO are incubated at room temperature for 30 min with 0–100 μM compound to allow equilibration.
2. Intrinsic tryptophan fluorescence spectra are acquired at 25°C, with λ_{exc} = 295 nm (5-nm slit) and emission from 300 to 400 nm (5-nm slit), and corrected for blank emission.
3. Fractional quenching values in the presence of the ligands are calculated from the decrease in fluorescence at 340 nm at increasing ligand concentrations relative to samples without ligand.
4. $C_{0.5}$ for each compound is estimated from fittings to a hyperbolic function, including a linear term for inner filter effects:

$$F = a \times [L] + ((F_{max} \times [L]) / (C_{0.5} + [L])) \tag{2}$$

5. Where F is the experimental fractional quenching, [L] is the total ligand concentration, F_{max} is the predicted maximal quenching due to specific ligand binding, and a is a constant that describes the ligand concentration dependence of the inner-filter effect.

3.5. Coupled In Vitro Transcription–Translation System

1. Coupled in vitro cell free transcription–translation systems have been very advantageous in previous investigations to prove the chaperone effect of compounds (*24, 25*). In this in vitro system, compounds can easily be added to the assay, allowing the synthesis of proteins in the presence of defined supplementations. The Rapid Translation System (RTS) is compatible with monitoring protein synthesis in short time scales in order to follow the kinetics of protein synthesis previous to saturation (occurring at about 60 min) and analyze whether the compounds accelerate protein synthesis (*13*).
2. In the case of PAH, selective synthesis of active [^{35}S]-labeled human enzyme is obtained at 37°C by RTS with the enhanced *E. coli* lysate (50 μl of the enhanced *E. coli* lysate), containing 50 μM ^{35}S-L-Met and the expression vector pcDNA3-WT-PAH (250 ng) in the absence and presence of hit compounds (100 μM) in 2.5% DMSO (13). Two and a half percent DMSO is also present in control synthesis assays. RTS is carried out at 30°C and 600 rpm.
3. Aliquots (7 μl) are taken from the synthesis assays at increasing times up to 2 h, for example, at 15, 22.5, 30, 60 min, and 2 h. Unlabeled L-Met (6 mM), ribonuclease A (1.4 mg/ml), and DNase (1.4 mg/ml) are added to stop the synthesis in each aliquot.

4. The aliquots from the RTS assays at different times and in the absence and presence of stabilizing compounds are boiled in SDS sample buffer; 1 part of 4× buffer is added to 3 parts RTS sample and 5 μl of the boiled mixture loaded on a denaturing polyacrylamide gel.
5. After SDS-PAGE, the gels are exposed to autoradiographic film for about 24 h, and the labeled protein in each lane analyzed and quantitized by phosphorimaging (see (13) and (17) (in this Series) for a careful description of the in vitro transcription–translation synthesis methodology and SDS-PAGE).

4. Notes

1. When Sypro Orange is used as an extrinsic fluorescence probe for HTP screening of stabilizing ligands for soluble proteins (Subheading 3.1), thermal denaturation is monitored at $\lambda_{exc} = 490$ nm and $\lambda_{em} = 575$ nm. Stock solution of Sypro Orange should be diluted (about 5×) to provide the working concentration.
2. As a maleimide-based thiol-reactive probe, CPM has an optimal pH between 6 and 8. Below pH 6, the rate of CPM-thiol adduct formation decreases while above pH 8, the chemical selectivity of the probe is reduced, and CPM may react with primary amines. Additionally, above pH 8, the rate of CPM-adduct hydrolysis increases substantially and this reaction can compete significantly with thiol modification. The slower rate of adduct formation can be partly compensated for by increasing the concentration of the CPM dye in the reaction.

Acknowledgements

The authors would like to thank their respective groups, especially to Angel L. Pey, Nunilo Cremades, Adrián Velazquez-Campoy, Michael A. Hanson and Mark T. Griffith for discussions and skilful execution of experiments shown in Figs. 2 and 3.

References

1. Sanchez-Ruiz JM (2007) Ligand effects on protein thermodynamic stability. Biophys Chem 126:43–49
2. Cremades N, Sancho J, Freire E (2006) The native-state ensemble of proteins provides clues for folding, misfolding and function. Trends Biochem Sci 31:494–496
3. Maclean DS, Qian Q, Middaugh CR (2002) Stabilization of proteins by low molecular weight multi-ions. J Pharm Sci 91:2220–2229
4. Meyer JD, Ho B, Manning MC (2002) Effects of conformation on the chemical stability of pharmaceutically relevant polypeptides. Pharm Biotechnol 13:85–107

5. Ericsson UB, Hallberg BM, Detitta GT, Dekker N, Nordlund P (2006) Thermofluor-based high-throughput stability optimization of proteins for structural studies. Anal Biochem 357:289–298
6. Niesen FH, Berglund H, Vedadi M (2007) The use of differential scanning fluorimetry to detect ligand interactions that promote protein stability. Nat Protoc 2:2212–2221
7. Alexandrov AI, Mileni M, Chien EY, Hanson MA, Stevens RC (2008) Microscale fluorescent thermal stability assay for membrane proteins. Structure 16:351–359
8. Nayar R, Manning MC (2002) High throughput formulation: strategies for rapid development of stable protein products. Pharm Biotechnol 13:177–198
9. Ulloa-Aguirre A, Janovick JA, Brothers SP, Conn PM (2004) Pharmacologic rescue of conformationally-defective proteins: implications for the treatment of human disease. Traffic 5:821–837
10. Loo TW, Clarke DM (2007) Chemical and pharmacological chaperones as new therapeutic agents. Expert Rev Mol Med 9:1–18
11. Pedemonte N, Lukacs GL, Du K, Caci E, Zegarra-Moran O, Galietta LJ, Verkman AS (2005) Small-molecule correctors of defective DeltaF508-CFTR cellular processing identified by high-throughput screening. J Clin Invest 115:2564–2571
12. Tropak MB, Blanchard JE, Withers SG, Brown ED, Mahuran D (2007) High-throughput screening for human lysosomal beta-N-Acetyl hexosaminidase inhibitors acting as pharmacological chaperones. Chem Biol 14:153–164
13. Pey AL, Ying M, Cremades N, Velazquez-Campoy A, Scherer T, Thony B, Sancho J, Martinez A (2008) Identification of pharmacological chaperones as potential therapeutic agents to treat phenylketonuria. J Clin Invest 118:2858–2867
14. McInnes C (2007) Virtual screening strategies in drug discovery. Curr Opin Chem Biol 11:494–502
15. Tanrikulu Y, Schneider G (2008) Pseudoreceptor models in drug design: bridging ligand- and receptor-based virtual screening. Nat Rev Drug Discov 7:667–677
16. Hanson MA, Cherezov V, Griffith MT, Roth CB, Jaakola VP, Chien EY, Velasquez J, Kuhn P, Stevens RC (2008) A specific cholesterol binding site is established by the 2.8 A structure of the human beta2-adrenergic receptor. Structure 16:897–905
17. Desviat LR, Perez B, Ugarte M (2003) Investigation of folding and degradation of in vitro synthesized mutant proteins in the cytosol. Methods Mol Biol 232:257–263
18. Martinez A, Calvo AC, Teigen K, Pey AL (2008) Chapter 3 Rescuing proteins of low kinetic stability by chaperones and natural ligands: phenylketonuria, a case study. Prog Nucleic Acid Res Mol Biol 83:89–134
19. Hanson MA, Stevens RC (2009) Discovery of new GPCR biology: one receptor structure at a time. Structure 17:8–14
20. Martínez A, Knappskog PM, Olafsdottir S, Døskeland AP, Eiken HG, Svebak RM, Bozzini M, Apold J, Flatmark T (1995) Expression of recombinant human phenylalanine hydroxylase as fusion protein in Escherichia coli circumvents proteolytic degradation by host cell proteases Isolation and characterization of the wild-type enzyme. Biochem J 306:589–597
21. Rosenbaum DM, Cherezov V, Hanson MA, Rasmussen SG, Thian FS, Kobilka TS, Choi HJ, Yao XJ, Weis WI, Stevens RC, Kobilka BK (2007) GPCR engineering yields high-resolution structural insights into beta2-adrenergic receptor function. Science 318:1266–1273
22. Irun MP, Maldonado S, Sancho J (2001) Stabilization of apoflavodoxin by replacing hydrogen-bonded charged Asp or Glu residues by the neutral isosteric Asn or Gln. Protein Eng 14:173–181
23. Senisterra GA, Soo Hong B, Park HW, Vedadi M (2008) Application of high-throughput isothermal denaturation to assess protein stability and screen for ligands. J Biomol Screen 13:337–342
24. Thony B, Calvo AC, Scherer T, Svebak RM, Haavik J, Blau N, Martinez A (2008) Tetrahydrobiopterin shows chaperone activity for tyrosine hydroxylase. J Neurochem 106:672–681
25. Thony B, Ding Z, Martinez A (2004) Tetrahydrobiopterin protects phenylalanine hydroxylase activity in vivo: implications for tetrahydrobiopterin-responsive hyperphenylalaninemia. FEBS Lett 577:507–511

INDEX

Peter Bross and Niels Gregersen (eds.), *Protein Misfolding and Cellular Stress in Disease and Aging: Concepts and Protocols*, Methods in Molecular Biology, vol. 648, DOI 10.1007/978-1-60761-756-3, © Springer Science+Business Media, LLC 2010

V

W

Y

MIX
Papier aus verantwortungsvollen Quellen
Paper from responsible sources
FSC® C105338

If you have any concerns about our products,
you can contact us on
ProductSafety@springernature.com

In case Publisher is established outside the EU,
the EU authorized representative is:
Springer Nature Customer Service Center GmbH
Europaplatz 3, 69115 Heidelberg, Germany

Printed by Libri Plureos GmbH
in Hamburg, Germany